AF361945

Nature-Based Solutions (NBS) in Cities and Their Interaction with Urban Land, Ecosystems, Built Environment and People

Nature-Based Solutions (NBS) in Cities and Their Interaction with Urban Land, Ecosystems, Built Environment and People

Debating Societal Implications

Editors

Dagmar Haase
Annegret Haase
Manuel Wolff
Diana Dushkova

MDPI • Basel • Beijing • Wuhan • Barcelona • Belgrade • Manchester • Tokyo • Cluj • Tianjin

Editors

Dagmar Haase
Lab of Landscape Ecology,
Department of Geography,
Humboldt University Berlin
Germany

Annegret Haase
Department of Environmental
and Urban Sociology, Helmholtz
Centre for Environmental
Research—UFZ
Germany

Manuel Wolff
Lab of Landscape Ecology,
Department of Geography,
Humboldt University Berlin
Germany

Diana Dushkova
Lab of Landscape Ecology,
Department of Geography,
Humboldt University Berlin
Germany

Editorial Office
MDPI
St. Alban-Anlage 66
4052 Basel, Switzerland

This is a reprint of articles from the Special Issue published online in the open access journal *Land* (ISSN 2073-445X) (available at: https://www.mdpi.com/journal/land/special_issues/NBS_cities).

For citation purposes, cite each article independently as indicated on the article page online and as indicated below:

LastName, A.A.; LastName, B.B.; LastName, C.C. Article Title. *Journal Name* **Year**, *Volume Number*, Page Range.

ISBN 978-3-0365-2143-5 (Hbk)
ISBN 978-3-0365-2144-2 (PDF)

Cover image courtesy of Dagmar Haase.

Contents

About the Editors

Dagmar Haase is Professor in Landscape Ecology Urban Ecology at the Geography Department, Humboldt-Universität zu Berlin in Berlin, and a Guest Scientist at the Helmholtz Centre for Environmental Research, UFZ, Leipzig, Germany. Her core expertise is in modelling urban land-use change and urban system dynamics and urban telecouplings. Dagmar's focus is on the quantification and assessment of ecosystem services and landscape functions using statistics and earth observation data.

Annegret Haase is a senior researcher at UFZ Leipzig, Department of Urban and Environmental Sociology. Her main research interests are: sustainable urban development, urban transformations, socio-environmental processes in cities and urban regions, land-use change and conflicts, use and development of urban green spaces, interaction between people and nature in cities, nature-based solutions and urban ecosystem services, engagement of urban society with nature (participation, stewardship, co-production), urban social–spatial inequalities and injustices, impacts of increasing diversity on human–nature interaction in cities, governance and participation.

Manuel Wolff is a research fellow at the Humboldt-Universität zu Berlin in the Department of Geography, Landscape Ecology Lab. He has worked in the integrated Project 'Urban Transformations' at the Helmholtz-Center for Environmental Research (UFZ), in the networking project 'CIRES—Cities Regrowing Smaller' and other projects at the Université Paris Sorbonne, University of Nottingham or Charles University Prague. Manuel's main interest is in land-use changes within the context of urban growth and shrinkage and the associated impacts on ecosystem services. He is working on a quantitative comparative analysis of human-environmental interactions and trends in European cities, including accessibility to green spaces and socio-environmental justice.

Diana Dushkova is a senior researcher at the Department Urban and Environmental Sociology of Helmholtz Centre for Environmental Research, UFZ, Leipzig, Germany and Associate Professor at the Peoples' Friendship University of Russia (RUDN), Agrarian Technological Institute. Her core expertise is in landscape research and spatio-temporal analyses of human–environment interfaces in cities using quantitative and qualitative methods, analysis and assessment of ecosystem services and their consequences for human health and well-being, sustainable development of cities and regions, and nature-based solutions. At presrnt, she is working on the H2020 projects: "RECONECT—Regenerating Ecosystems with Nature-based solutions", "CONNECTING Nature" and co-leading the International Summer school HU/MGU "Urban ecosystem services and human well-being: exchange of experience between Russia and Germany".

Editorial

Editorial for Special Issue "Nature-Based Solutions (NBS) in Cities and Their Interactions with Urban Land, Ecosystems, Built Environments and People: Debating Societal Implications"

Diana Dushkova [1,2,*], **Annegret Haase** [1], **Manuel Wolff** [3] **and Dagmar Haase** [3,4]

[1] Department of Urban and Environmental Sociology, Helmholtz Centre for Environmental Research—UFZ, Permoser Str. 15, 04318 Leipzig, Germany; annegret.haase@ufz.de

[2] Landscape Design and Sustainable Ecosystems, Agrarian and Technological Institute, Peoples' Friendship University of Russia (RUDN), Miklukho-Maklaya Str. 8, 117198 Moscow, Russia

[3] Lab of Landscape Ecology, Department of Geography, Humboldt University Berlin, Rudower Chaussee 16, 12489 Berlin, Germany; manuel.wolff@geo.hu-berlin.de (M.W.); dagmar.haase@geo.hu-berlin.de or dagmar.haase@ufz.de (D.H.)

[4] Department of Computational Landscape Ecology, Helmholtz Centre for Environmental Research—UFZ, Permoser Str. 15, 04318 Leipzig, Germany

* Correspondence: diana.dushkova@ufz.de or kodiana@mail.ru

Citation: Dushkova, D.; Haase, A.; Wolff, M.; Haase, D. Editorial for Special Issue "Nature-Based Solutions (NBS) in Cities and Their Interactions with Urban Land, Ecosystems, Built Environments and People: Debating Societal Implications". *Land* **2021**, *10*, 937. https://doi.org/10.3390/land10090937

Received: 16 August 2021
Accepted: 30 August 2021
Published: 6 September 2021

Publisher's Note: MDPI stays neutral with regard to jurisdictional claims in published maps and institutional affiliations.

Today's cities increasingly serve as the nexus between nature and people in times of strong urban growth and, in some cases, urban decline. There is no doubt that today's most major and urgent challenges occur in cities. Among them are challenges such as rapid climate and environmental change, complex water and waste management issues, adverse health and well-being as well as changes in social cohesion, land use and migration patterns. The increasing concentration of people in cities and the fact that cities are strongly tied to non-urban areas in relation to economics, consumption and power reveals the considerable significance of cities in terms of global challenges. This poses new tasks for the cities of the future, which should be designed as sustainable and liveable to serve the health and well-being of the population, on the one hand, and to support biodiversity and healthy ecosystems, on the other. In this regard, nature-based solutions (NBS) can provide an entry point for addressing these challenges, as they involve integrating the ecological dimension within spatial planning policies and practices in cities.

Defined by the European Commission as "actions [...] and solutions to societal challenges [...] which are inspired by, supported by, or copied from nature", NBS simultaneously provide multiple environmental, social and economic co-benefits [1]. For example, they can improve both air quality and a location's physical attractiveness and have positive impacts on public health and quality of life while also allowing for more biodiversity, or create green jobs through the greening of cities or planting of trees in former brownfields [2,3]. In comparison to ecosystem services aimed at the assessment and valuation of the immediate benefits to human well-being and the economy, NBS focus on the benefits to people and the environment itself. They allow for sustainable solutions that are able to respond to environmental changes and hazards both in the short and long term [4]. In this sense, NBS go beyond the traditional biodiversity conservation and management principles by "re-focusing" the debate on humans and specifically integrating societal factors such as human well-being, poverty reduction, socio-economic development and governance principles.

Nowadays, NBS are on their way to becoming mainstream in national and international policies. A great number of ongoing European research projects endeavour to explore how NBS work in different urban contexts in relation to the political, social, cultural, institutional, environmental and economic background, and how to successfully

1

implement NBS in Europe and worldwide. In most cases, these projects aim to analyse how the NBS concept could help link research and innovation in the areas of biodiversity, ecosystem services, economic demands and societal challenges. Researchers have explored what actions are needed to further support the knowledge base for NBS and presented key recommendations for identifying the drivers of NBS success and for overcoming barriers and bridging gaps to boost the promotion and uptake of NBS worldwide. This is reflected by an increasing number of relevant scientific publications [5–8].

Although scholarly work on NBS has since entered the transdisciplinary arena and urban practice reality, there are still many open questions and challenges awaiting a response. This includes, among others, methodological, planning and resilience issues, not to mention challenges related to the question of how NBS can respond to complex societal, political, economic and environmental challenges and how they may contribute to more socially sustainable and responsible futures. These questions are particularly relevant in cities because urban areas are often characterised by high inequality in terms of access to green space as well as demographic, socio-economic, environmental and power-related factors.

As a result, the notion of "nature-based" usually describes heterogeneous phenomena that operate differently according to different contexts, even when they are based on the same or a similar conceptual model.

It is acknowledged that when NBS are locally set up, they interact strongly with the local conditions such as the built environment, the local natural resources and ecosystems, the socio-demographic potential and the land use, in addition to the way urban policy and planning processes are organised and, not least, the makeup of the urban population, the society and its actors. Thus, by and large, if we want to better understand how NBS may operate successfully at the local level and which challenges have to be overcome, we require a great deal of knowledge about the interactions between the multiple contextual conditions and drivers of NBS and their impacts [3].

The interactions between NBS, cities and urban populations are complex and multidimensional. Conceptual knowledge is required to better understand urban transformations and their consequences for cities, which are, of course, highly complex systems. Models or principles are used to analyse the outcomes of different local responses, and tools serve to resolve specific problems. NBS implementation also enters the field of urban policymaking, governance and participation, for instance, when new models of cooperation are set up or when new local business associations or civic society groups are involved in strengthening sustainability or environmental stewardship [4,6,9–11]. The assessment report by Wild et al. [8] shows that the societal implications of NBS have become an increasingly important topic, and there are still many challenges for future research related to the acceptance of NBS, their methods of implementation and impacts or trade-offs with regard to social inequalities, diverging interests and conflicting goals.

At the same time, the urban green scholarly debate has become more attentive to topics such as the interaction between urban ecosystem services or NBS and the unequal distribution of their benefits and burdens in cities [10,12,13], as well as the impact of power relations and imbalances [14]. Furthermore, within recent years, the connections between greening programmes/policies and issues of inclusiveness, justice and inequalities have also come into focus [15–19].

Set against this background, this Special Issue seeks to provide an overview of the current state of knowledge about the interactions between NBS and urban land, built environments, ecosystems and people in cities. In particular, it looks at those interactions through the lens of societal challenges which NBS are aimed to address. With the help of a number of conceptual and empirical studies on NBS development and implementation in different cities, we discuss the interactions between NBS and their urban context, the resulting benefits and trade-offs, as well as the consequences for policy, planning, maintenance, stewardship and governance.

This Special Issue of *Land* brings together a collection of diverse papers that debate the societal implications of NBS in cities and their interactions with urban land, ecosystems, the built environment and people. In inviting papers, we were particularly interested in studies that could assist in answering some of the following questions:

- What types of NBS based on green and blue infrastructure (GBI) are being implemented in cities across the globe?
- Which properties of urban nature and/or urban ecosystems do they make use of, and how do NBS themselves influence urban ecosystems and ecosystem services flows in cities?
- What are the trade-offs of NBS compared to other ecosystem services and urban biodiversity?
- What are the typical types of land and land units where NBS are implemented?
- How do NBS and their implementation interrogate/interact with the social environment and issues of social cohesion and justice?
- What are land governance and policy schemes for NBS in cities? Do they differ from the prevailing land management and governance policies implemented so far in our cities?
- How does the implementation of NBS correspond to and interact with general directions and priorities of urban development?

We have compiled a range of articles from research undertaken in different countries, continents and hemispheres (Netherlands, Germany, USA, Bangladesh, India and Australia) to show that the interconnections between NBS and urban societies are a global challenge and how contextual factors can impact NBS design, implementation, acceptance and effects.

Looking more closely, we can see that a number of studies demonstrate the value of NBS from the perspective of urban GBI networks [3,20,21]. Several studies analysed linkages between urban green spaces, the ecosystem services they provide and public health and well-being through a range of benefits such as the mitigation of climate change, improvement of mental health and well-being through contact with nature, stormwater management and biodiversity conservation [20–22]. This Special Issue considers two sides of GBI development with NBS. On the one hand, we have included a range of papers that address the perspective of ecosystem services provision and its benefits, such as carbon storage and sequestration, pollution removal, food production, noise reduction and recreational and cultural values (see papers [21–23]), while other papers deal with the undesired effects and trade-offs of NBS implementation, such as green gentrification, negative effects on neighbourhoods/residential development and housing prices, as well as increases in social disparities and disintegration [24,25]. This underlines that, firstly, NBS are a complex response to the need for greener and more sustainable cities and include multiple impacts that bring about very different results for different actors, people, structures and spaces in the city. Secondly, the authors of most of these papers demand a more serious consideration of the multiple impacts of NBS (implementation), including existing trade-offs.

The role of NBS for spatial planning and landscape-based visions in Dutch cities was analysed by Van Roiij et al. (2021) [23]. They applied a landscape-based and co-creation-based planning approach to regional spatial policy challenges, paving the way for a paradigm shift towards a future land management system that is resilient to external pressures.

The value of co-creation in the process of successful NBS development and implementation is also highlighted in the paper by Dushkova and Haase (2020) [3]. The authors use the city of Leipzig (Germany) as an example to discuss the main drivers behind NBS, possible design options and the involved governance actors. By discussing these drivers and governance strategies, the authors introduce a framework for assessing the co-benefits, opportunities and challenges of NBS in urban areas. They also provide examples of best practices that demonstrate the multiple co-benefits provided by NBS.

The types and quality of GBI-based NBS implemented in cities were studied by Lahoti et al. (2020) [22] and Ahmed et al. (2019) [20]. They analysed the existing spatial morphology to understand the potential for GBI development and its challenges. Using

Dhaka (Bangladesh) as a case study, the paper by Ahmed et al. (2019) [20] explores how urban growth planning can be guided by a GBI network that combines blue, green and grey elements to provide a multifunctional urban form. The authors highlight the meaning of the spatial morphology for potential locations of NBS development and the types of solutions necessary for different typologies of urban densities. The proposed network takes on different forms at different scales and locations and offers different types of climate mitigation actions, controls and management options. The paper also provides some practical implications and challenges for implementing BGI at different urban scales.

Ahmed et al. (2019) [20] consider the challenges addressed by GBI, such as flood mitigation and water sensitive design, while Shade and Kremer (2019) [26] focus on green infrastructure implementation as one of the important measures for climate adaptation. Using a combination of cellular automata, machine learning and Markov chain analysis, the authors demonstrate that land use and land cover modelling (such as the modelling conducted for Philadelphia, USA) is an important tool for city officials planning future land usage.

The papers by Ali et al. (2020) [24] and Schwarz et al. (2021) [25] introduce green regeneration NBS as strategies for tackling land abandonment and improving the quality of life in disadvantaged neighbourhoods in shrinking cities. A public park was found to operate as a trigger for structural, social and symbolic upgrades in the formerly shrinking city of Leipzig, but only in combination with dynamic real estate market developments, which are the main drivers of change. Ali et al. (2020) [24] identify various facets of green gentrification. Schwarz et al. (2021) [25] critically examine the positive and negative immediate impacts of green space NBS on residents' well-being, residential location choice and housing and land markets. The paper directly addresses questions posed by this Special Issue, arguing that social settings, such as property constellation and real estate agents, benefit from higher income clients' preferences to live close to high-quality urban green spaces and thus foster the green gentrification process discussed in the Ali et al. (2020) [24] article. The paper by Rink and Schmidt (2021) [27] adds to this topic by describing the use of pocket forests and larger urban forests on inner-city brownfields as multifunctional NBS for shrinking cities. Even though urban forests do not constitute an independent or new type of NBS, they create new ecosystems from existing abandoned, brownfield or neglected areas. These forests were found to be multifunctional in terms of urban climate alleviation and air quality improvement, as they simultaneously enhance the value of adjacent neighbourhood areas while creating new recreational opportunities and supporting local biodiversity. Moreover, the afforestation of brownfields was revealed as the cheapest way to create greenery, which not only fulfils the main objectives but also was accepted and used by the local population.

Taken together, the articles in this Special Issue indicate that NBS provide clear benefits for urban societies responding to social–environmental challenges. NBS can help achieve strategic planning goals such as climate adaptation, biodiversity conservation and the improvement of recreational facilities and public health. In terms of costs, NBS such as lawns or afforestation were found to be the cheapest ways to create urban greenery. At the same time, the papers report that NBS often involve trade-offs. For instance, greening can cause changes and new imbalances in the real estate market that limit the aforementioned recreational benefits, particularly for low-income households. Thus, and in line with recent arguments in urban social–ecological–technological studies [28,29], the societal implications we are examining in this Special Issue are ambivalent in the best sense.

An interdisciplinary approach is vital for land-related studies, and the contributions to this Special Issue represent a robust and broad panorama of disciplines, approaches and research traditions. Perhaps this is the best evidence for the fact that NBS, both as real-world tools and an area of research, have become increasingly relevant for the transformation of cities towards greater sustainability. At the same time, their implementation is increasingly controversial and critically debated, for instance, with respect to issues of justice.

While this collection may not provide a definitive summary of the NBS phenomenon, we are convinced that it will at least contribute to a better understanding of all those processes and tendencies which take place in the urban environment when NBS are put into action.

Funding: This research was funded by the Horizon 2020 Framework Programme of the European Union, research and innovation project "—Regenerating ECOsystems with Nature-based solutions for hydro-meteorological risk rEduCTion", grant Agreement No. 776866.

Conflicts of Interest: The authors declare no conflict of interest.

References

1. European Commission. Policy Topics: Nature-Based Solutions 2016. Available online: https://ec.europa.eu/research/environment/index.cfm?pg=nbs (accessed on 15 June 2021).
2. Albert, C.; Schröter, B.; Haase, D.; Brillinger, M.; Henze, J.; Herrmann, S.; Gottwald, S.; Guerrero, P.; Nicolas, C.; Matzdorf, B. Addressing societal challenges through nature-based solutions: How can landscape planning and governance research contribute? *Landsc. Urban Plan.* **2019**, *182*, 12–21. [CrossRef]
3. Dushkova, D.; Haase, D. Not Simply Green: Nature-Based Solutions as a Concept and Practical Approach for Sustainability Studies and Planning Agendas in Cities. *Land* **2020**, *9*, 19. [CrossRef]
4. Almenar, J.B.; Elliot, T.; Rugani, B.; Philippe, B.; Gutierrez, T.N.; Sonnemann, G.; Geneletti, D. Nexus between nature-based solutions, ecosystem services and urban challenges. *Land Use Policy* **2021**, *100*, 104898. [CrossRef]
5. Dushkova, D.; Haase, D. Methodology for development of a data and knowledge base for learning from existing nature-based solutions in Europe: The CONNECTING Nature project. *MethodsX* **2020**, *7*, 101096. [CrossRef] [PubMed]
6. Kabisch, N.; Korn, H.; Stadler, J.; Bonn, A. Nature-based Solutions to Climate change in Urban Areas–Linkages of science, policy and practice. In *Theory and Practice of Urban Sustainability Transitions*; Springer: Cham, Switzerland, 2017.
7. Raymond, C.M.; Berry, P.; Frantzeskaki, N.; Kabisch, N.; Breil, M.; Nita, M.R.; Geneletti, D.; Calfapietra, C. A framework for assessing and implementing the co-benefits of NBS in urban areas. *Environ. Sci. Policy* **2017**, *77*, 15–24. [CrossRef]
8. Wild, T. (Ed.) *Nature-Based Solutions Improving Water Quality & Waterbody Conditions. Analysis of EU-funded Projects (Report)*; Publications Office of the European Union: Luxembourg, 2020. [CrossRef]
9. Eggermont, H.; Balian, E.; Azevedo, J.M.N.; Beumer, V.; Brodin, T.; Claudet, J.; Fady, B.; Grube, M.; Keune, H. Nature-based Solutions: New Influence for Environmental Management and Research in Europe. *Gaia–Ecol. Perspect. Sci. Soc.* **2015**, *24*, 243–248. [CrossRef]
10. Kabisch, N.; Frantzeskaki, N.; Pauleit, S.; Naumann, S.; Davis, M.; Artmann, M.; Haase, D.; Knapp, S.; Korn, H.; Stadler, J.; et al. Nature-based solutions to climate change mitigation and adaptation in urban areas: Perspectives on indicators, knowledge gaps, barriers, and opportunities for action. *Ecol. Soc.* **2016**, *21*, 39. [CrossRef]
11. Scott, M.; Lennon, M.; Haase, D.; Kazmierczak, A.; Clabby, G.; Beatley, T. Nature-based solutions for the contemporary city. *Plan. Theory Pract.* **2016**, *17*, 267–300. [CrossRef]
12. Keeler, B.L.; Hamel, P.; McPhearson, T.; Hamann, M.H.; Donahue, M.L.; Meza Prado, K.A.; Arkema, K.K.; Bratman, G.N.; Brauman, K.A.; Finlay, J.C.; et al. Social-ecological and technological factors moderate the value of urban nature. *Nat. Sustain.* **2019**, *2*, 29–38. [CrossRef]
13. Neßhöver, C.; Vandewalle, M.; Wittmer, H.; Balian, E.V.; Carmen, E.; Geijzendorffer, I.R.; Görg, C.; Jongman, R.; Livoreil, B.; Santamaria, L.; et al. KNEU Project Team. The network of knowledge approach: Improving the science and society dialogue on biodiversity and ecosystem services in Europe. *Biodivers. Conserv.* **2016**, *25*, 1215–1234. [CrossRef]
14. Berbés-Blázquez, M.; González, J.A.; Pascual, U. Towards an ecosystem services approach that addresses social power relations. *Curr. Opin. Environ. Sustain.* **2016**, *19*, 134–143. [CrossRef]
15. Cousins, J.J. Justice in nature-based solutions: Research and pathways. *Ecol. Econ.* **2021**, *180*, 106874. [CrossRef]
16. Haase, D.; Kabisch, S.; Haase, A.; Andersson, E.; Banzhaf, E.; Baró, F.; Brenck, M.; Fischer, L.K.; Frantzeskaki, N.; Kabisch, N.; et al. Greening cities–To be socially inclusive? About the alleged paradox of society and ecology in cities. *Habitat Int.* **2017**, *64*, 41–48. [CrossRef]
17. Kremer, P.; Haase, A.; Haase, D. The future of urban sustainability: Smart, efficient, green or just? Introduction to the Special Issue. *Sustain. Cities Soc.* **2019**, *51*, 101761. [CrossRef]
18. Langemeyer, J.; Connolly, J.J.T. Weaving notions of justice into urban ecosystem services research and practice. *Environ. Sci. Policy* **2020**, *109*, 1–14. [CrossRef]
19. Pineda-Pinto, M.; Frantzeskaki, N.; Nygaard, C.A. The potential of nature-based solutions to deliver ecologically just cities: Lessons for research and urban planning from a systematic literature review. *Ambio* **2021**, in press. [CrossRef]
20. Ahmed, S.; Meenar, M.; Alam, A. Designing a Blue-Green Infrastructure (BGI) Network: Toward Water-Sensitive Urban Growth Planning in Dhaka, Bangladesh. *Land* **2019**, *8*, 138. [CrossRef]
21. Ignatieva, M.; Haase, D.; Dushkova, D.; Haase, A. Lawns in Cities: From a Globalised Urban Green Space Phenomenon to Sustainable Nature-Based Solutions. *Land* **2020**, *9*, 73. [CrossRef]

22. Lahoti, S.; Lahoti, A.; Kumar Joshi, R.; Saito, O. Vegetation Structure, Species Composition, and Carbon Sink Potential of Urban Green Spaces in Nagpur City, India. *Land* **2020**, *9*, 107. [CrossRef]

23. van Rooij, S.; Timmermans, W.; Roosenschoon, O.; Keesstra, S.; Sterk, M.; Pedroli, B. Landscape-Based Visions as Powerful Boundary Objects in Spatial Planning: Lessons from Three Dutch Projects. *Land* **2021**, *10*, 16. [CrossRef]

24. Ali, L.; Haase, A.; Heiland, S. Gentrification through Green Regeneration? Analyzing the Interaction between Inner-City Green Space Development and Neighborhood Change in the Context of Regrowth: The Case of Lene-Voigt-Park in Leipzig, Eastern Germany. *Land* **2020**, *9*, 24. [CrossRef]

25. Schwarz, N.; Haase, A.; Haase, D.; Kabisch, N.; Kabisch, S.; Liebelt, V.; Rink, D.; Strohbach, M.W.; Welz, J.; Wolff, M. How Are Urban Green Spaces and Residential Development Related? A Synopsis of Multi-Perspective Analyses for Leipzig, Germany. *Land* **2021**, *10*, 630. [CrossRef]

26. Shade, C.; Kremer, P. Predicting Land Use Changes in Philadelphia Following Green Infrastructure Policies. *Land* **2019**, *8*, 28. [CrossRef]

27. Rink, D.; Schmidt, C. Afforestation of Urban Brownfields as a Nature-Based Solution. Experiences from a Project in Leipzig (Germany). *Land* **2021**, *10*, 893. [CrossRef]

28. Lin, B.; Ossola, A.; Ripple, W.; Alberti, M.; Andersson, E.; Bai, X.; Dobbs, C.; Elmqvist, T.; Evans, K.L.; Frantzeskaki, N.; et al. Cities and the "new climate normal": Ways forward to address the growing climate challenge. *Lancet Planet. Health* **2021**, *5*, e479–e486. Available online: http://www.thelancet.com/planetary-health (accessed on 15 June 2021). [CrossRef]

29. Egerer, M.; Haase, D.; McPhearson, T.; Frantzeskaki, N.; Andersson, E.; Nagendra, H.; Ossola, A. Urban change as an untapped opportunity for climate adaptation. *NPJ Urban Sustain.* **2021**, *1*, 22. [CrossRef]

Article

Gentrification through Green Regeneration? Analyzing the Interaction between Inner-City Green Space Development and Neighborhood Change in the Context of Regrowth: The Case of Lene-Voigt-Park in Leipzig, Eastern Germany

Lena Ali [1],*, Annegret Haase [2] and Stefan Heiland [1]

[1] TU Berlin, Department of Landscape Planning and Development, 10623 Berlin, Germany;
    stefan.heiland@tu-berlin.de
[2] Helmholtz Centre for Environmental Research-UFZ, Department of Urban and Environmental Sociology,
    04318 Leipzig, Germany; annegret.haase@ufz.de
* Correspondence: lena.r.ali@campus.tu-berlin.de; Tel.: +49-341-2351735

Received: 4 November 2019; Accepted: 13 January 2020; Published: 16 January 2020

Abstract: Green regeneration has become a common strategy for improving quality of life in disadvantaged neighborhoods in shrinking cities. The role and function of new green spaces may change, however, when cities experience new growth. Set against this context, this paper analyzes a case study, the Lene-Voigt-Park in Leipzig, which was established on a former brownfield site. Using a combination of methods which include an analysis of housing advertisements and interviews, the paper explores the changing role of the park in the context of urban regeneration after the city's turn from shrinkage towards new growth. It discusses whether the concept of green gentrification may help to explain this role. As a result of our analysis, we argue that Lene-Voigt-Park has indeed operated as a trigger for structural, social, and symbolic upgrades in the growing city of Leipzig, but only in combination with real estate market developments, which are the main drivers of change. The concept of green gentrification does help to better understand the role of different factors—first and foremost that of green space. We also discovered some specifics of our case that may enrich the green gentrification debate. Leipzig serves as an example for a number of regrowing cities across Europe where green gentrification might represent a challenge.

Keywords: green gentrification; regeneration; urban green space; neighborhood change; housing market; regrowth; Leipzig

---

1. Introduction

Following a period of massive shrinkage in the 1990s and having faced a subsequent outflow of people and housing vacancies, larger cities in eastern Germany (e.g., Leipzig, Dresden, Potsdam) have seen new growth since around the year 2000. In the period of shrinkage, including the first years after shrinkage had stopped, several regeneration measures aimed at improving quality of life [1] had been introduced and financed by large-scale state funding programs such as Stadtumbau Ost (Urban Restructuring East). Since 2001, Leipzig has witnessed the physical regeneration of housing areas, improvement of streetscapes and urban green spaces, as well as the reuse of vacant lots [2]. In this context, the reuse of urban brownfields and demolished former industrial and residential buildings made the expansion of urban greenery a key measure in sustainable urban and neighborhood planning [3].

The city of Leipzig is an outstanding example for urban regrowth. In the last ten years, Leipzig has been hyped as a great place to work, study, and live, and the city was even nicknamed "Hypezig".

New growth can be observed throughout the city, but especially in districts with Wilhelminian-style buildings[1] which were previously rundown and unrenovated, and where urban development funding plays an important role [4,5]. Presently, these districts are characterized by a housing market saturation, exclusive building projects, and rising apartment rents. Reudnitz-Thonberg, one such district in the eastern part of the city, is home to the Lene-Voigt-Park (hereinafter abbreviated as LVP). The park was created on a former railway industrial area during the post-shrinkage period around the year 2000, when housing vacancy was high. The aim was to encourage residents not to move away and to improve the quality of life in the neighborhood. Since then population growth, socio-structural dynamics within and between districts, rising apartment rents, and the city's improved image have made gentrification an increasingly hot topic in Leipzig.

Neighborhoods experiencing such structural, social, and symbolic upgrading, which results in residents being forced to move away, are referred to as gentrified [6]. Yet the fundamental determinants that cause or trigger gentrification have not yet been identified, hence gentrification is considered a process that can take various forms. The green gentrification[2] discourse, which emerged more recently, analyzes the link between sustainable urban planning and "green urban developments", and their effects on the housing market as well as their social implications (renovations, rising rents, displacement, segregation). Urban upgrading by establishing new green or blue qualities (e.g., a high-quality, planned green space or waterfront) leads to the displacement of low-income residents, because richer households move to these newly developed areas (for an initial study see [7]).

The aim of this paper is to examine the role of urban green spaces for gentrification in the context of new growth after shrinkage and urban regeneration, using Lene-Voigt-Park (LVP) in Leipzig as a case study. We discuss whether LVP might have operated as a trigger of residential change and displacement under the new conditions of growth since 2010 and if, consequently, evidence for green gentrification can be found. Moreover, we examine the extent to which this approach can be used to analyze similar cases. Our pivotal question is: What role does green space development play in a context of urban regeneration, when a city is experiencing new growth after shrinkage? Can the concept of green gentrification help to explain this role?

The paper is structured as follows: After the introduction, Section 2 expands on the debates about greening, regeneration, and green gentrification in the specific context of shrinking and regrowing cities. Section 3 introduces Leipzig and particularly Leipzig's inner east and LVP as a case study, and describes the methods used for the study. Section 4 presents the key results, which are discussed in Section 5 in relation to the research question and the debates introduced in Section 2. We finish with some concluding remarks in Section 6.

2. Interrogating Debates: Greening, Regeneration, and Green Gentrification in the Context of Urban Shrinkage and Regrowth

In this section, we cross-reference debates on urban regeneration, greening (policies) and green gentrification. The first part briefly describes urban development in eastern Germany since the early 1990s with special attention to the role of greening (policies) and the general debate on gentrification. In the second part, we introduce the arguments of the green gentrification debate that look critically at interactions between "green regeneration" and the (re)production of socio-spatial inequalities and inequities. We focus on processes in cities that have turned shrinkage towards regrowth.

---

[1]   Wilhelminian-style building stock means buildings erected in the period between the 1870 and 1914.

[2]   In this study, which focuses on greening (strategies) and urban green spaces, we find the term "green gentrification" to be the most appropriate, but use it in line with other terms such as eco-/ecological gentrification and environmental gentrification.

2.1. Greening Strategies in the Context of Shrinkage and Regrowth

Many cities in eastern Germany experienced a period of shrinkage after the fall of the Berlin Wall and German Reunification in 1989/90. Large cities like Dresden and Leipzig were characterized by unemployment, out-migration, decay, and vacancy of buildings. Particularly inner-city neighborhoods with Wilhelminian architecture suffered a loss of function and value [8]. Following a huge wave of out-migration to western German cities, suburbanization became the second major reason for people to leave eastern cities in 1996/97, as the suburbs promised a better quality of life for many families.

City authorities were faced with the challenge of successfully transforming brownfields and derelict areas into green spaces with a positive appearance and social functions, given that vacancy and abandoned spaces can easily be associated with decline and a lack of prospects. Integrating the development of urban green areas into the comprehensive set of urban development support and funding programs (from municipality to EU level), cities took the opportunity to restructure neighborhoods and promote less dense residential areas with newly designed green spaces (cf. e.g., [9,10]). However, due to modernization, new building projects and persistent suburbanization, the vacancy rates often still exceeded 20% in the city centers (cf. e.g., [11]).

In around 2000, urban shrinkage changed (first moderately, from 2010 onwards more dynamically) to urban regrowth with the beginning of a new population influx to inner-city districts prompted by attractive, newly renovated housing stock and increased green space. This has pushed forward revitalization processes also in the areas with high vacancy rates [12]. Reurbanization describes the renewed in-migration of various household types and their lifestyles to the city centers, including their intention to stay (cf. [2,13,14]). Green spaces have played an important role in this process, as they have contributed to the revaluation of many neighborhoods. Projects on a local scale that are aimed at improving the living conditions in disadvantaged neighborhoods are particularly useful for improving the image of residential areas, if not entire districts (cf. e.g., [15–17]). In Leipzig, the long-term establishment of green spaces is considered a key measure in the regeneration process, as building stock redevelopment increased at the edges of large and attractive green spaces [16].

The benefits humans derive from urban green spaces are well documented and beyond question (cf. e.g., [18–20]), as access to and use of urban green spaces is crucial for people's wellbeing and both physical and mental health [21,22]. Consequently, greening has become increasingly important as a strategy to improve quality of life and sustainability of cities throughout the last decades. Hence, in the real estate sector, green spaces act as a soft location factor, potentially increasing the value of nearby properties. While establishing new green spaces during urban shrinkage can help cities avoid total decline, in times of regrowth, green spaces can contribute to gentrification processes. Starting with the in-migration of so-called "pioneers"—mostly artists and students taking advantage of available and cheap space—such neighborhoods soon develop further, showing the typical features of a gentrification process, such as changes of building stock, apartment rents, and residents, as well as a functional and image change (cf. [23,24]).

Such developments can be observed in regrowing eastern German cities. There are a few main differences to the gentrification that has taken place in steadily growing western German cities, like Munich or Hamburg. In the east, home ownership has not increased very much, rental costs long stayed at a relatively low level, and people have still had relatively great freedom of choice while looking for their preferred neighborhood [25]. The displacement of residents has not been a typical characteristic of this development [26], implying that in-migration to neighborhoods has been driven by housing preferences and the image of the different areas rather than by rental costs (ibid.). However, since around 2010, dynamic growth has been taking place in some large East German cities—upgrading now includes high-end renovations and new upmarket constructions as well [27]. At the same time, the concept of gentrification attracted more attention in public debate and scientific discourse [28], and led to the eastern German development being called "new-build gentrification" (i.e., a process that contributes to a small but distinctive segment of the housing market [29]). It is also referred to as "soft gentrification", which emphasizes the slow speed of the development process [30].

Ongoing in-migration during a housing shortage results in competing land use claims. Consequently, green spaces often have to make space for new construction and building density increases again. Diminishing urban green increases the value of remaining or newly developed individual green spaces. As a result, this leads to an extra boost in value for the residential areas near those spaces, which in turn leads to higher rental costs.

According to Marcuse [31], exclusionary displacement is the consequence of high rents that do not allow poorer households to move to a certain area. This indirect displacement, combined with direct displacement (when residents are forced to move out), results in (higher) segregation within the city. This segregation is partly a reflection of the (lack of) high-quality green space: While better-off households often live in areas with a good provision of urban green spaces, poorer households more often live in densely built areas with a worse or even under-provision of urban green (cf. e.g., [32,33]). Due to this insufficient supply, the accessibility and quality of remaining green spaces are of major importance. Generally, a spatially uneven distribution of green spaces is an effect of limited development regulations, and the basis for the question of environmental justice.

2.2. Green Gentrification: A Critical Perspective on the Impacts of Green Urban Regeneration

Greening under market conditions may cause negative effects on housing costs and lead to a (re)production of inequalities and injustices. The value-adding impact of green spaces on real estate objects is provable in an economic sense, as shown by several studies (cf. e.g., [34,35]). Depending on their function and amenities, green spaces may increase the standard ground value up to 20% [34]. Generally, the awareness of this interaction between greening and real estate development, and resulting social injustices, has slowly been increasing within recent years (e.g., [36–38]). This is supported by a recent study by Rigolon and Németh [39] that investigated predictors for gentrification using the example of parks in US cities: It has shown that both function and location of parks are good predictors, whereas size is not.

In this vein, the concept of green gentrification emerged in the scientific discourse; this approach is used to critically assess the impacts of neighborhood upgrading due to urban green, which results in the displacement of economically vulnerable people as stated by one of the inaugural papers by Dooling in 2009 [7]. Later works also describe green gentrification as a strategy to upgrade neighborhoods and taking into account displacement if not intending it (cf. e.g., [40]). Generally, green spaces can operate as triggers for gentrification in different ways. Either they unintentionally lead to an increase in property prices and housing costs, because property owners and real estate agents regard them as a factor that increases property value, or they are intentionally implemented for economic gains that benefit high-income households, regardless of the consequence that low-income residents are excluded from the advantages of newly designed greenery (cf. e.g., [7,41,42]). Checker ([43], p. 212) explains this targeted strategy as follows: "Operating under the seemingly a-political rubric of sustainability, environmental gentrification builds on the material and discursive successes of the urban environmental justice movement and appropriates them to serve high-end redevelopment that displaces low-income residents". This means that the development is technically profit-oriented and disregards the social dimension of sustainability, (re)producing social inequality (cf. e.g., [44–46]).

Green gentrification is regarded as related to greening strategies in the context of urban renewal and sustainability initiatives in the neoliberal era [47], or to the revitalization of old industrial brownfield sites [48,49] which is especially important for (post)industrial cities such as Leipzig. Expressions such as "cleaning up and clearing out" [50] and "from toxic wreck to crunchy chic" [51] highlight the exclusive character of the newly developed green spaces and the surrounding residential areas. Gould and Lewis [40] describe it as the transformation of a low-value environmental site with potential into a high-value environmental site, which is followed by a population shift. Curran and Hamilton [45], as well as Wolch et al. [52], ask when a neighborhood is "just green enough" to mitigate or avoid effects like rising housing costs and displacement, but still provide good quality of life. "Just green enough"-approaches represent a means in which to tackle the seemingly omnipresent logics of improvement of residential quality (here: through greening) and the unavoidable concurrent increase in prices and rents and related social consequences (here: direct or indirect displacement) [42,46,53]. Especially endangered by potential gentrification are neighborhoods in good location with multi-functional green spaces—caused by either greening or by the conditions of context change (e.g., through a change from supply to demand-driven housing markets) as is the case in many regrowing cities.

Different solutions have been proposed to address the negative outcomes of greening strategies. They have a clear focus on incentives designed to regulate housing market dynamics. Profit-oriented development is to be restricted, for instance by means of social housing programs or rent control (cf. e.g., [43,54]). Another approach appeals to the residents, who are encouraged to oppose high-end redevelopment and enforce small-scale greening initiatives in the form of a bottom-up or grassroots movement (cf. e.g., [55,56]). Urban gardening is one example of such a "just green enough" strategy that involves civic participation, thus accounting for the real needs of the residents [57], although other scholars question whether bottom-up greening strategies can actually prevent gentrification as long as they happen under market conditions [40]. Other studies analyze the conflicting interests of local actors when it comes to greening with an unequal distribution of benefits and losses at the neighborhood scale [58]. There are a growing number of studies dealing with marginalization and exclusion related to greening projects, including strategies aimed at contesting or resisting gentrification (cf. e.g., [59,60]). However, studies that look at greening from the perspective of housing market development and gentrification theory are so far exceptions (e.g., Holm [53] who calls green gentrification the "ecology of upgrading").

3. Case Study, Materials, and Methods

3.1. Leipzig: The Shift from Shrinkage Towards New Growth

We have chosen to focus on the German city of Leipzig as it is one of the most prominent examples of urban shrinkage and regrowth across Germany and Europe, and exemplifies a larger group of cities with similar development pathways and features. Leipzig was recently dubbed a "city of extremes" [28], as the city went from massive shrinkage towards dynamic regrowth in only 20 years [61].

The city's period of severe shrinkage started in the 1960s but saw its most dramatic phase in the 1990s when the city lost about 20% of its inhabitants (approximately 100,000 people) in only 10 years [62]. At that time, this exodus not only led to massive job losses and high unemployment rates, but also to high rates of vacant housing and a lot of abandoned space throughout the city. Greening these places to improve the quality of life in the residential areas, therefore, became a key strategy for counteracting shrinkage [1]. Greening strategies operated together with the demolition of surplus housing. They included the creation of new green spaces such as parks and urban gardens and the expansion of existing ones, interim greening (particularly in those areas that were unlikely to be rebuilt in the near future), new street greenery, and the refurbishment of urban waterways [1,63]. At the same time, Leipzig's housing market was characterized by high vacancy rates, which were highest in built-up, inner-city areas with Wilhelminian architecture; the socio-spatial segregation patterns re-configured after the first half of the 1990s [64].

When shrinkage came to a halt around the year 2000, Leipzig had approximately 70,000 vacant apartments and 3000 brownfield sites [62]. During the 2000s, Leipzig experienced reurbanization [13], mostly in the inner city, decreasing vacancy rates and modest annual gains in population numbers (2000–4000 people). From 2010 onwards, the city entered a new phase of dynamic regrowth, leading to a population increase of almost 100,000 by 2018 with growth rates of more than 2% per year [5]. This growth was facilitated by new, large-scale investments by major corporations such as the Deutsche Post DHL Group, BMW, and Porsche, and the creation of over 70,000 new jobs in the industrial and service sector since the mid-2000s. Investment in housing renovations, new construction, and urban land in good locations increased. This has not only encouraged (young) people to come to Leipzig, but also to stay and to start families. Unemployment rates have fallen from 18% in 2003 to 6% in 2018. As a consequence, housing vacancies quickly decreased to less than 4% overall and 2% of the housing available on the market by the end of 2016 [27]. Both the number of transactions and the amounts of turnover in the real estate market have been continuously increasing since 2010; the same applies to real estate prices and rents for new buildings and existing stock [65]. The supply of low-cost housing has dramatically decreased and does not correspond with the rising demand (ibid.). Given this context, housing construction, which was marginal before 2010, has experienced a new boom, and the pressure on vacant land has been increasing. Inner-city areas have seen a dynamic re-densification, and urban land prices have skyrocketed in some areas [27]. Patterns of segregation reported for the 2000s have consolidated, but the levels have considerably increased [27,61]. Between 2013 and 2017, the average net rent increased by 10.6% in Leipzig—with rents of new contracts increasing up to 25%. Looking at flat sales, the market for owner-occupied housing is dominated by purchases of renovated built-up flats [66]. While 29% of those flats had been sold in first sale, 71% had been sold in resale. Ninety-five percent of those having been sold in first sale were purchased by people coming from outside Leipzig [67].

In line with the "shift in thinking from shrinkage to growth", the city adapted its development strategies to the new conditions of growth, and new green space strategies were set up [5,68]). City authorities focused on the construction of housing and infrastructure, and on questions of how to maintain existing green spaces and to create new ones under the prevailing conditions. The issues of socially responsible living and housing conditions and social cohesion have also received increased attention. In reality, green and open spaces in inner-city areas have been under increasing pressure; some have already disappeared and been replaced by new constructions. This is not only true for private properties, but also for public properties that had been developed as interim green spaces or abandoned after 1989.

3.2. Lene-Voigt-Park in Leipzig

After 1990, Leipzig's eastern districts of Reudnitz-Thonberg and Anger-Crottendorf (cf. Figure 1) were not very popular among residents, as they were known for having urban development deficiencies, highly polluted residential areas, housing vacancies, traditionally weak social structures, and a lack of attractive open spaces. Consequently, the districts remained focus areas of urban development funding in the 2000s [69]. This situation has changed: With the inner-city reurbanization in the 2000s and the dynamic overall growth in the 2010s, the repopulation of Leipzig's inner east recently turned into dynamic yearly growth. At the beginning of the 2000s, when LVP was being established, the area experienced a coexistence of vacancies, cheap rents, refurbished buildings, and flows of incoming and outgoing residents. In other words: Leipzig's inner east faced a lot of challenges, but also experienced an incipient upswing and grew in popularity as a destination for mostly younger households ([13,70]).

The overlapping of multiple measures for urban renewal, and structural funding aimed at combatting economic, ecological, climatic, demographic, and social disadvantages in those neighborhoods, has significantly contributed to this upward trend. In the formally defined Leipzig-Reudnitz redevelopment area (Sanierungsgebiet), 79% of the public spaces were being rehabilitated and 73% of the properties were either completely new or modernized [71]. The creation of LVP in the former Eilenburger Bahnhof area (cf. Figure 2) was one of the first development goals completed in the eastern districts. Due to these kinds of successes, the redevelopment area is now set to be repealed.

After a long participatory process, LVP was inaugurated in 2001 and its establishment finalized in 2004. Due to its varied usage structure with sport and relaxation areas, the park is able to meet the different needs of its visitors. At 800 m in length and 80 to 130 m wide, the park is located in the middle of Reudnitz (see Figure 1), but also forms a green axis from the city center into the landscape of the city's surrounding area. This connecting function will become even more important once LVP is further developed as part of the Parkbogen Ost concept: An approximately five-kilometer-long pedestrian and bicycle path is to be created along former railway lines through green spaces that connect the city districts. The path will lead from the central station through eastern Leipzig and back to the city center. The project is intended to provide further impetus for the revaluation of the residential and business districts of Leipzig East and is part of the district-based regeneration strategy STEK LeO 2013 [70] and the city-wide masterplans (SEKo 2009 [4], INSEK 2018 [5]), as well as other municipal concepts [73]. In 2017, the masterplan for Parkbogen Ost was approved by the city council. The original idea for Parkbogen Ost was driven by civic engagement and the project is also being developed with public participation.

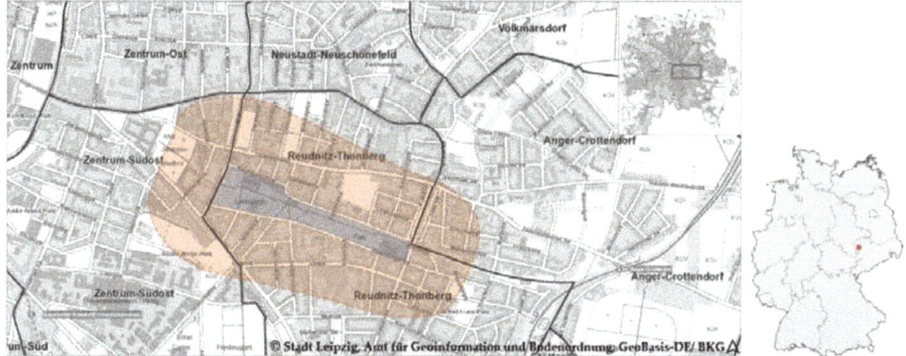

Figure 1. Study area Lene-Voigt-Park (LVP; blue) with its surroundings as focus area (orange), district Reudnitz-Thonberg (City of Leipzig, Germany).

(a)   (b)

Figure 2. (a) The railway industrial area in Reudnitz [72] has been transformed into (b) the Lene-Voigt-Park (LVP) (Photo: Annegret Haase).

According to the Federal Office for Building and Regional Planning, "Leipzig East receives a new attraction through the transformation of the former railway station, which gives the district a new identity and a positive appearance. The project is an exemplary response to structural changes in the city" [74]. In fact, since LPV opened, new types of residents have been moving into the area, the housing market offers have evolved and Reudnitz is spoken about as a trendy district. Whereas in 2003 students and creatives were the most important newcomers, today especially the Lene-Voigt-Quartier north of LVP is dominated by economically and socially established households (mostly young families with average to above-average incomes). The building stock has been almost completely renovated and the vacancy rate fell to below 2% by as early as 2010. New construction projects are being implemented on an ongoing basis and increased ownership of property can also be observed [75].

3.3. Methods

The empirical analysis consists of complementary methods, each contributing to the characterization of the LVP neighborhood and the description of its development. As gentrification is a manifold process, and the effects of the LVP green space might only be one of its influencing factors, the case study required a thorough analysis of different variables in order to identify the role of LVP in this framework. Such factors might be changes of the residential population, the housing market and the neighborhood's image, as well as their interdependencies and mutual reinforcement. The methods and the variables are listed in Table 1, which also outlines the sources/ approaches and the rationale for using each particular method. The approach was exploratory and focused on the advantages of combining quantitative and qualitative methods (see Table 1). As this paper grew out of a bachelor's thesis submitted in 2017, it must be noted that some of the results were obtained in 2016/17. This refers to the interviews, the site visit, and the analysis of the housing advertisements in particular. Despite being slightly older, these results remain relevant as they simply reflect another stage in the gentrification process. We also included more recent data in the statistical analysis and elsewhere.

The LVP neighborhood in Reudnitz was our focus area, but the districts Reudnitz-Thonberg and Anger-Crottendorf were included as well, particularly because the latter is likely to gain relevance through the development of Parkbogen Ost. The statistical data (available at the district level only) were analyzed from the year 2000 onwards in order to identify whether the opening of LVP in 2004 influenced the development of the analyzed variables. The annual growth rate was therefore calculated for two periods, from 2000 to 2004 and from 2004 to 2017/18. The years had to be selected according to the given data base, so the periods were not always consistent. Applicable data from before 2000 was not available.

The interviews with key stakeholders (in the following marked by "Int." and the respective number) were conducted in January and February 2017. By asking people working in different fields (green space development, housing market, neighborhood development and civic society) we tried to draw a comprehensive picture of our case study. All interviewees had or still have extensive professional or stakeholder involvement with the development of LVP and its neighborhood, the respective political processes, as well as many years of insight into the social structure of the district Reudnitz-Thonberg. Each interview comprised around ten questions and took approximately 30 mi. First, all interviewees were asked to describe the neighborhood and its general development, later on with explicit reference to the establishment of LVP. The following questions referred to each interviewee's special field of expertise (cf. Table 1), focusing on (1) strategies and measures regarding urban regeneration, greening, and development goals, as well as conflicts from the municipality's perspective; (2) the planning process, functions and role of LVP in neighborhood changes; and (3) urban development across scales and the question of whether the (green) gentrification concept helps to explain the recent development in LVP's neighborhood. During the interviews, neither the questions nor the order of the questions were mandatory. This allowed to gain further insights into how the experts assess the role of LVP. In Section 4 some interview statements are cited in order to illustrate the narrative and to reflect the different perspectives. The interviews were very valuable, as the gentrification process can be very small-scale and/or start with a change in the way a neighborhood is perceived. In such cases, developments that are part of the process might only be visible in the statistical data later on. This highlights the importance of using complementary methods.

Table 1. Overview of the methods used in the case study ([1] Leipzig Informationssystem LIS: https://statistik.leipzig.de/statdist/index.aspx).

Method	Main Emphasis	Source/Approach	Rationale
Literature and document analysis (qualitative)	Revitalization, reurbanization, green gentrification	Technical literature	General understanding of the debates and their interlinkages
	Urban/neighborhood development, green space development, population development, housing market	City policy and planning documents (e.g., concepts which include development challenges, goals)	Evaluation of documents on strategies and developments in the city/ LVP area (e.g., renovations, upgrading, densification and greening), and their meaning in the context of the research question
	Urban growth, "Hypezig", gentrification as addressed by society and the media	Newspaper and magazine archives	Media coverage of the debates and societal awareness; recognizing (future) challenges
Secondary analysis of statistical data (quantitative)	Total population, population movement, average age, unemployment, housing market, income and rent	Data from the Leipzig Office for Statistics and Elections (city and district data [1], reports, citizen surveys): Calculation of total growth rates and average annual growth rates	Identification and evaluation of trends (before vs. after LVP opening) and interrelations of variables such as population movement/characteristics or rental prices and comparison to gentrification theory
	Rental price development	Online real estate portals, rent index	Analyzing housing market data in order to detect price trends and to assess and interpret the results of the housing advertisement analysis
Analysis of housing advertisements (quantitative, qualitative)	Price development and the role of LVP/urban green	GIS-based analysis of rental housing market advertisements on the online portal ImmobilienScout24 (300 m buffer around LVP, cf. Figure 1)	Evaluation of the role of green spaces/LVP for the value of apartments and its impact on rental prices and thus also for the gentrification process in general
Site visit (qualitative)	Characterization and image of the neighborhood	(Photo) documentation of the building stock quality, type of residents, gastronomy facilities (300 m buffer around LVP, cf. Figure 1)	Identification and exploration of structural, social and symbolic upgrading as signs for gentrification around LVP and comparison to such signs in the wider area
Semi-structured interviews with experts (qualitative)	Guidelines with focus on: 1. Urban planning/development 2. Park design and function 3. Urban research/impact analysis	Interviewees: Int.1 landscape architect, Int.2 city official (urban renewal and housing), Int.3 city official (urban green and water), Int.4 scientist (environmental economy), Int.5 scientist (urban sociology), Int.6 and Int.7 civic associations/locals	Interviews allow for in-depth exploration and an understanding of participants' experiences and perceptions which is of major importance given the topic of gentrification and the role LVP plays in it. Interviewees were selected so as to cover a wide range of perspectives and expertise across different working fields.

The analysis of housing advertisements was limited to an area of no more than 300 m from the park boundaries, which is a figure commonly used to evaluate the area of pedestrian recreation around inner-city green spaces (cf. e.g., [3,76,77]). Online advertisements were viewed daily between 16 November and 16 December 2016. The analysis documented both the base rents (later given in average per m^2) and any references to urban green or specifically to LVP as a significant benefit of the residential area. The rents were then compared to those listed on the real estate portal for Reudnitz-Thonberg, Anger-Crottendorf, and all of Leipzig. The analysis focused on rents (and not property prices), because the percentage of home ownership in the city is as low as 14% (Reudnitz-Thonberg: 5%; 2017) [78].

The results need to be handled with caution since the sample of the housing advertisements was rather limited in both the size and the time period of documentation. Nevertheless, it is a promising approach to identify the effects of green spaces on rental values, and to assess the role of green spaces, such as LVP, as a part of the gentrification process in the respective neighborhoods. The analysis of our sample showed a significant result (see Section 4.4), but it cannot simply be generalized. The same applies to the site visit and the analysis of statistical data, due to its limited availability in some cases. Considering the limitations of the single methods as well as the complexity of gentrification processes, the focus of our methodological approach was put on combining different, independent methods which support each other (method triangulation). Only this allowed for a thorough analysis (e.g., identifying the forms and causes of urban regeneration and upgrading by creating an overall picture of the case study), interpretation and validation of the results. The following section presents the key findings of the analyses.

4. Results: Green Gentrification in the Context of Regrowth? The Case of Lene-Voigt-Park

4.1. Population Growth and Population Change

Since 2002 there has been a continuous increase in the number of residents in Reudnitz-Thonberg. From the opening of LVP in 2004 to the year 2018, the average annual growth rate was 2.9%, whereas from 2000 to 2004 it was still as low as 0.1%. Looking at the migration dynamics only, it can be shown that from 2006 onwards, there is a clear upward trend of people moving to the district from outside Leipzig. The number of people in-migrating increased most between 2007 and 2012. In general, there is a noticeable trend of population growth caused by people moving to Leipzig from other cities, as well as an increase in the number of those people moving to the districts Reudnitz-Thonberg and Anger-Crottendorf. Reudnitz-Thonberg had a higher growth rate (157% from 2000 to 2017) than Anger-Crottendorf (127%) and Leipzig as a whole (70%)[3]. Leipzig's inner east had started to profit from migration as the last of all the areas with old building stock featuring Wilhelminian architecture in Leipzig, Int.5 says[4]. But the LVP neighborhood already profited from migration in the early and mid-2000s, when the park opened (ibid.).

This trend is accompanied by a significant decrease in the average age of the residents (from 41 years in 2000 to 37 years in 2018) and the unemployment rate (14.3% to 4.8%). During the same time, the average age across the whole city went from 43 years to 42 years and Leipzig's unemployment rate dropped from 12.4% to 5.4%. When considering the park opening in 2004, the trend for unemployment rates is particularly striking: From 2004 to 2017 the rates decreased by 10.1%, whereas from 2000 to 2004 it had still increased by 2.0%. For the whole period (2000–2017) this equates to a total decrease in unemployment of 66%. Two factors played an important role in attracting students to the area: The previously moderate rents and the fact that the park serves as a meeting point. Once the students have moved in, more and more of their friends follow, and that leads to changes in the community (Int.3). The site visit showed that besides the growing number of students, many young families can

---

[3]   For the data sources used for these and all other calculations see Leipzig Informationssystem LIS: https://statistik.leipzig.de/statdist/index.aspx (see also chapter methods, Table 1).

[4]   All interview quotes are translated by the authors.

be found in and around the LVP area and meeting in cafés after walking through the park. Areas in Leipzig East that are close to the city center have become more relevant for families in search of an apartment, because the housing markets in popular districts, such as the Südvorstadt, have already been saturated. Combined with the new park, this market saturation has caused a shift in migration dynamics (Int.4).

4.2. Housing Market and Income

Vacancy rates have always been lower in the LVP area compared to those in adjacent neighborhoods, and the buildings were already renovated around the year 2000 (Int.5). Today, a growing number of construction projects can be allocated to a more costly housing segment (Int.4, Int.5). In Anger-Crottendorf, new loft apartments and prestigious urban villas are planned as opportunities for profitable investments. Such projects address a very specific type of clientele that is willing to invest in housing after luxury renovations have been completed (Int.4, Int.5, Int.7). The city of Leipzig highlights that the LVP area is important for the generation of home ownership and increasing the value of residential areas, as well as for competitiveness in terms of location advantages [10]. Another building project has been completed by a company from outside Leipzig, just one street south of LVP in Josephinenstraße (Figure 3). Its marketing strategy framed the apartment complex as an intelligent financial investment in an excellent location [79].

Figure 3. Billboard in Josephinenstraße: "We're building 82 iQ apartments for students" (Photo: Lena Ali).

In sharp contrast to Figure 3 is Figure 4, both photos taken in February 2017 during the site visit. In 2016, a banner was hung on the former engine shed of the Eilenburger Bahnhof inside the LVP area, which read: "What happens when rents are rising, but wages are not?" (Figure 4). According to Int.3, new construction projects are usually accompanied by a rent increase, which is not something remarkable in a market economy. The real estate section of the magazine Capital [80] notes that the marketing period for apartments in Reudnitz-Thonberg has considerably shortened: Advertisements are usually only online for a few days. The magazine also reports that rental apartments in existing stock are offered for around €6.34/m^2 (2019: 6.93), while new apartments cost €10.73/m^2 (2019: 14.22). Newly renovated building complexes and newly built apartments are driving up average prices in the district (Int.5, [81]). The Immaxi Immobilien agency [82] forecasts that property prices for rented apartments and condominiums will rise significantly in the medium term. The 2019 prognosis adds that there are hardly any apartments for low-income earners on the free market at present. The only apartments available for this housing segment are those offered by municipal real estate companies and housing cooperatives [83]. According to real estate portals, rents in Reudnitz-Thonberg are rising due to a shortage of supply (cf. e.g., [80,84]). Int.6, who lives in the district, points out that from 2014 to 2016, he was asked twice to pay a rent increase. The 2017 municipal citizen survey indicates that 44% of all households living in the district have had a rent increase during the last four years [85].

Figure 4. Protest banner inside Lene-Voigt-Park (LVP): "What happens when rents are rising, but wages are not?" (Photo: Lena Ali).

Table 2 shows the development of both rent and income between 2008 and 2017. While in Reudnitz-Thonberg the average net household income has increased by 55%, in Leipzig it has only risen by 28%. In contrast, rental prices have risen by 13% (total rent: 8%)[5] in the district, and 13% (12%) in Leipzig[6]. Those numbers support the general supposition that gentrification is occurring in Reudnitz-Thonberg, as the increase in rent cannot be explained by a general raised level of income. Well-educated young people are gentrifiers settling there, states Int.4, hence supporting this conclusion as well. At some point in this process, rents simply rise too fast and once a certain tipping point is reached, people can no longer afford their housing, he says. But so far, displacement of residents has occurred almost exclusively in the form of indirect displacement (Int.5, Int.6). When people moved there in the early 2000s, vacancy rates were still high. They were the first ones to move in after renovations and did not directly displace others. This is specific to urban regrowth situations (Int.5).

Table 2. Income and rent development in Reudnitz-Thonberg and Leipzig (Data sources: [86]).

	Reudnitz-Thonberg			Leipzig		
	2008	**2017**	**Growth Rate**	**2008**	**2017**	**Growth Rate**
Net household income (median, €/month)	1317.00	2038.00	54.75%	1379.00	1767.00	28.14%
Base rent (median, €/m²/month)	4.98	5.61	12.65%	4.98	5.62	12.85%
Total rent (median, €/m²/month)	6.93	7.49	8.08%	6.92	7.77	12.28%

4.3. Neighborhood Change

Leipzig's new image has been discussed in local, national, and foreign newspapers, often in relation to gentrification, for example: "Hypezig-Leipzig instead of Berlin" [87], "From Leipzig to Hypezig-hipster's eye new playground" [88]. Such media coverage refers to the districts close to the city center in the west and south, as well as in Leipzig East. Reudnitz-Thonberg now ranks among the city's most popular districts and LVP is known citywide for its rich cultural life, thanks to its diverse usage that includes social events (Int.4, Int.6). Being frequently used throughout all the year, LVP suffers especially in summer from overuse (Int.1–Int.7). This is connected to the fact that it is the only green space in the area where people meet and like to spend time, and where there is a cultural

---

5. The base rent is the net rent for an apartment without additional costs such as heating or water, whereas total rent includes these additional costs.
6. The housing advertisement analysis provides rent data on a smaller scale, but as they explicitly aim to point out the role of LVP in the neighborhood development, they are not shown here but in Section 4.4.

exchange because of the different people who gather there, Int.6 says and adds that other green spaces in the area are less inviting for spending longer periods of time, because they are located close to streets, partly covered by bushes, or "occupied" by groups such as people with dogs or homeless people.

The LVP neighborhood has developed a new urban flair that is expressed through new businesses, such as the vegan-vegetarian restaurant and espresso bar across the street from the park, which both opened in 2015. In the wider area, there is now a wholefood store and several Spätis (late-night corner shops), which are especially popular among students. With the shift in supply and services, the character of a neighborhood changes, in some cases leading to an identification loss with one's surroundings. Such neighborhood changes have made certain residents even more aware of gentrification, especially when attention is drawn to income on the one hand, and to rents on the other hand (Int.6), as is reflected by the protest banner (Figure 4).

Table 3 summarizes how the LVP neighborhood has developed in different aspects and draws a parallel to the depiction of the transformation process in gentrification studies. The next chapter highlights the role of LVP for this development.

Table 3. Overview of the Lene-Voigt-Park (LVP) neighborhood development since 1990, based on the empirical findings.

In the 1990s/Beginning of the 2000s	Starting from the Mid 2000s
Population loss: Decline in the birth rate and emigration	Population growth: Increase in the birth rate and immigration; most immigration from outside Leipzig
Vacancy (rate up to 20%) and decay of buildings; first renovations in the area surrounding the park	Renovated building stock (vacancy rate of less than 2% since 2010); new construction projects; growing and target group-oriented gastronomic offers
Low rents (around €4.50/m^2 base rent in 2002, new rentals)	Moderate but rising rents (around €7.50/m^2 base rent in 2018, new rentals) and increased home ownership
Former industrial working-class district with high population density; low-income households	Socially mixed population, but with increasing average income; rejuvenation and studentification
Many brownfield sites, only a few designed green spaces	Brownfield redevelopment: Creation of interim green areas, urban gardening, new parks, Parkbogen Ost green belt concept, new construction sites
Rather unknown, unpopular neighborhood	Attractive residential area with vastly improved image (particularly close to the park); especially popular among students and young families

4.4. The Role of Lene-Voigt-Park for the Neighborhood Development and Opinions on Future Challenges and Opportunities

In all Leipzig residential areas where urban green has been created or improved—particularly in those neighborhoods with Wilhelminian architecture—renovation and in-migration have followed at about the same time. The real estate industry openly inquires in advance about when new areas will be developed, or how long certain construction measures will take, and for private investors "green" is always a must (Int.3). The interviewees agree on the role LVP has played in the development of its surroundings and argue that the park has been the central regeneration element in the district. It has even had an impact far beyond this area, as it functions as a pedestrian and bicycle pathway (Int.3) as well as having additional spill-over effects (Int.5). This means that the park is perceived as a high-quality amenity and it is likely that it has impacted gentrification dynamics: With respect to residential changes (cf. also Section 4.1), the park sped up the process by which the neighborhood became more desirable, as families moved to the area since the park offers good playing opportunities for kids (Int.7). Beyond that, the park has triggered upgrading, as its creation marks the beginning of an entire (re)development process; buildings right next to LVP were the first to be renovated (Int.2).

The statistics shows that the dynamics of all analyzed variables magnifies from 2004 onwards, implying that the influence of the park can be demonstrated on the temporal scale.

Its influence can be observed on the spatial scale as well: There are still some vacant buildings and construction sites in the area two or three streets away from LVP. Sometimes the area known as the LVP neighborhood only extends one street away from the park (Int.4). Since in the wider area some spots have not experienced any upgrading, the park has a high local impact. According to Int.3, there is a difference between people saying "I live in Leipzig East" and people saying "I live close to LVP". When people said the latter, the words sounded very deliberately chosen, so as to emphasize their neighborhood's good reputation. The view that LVP serves as a positive location factor is supported by the advertisement analysis. Within a radius of 300 m of the park boundaries, LVP is mentioned in 59% of 39 advertisements for rental apartments that were on the internet within a one-month period (see Figure 5). The description of the neighborhood usually stated that the nearby LVP offers tranquility and relaxation, leisure activities, walking and cycling opportunities. 10% of the advertisements do not mention the park by its name, but refer to urban green in the area. In some cases both LVP and the Friedenspark are mentioned, while four advertisements only refer to the Friedenspark, and none of the advertisements mention any other green spaces.

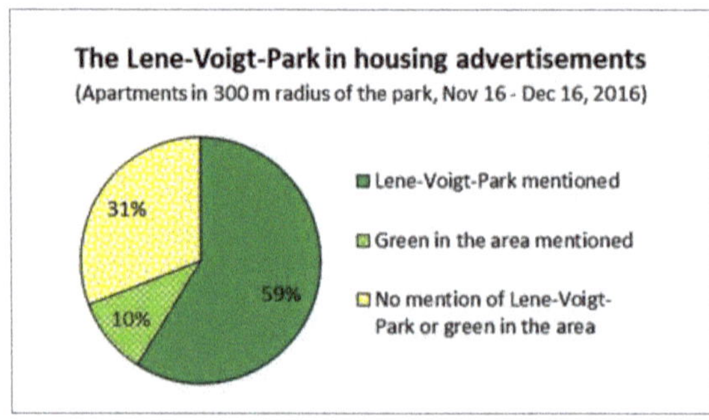

Figure 5. Mentioning of the Lene-Voigt-Park (LVP) in housing advertisements (data sources: [89]).

Figure 6 shows the rents (base rent) listed for the advertised apartments mentioned above and compares them to district and city prices. The prices on online portals are for new rental contracts and are therefore higher than the rents for existing rental contracts as shown in Table 2[7]. Based on our advertisement analysis, the average price within 300 m of LVP is €7.38/m^2 (with a standard deviation of 1.77; sample ranging from €5.49/m^2 to €12.86/m^2). In contrast, the average price in Reudnitz-Thonberg (as well as in Anger-Crottendorf and Leipzig) in the same time period did not exceed €7.00/m^2. This is true for the prices in Reudnitz-Thonberg published by Capital [80] and PWIB Wohnungs-Infobörse [90] as well: €6.34/m^2 and €6.83/m^2, respectively (Immobilien Scout, Figure 6: 6.76). In the first quarter of 2019, the average price increased to around €7.50/m^2 (Figure 6).

Considering the future development of LVP as part of Parkbogen Ost, the city will have to face the challenges resulting from property being increasingly in private hands—especially when it comes to big, sometimes foreign, real estate companies (as is already the case in parts of the eastern districts) (Int.2, Int.6). In Leipzig, a large group of people is still very vulnerable to rising rents, but the social mix is mostly in danger due to indirect displacement. As low-income residents become increasingly restricted in their choice of residence, they will be pushed further to the outskirts (Int.5). Many properties adjacent to the future Parkbogen Ost will "awaken from a deep sleep" by catching the attention of

---

7   Given rents across the article sections may also vary according to the data sources and their survey method (e.g., data from the Leipzig Office for Statistics and Elections vs. data from real estate portals).

investors (Int.2). The city's intention is to attract companies and stimulate new employment, using Parkbogen Ost as a brand for city marketing and tourism [73].

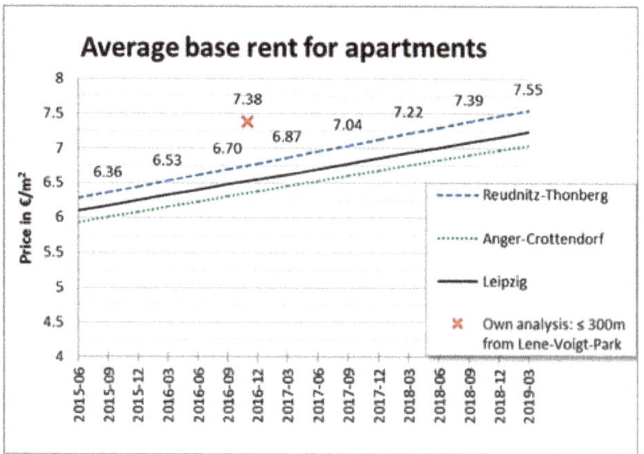

Figure 6. Rental prices for advertised apartments at the city, district, and neighborhood levels. (Data labels shown for Reudnitz-Thonberg and neighborhood level only; data sources: [89,91]).

The interviewees are rather skeptical about green gentrification and the new role of the park within the context of Leipzig East's changing neighborhood character, even though they do not deny the general impact of the park as a "catalyst" or "ingredient" of change or gentrification. More specifically, the interviewees said that LVP is having an impact, but not developing urban green is no solution either. If there was no park, one would still notice an increase in prices in areas with nice cafés and other amenities (Int.4). Social and milieu cohesion must also be seen as key factors, rather than green spaces alone (ibid.). A new park or a new pedestrian and bicycle path might only tip the scales in some cases (Int.3). Consequently, the interviewees declared that other factors (such as real estate sales, speculation etc.; Int.3) and parallel developments (such as renovations and in-migration; Int.5) are also decisive for upgrading and residential change. LVP only serves as a trigger for change in combination with these other factors and developments, which suggests that there is no simple cause–effect relationship between urban green and the social structure (Int.5).

However, social inclusion through affordable housing and sustainable green space development for everybody remain the two main challenges (Int.1, Int.3, Int.4, Int.6). Up until now, LVP has been important for social cohesion as well as for environmental justice in Leipzig East, but this is threatened by the pressure on open spaces and diverging local interests. The wishes of the population concerning urban renewal no longer match the availability of land or the city's financial resources (Int.3, Int.5). Land prices have risen drastically, which limits the city's ability to buy plots of land in order to preserve urban green, for example (Int.3). In the future, the city will require policy instruments for regulating the housing market and defending urban green against financially more profitable land use. Int.6 argues that a key measure would be to limit the privatization of real estate, so that more properties and apartments are owned by the city and not by big real estate companies.

While there is no doubt that developments like Parkbogen Ost are desirable, people must, nonetheless, demand respect for their rights if they are negatively affected by the consequences of such developments. For instance, residents should question the legitimacy of rent increases and oppose illegal increases imposed by their landlords (Int.1, Int.6). One idea proposed by the interviewees to prevent (even if only to a limited extent) further upgrading of the LVP neighborhood is to open a beer garden inside the former engine shed instead of an art gallery or high-class restaurant, as these would only target a specific clientele (Int.1, Int.3). Public participation in the planning phase is also not necessarily a solution, as the LVP case shows: Gentrification was not mitigated by taking into account the opinion of different groups of residents, among them children and stereotypical working class men discussing the plans with a beer bottle in their hands (Int.1).

5. Discussion

In this chapter we come back to the research questions posed at the beginning of this paper about the role of green space development in the context of urban regeneration and new growth, and discuss whether or not the concept of green gentrification is applicable to the results of the case study.

5.1. The role of Green Spaces in the Context of Urban Regeneration and Urban Regrowth

The results of the case study point first and foremost to the fact that the context for the development of the LVP neighborhood has changed considerably since the opening of the park, due to the shift from urban shrinkage to urban regrowth. Furthermore, in the course of these developments the role of the park has changed as well. Under the conditions of shrinkage, green spaces or greening strategies had been used to stabilize urban neighborhoods by cleaning up and reusing brownfields. The interviewees highlighted that the revitalization of the former train station area by the creation of LVP has been a decisive factor in the regeneration of the entire inner east of Leipzig. Its design and effective integration into the neighborhood's road infrastructure make LVP a valuable inner-city park. However, we can assume that LVP was only able to have such a strong effect on the influx of residents because the entire city of Leipzig had seen considerable in-migration, new growth, and reurbanization due to its rising attractiveness after 2000. Only within this context did the role of LVP change to become a trigger or catalyst for further residential change and upgrading.

LVP was created at a time when urban regeneration was promoted by a bundle of factors such as the municipal prioritization of housing and streetscape renovations, the creation of new jobs, and the influx of young people to Leipzig. In the 2000s, however, Leipzig's inner east was still not among the attractive areas that benefitted most from new residents and jobs, although reurbanization and rejuvenation could also be observed there [13]. For LVP it can be stated that there is no simple causality between the opening of the park on the one hand, and the changes in the residential milieu, housing market, and neighborhood image on the other hand (cf. also Int.5). The park, after its establishment, did have an impact on the development of residential composition, in- and outflows of residents, the development of housing costs, the housing market and housing vacancies, as well as the area's reputation. This assumption is supported by the development around LVP since 2004 and its comparison to respective tendencies on a Leipzig East or a whole city scale (see particularly the statistical data on the population, renovations and the interview quotations, e.g., on the neighborhood's image). Yet we can only hypothesize about the extent to which this impact is direct or mediated, and about the precise ways in which the green space operated as a trigger, catalyst or accelerating factor for an ongoing process.

Without any doubt, the housing market development has fundamentally changed through reurbanization and regrowth since 2000 and especially since 2010. Initially, only the streets directly around LVP developed a reputation as a "prestigious residential area" and were affected by in-migration and investment in the building stock. Since the rental prices are correspondingly higher there (cf. Figure 6), we can assume that the existence of and vicinity to LVP had an additional "trigger function" for local upgrading, although real estate and housing market processes certainly play the most important role. Even if we cannot prove any "statistical" causality, this impact of LVP is clearly detectable (e.g., by the development of rents, the analysis of housing advertisements, and the interviews' results as well). For our case study and the whole debate on interactions between green(ing) and upgrading, it is crucial to show this context-dependent catalyst function under changed supply-demand and cost development conditions. The study thus provides good evidence to suggest that the trigger function of green spaces is one element within a complex and multifaceted transformation process, which has been too often neglected or not looked at in theoretical works on gentrification in German cities (see, for example, [23] or [24]) or at neighborhood scale. In our perspective, this result seems to be much more striking than the proof of any causality, especially in situations where context conditions change as we described for Leipzig, a city which can be taken as an example for many regrowing cities across Europe.

The role of LVP for urban and neighborhood development can be divided into three phases. Firstly, as an urban regeneration project, the park had a major influence on the revival of the neighborhood at a time when Leipzig was suffering from shrinkage, high out-migration, and housing vacancies. Secondly, the newly created park triggered an influx of new residents and operated as a factor for attractiveness, accelerated residential change, and decreasing vacancies. Thirdly, in the context of new growth from 2010 onwards, LVP is referred to as a location factor providing exclusive quality in an inner-city area that is becoming more densely populated, and can be regarded as a driver of rising rents and further residential change, including exclusionary displacement.

Given the context of dynamic growth in Leipzig since 2010, green space development might play a larger role in upgrading and gentrification over the next few years, particularly given that the Parkbogen Ost green belt project will be fully realized during that time. If newly built housing or upmarket renovations concentrate along the green belt, the expected revenues through rents will be potentially even much higher than at present (fall 2019) when population growth and growing demand for housing continue within the next years. Today, decisions about new housing and home ownership are already strongly influenced by the planned Parkbogen Ost (Int.3, based on respective inquiries from the real estate industry). The municipality has limited capacity to prevent increases in housing costs in areas close to the existing and planned green spaces, given that financial austerity and market forces are the main drivers of housing and real estate market development in Leipzig (cf. Int.3).

We can assume that LVP represents a typical "change of function" of green spaces in inner-city areas that are experiencing new growth after decline, and conclude: In times of decline, green spaces are established or operate as elements to encourage people to stay, and to make neighborhoods more attractive. So, while they initially act as a stabilizing factor in shrinking cities, green spaces contribute to selective upgrading and the localization of "better" residential areas within the framework of reurbanization or growth.

5.2. Lene-Voigt-Park in Leipzig: A Case of Green Gentrification?

In the concept of green gentrification, the role of urban green or blue structures is, in some cases, clearly defined (particularly in early studies on the subject). Such structures represent exquisite quality developed within the sustainability paradigm, they facilitate the upgrading of neighborhoods and consequently cause gentrification as stated by the green gentrification literature introduced above (cf. for example [43,49,51]). The interviewees' narratives provide strong evidence that the park is an outstanding amenity with a high local impact, and it clearly indicates that gentrification is a concern in the neighborhood. The LVP case study can, subsequently, be evaluated using this concept. In the course of its redevelopment into a high-quality park, the former brownfield site was able, due to its location and form, to successfully add value to its surroundings and act as a trigger for upgrading. In this sense, our study is in line with the findings of Rigolon and Németh (see above; [39]). The opening of LVP was followed by the in-migration of better-off households, while the increase in rents put more and more pressure on lower-income households and increasingly excluded them from entering the area. Residents living close to the park now belong to social milieus such as the "urban young professionals" or environmentally conscious "middle-class professionals" [92], who value and demand good quality of life and can afford rising property prices and rents. The LVP case can consequently be integrated into the green gentrification scheme developed by Gould and Lewis [40], as well as the pathway described by Kern [51] (p. 70): "industry—pollution—disinvestment—cleanup—reinvestment—gentrification—displacement"; even though the last step is currently still limited to indirect displacement (cf. Int.5, Int.6).

However, our study also deviates from the original concept of green gentrification in some respects: The main incentive for transforming the brownfield site into LVP was to "save" the neighborhood from social and structural decay, or to end the "downward spiral" of existing deficiencies (cf. Int.2, [93]). Having been constructed in a bottom-up, integrative process with a very small budget, it resembles a robust design rather than a "bourgeois aesthetic" (as called in the inaugural study by Dooling; cf. [7]). The focus on urban regeneration through greening in Leipzig's inner east was largely initiated by civic engagement for sustainable, socio-ecological urban development. Consequently, the planning and realization of LVP did not involve capital accumulation through eco-branding (cf. [42]), or through stakeholders operating under a "seemingly a-political rubric of sustainability", as Checker [43] (p. 212) puts it. LVP represents a case where gentrification follows the dynamics of the overall urban development context, of which the park is an intrinsic part. This shows that it is not green space development per se that should be questioned, but primarily the broader political and economic context which encourages rather than confronts inequality. A clear distinction between these two different green gentrification pathways (gentrification triggered either intentionally or unintentionally; cf. also [94]) has not been discussed much in the literature, though it is important for evaluating upgrading, rising costs and (the threat of) displacement as a result of green(ing).

To date, LVP has significantly improved livability in the neighborhood and provided important impetus for the development of eastern Leipzig. While the positive effects must be emphasized, it is also vital to question how long the status quo will prevail. Retracing the transformation of the LVP neighborhood from the 2000s onwards (cf. Table 3) revealed that gentrification dynamics are intensifying. Even though rental prices in the area are still moderate (for an inner-city district), long-term residents are increasingly under pressure to defend their homes—including those who were involved in creating LVP (cf. for example Int.1, Int.6). This results from the rent expenditures, which are high when incomes are low, a situation that is typical for Leipzig's inner east (cf. [95]). In addition, it can already be observed that not all people have equal access to the apartments surrounding the park; better-off households dominate this market. The large-scale Parkbogen Ost green belt project is likely to further fuel such processes. This means that positive effects, including social and environmental justice implications, are increasingly threatened. In order to avoid the tipping point, which is reached when environmental improvement no longer benefits the residents living there, it is crucial for the different sectors influencing urban development to work hand in hand.

Solutions for avoiding social inequities fostered by green space development need to be assessed in-depth, while accounting for case-related differences. As the LVP case shows, gentrification cannot be prevented simply by including public participation in the planning phase and implementing a bottom-up process (cf. [40]). The LVP project strongly emphasized a design shaped by community concerns and needs (cf. Int.1), instead of being an upmarket prestige project. The conditions in Leipzig and in this case study are different to those in "hot spot cities" of gentrification, where "just green enough" strategies are discussed and proposed as solutions (cf. [45,52]). Consequently, different kinds of solutions are required. They must account for the specific urban context, including its broader political and economic processes, which are vital for assessing whether green spaces may increase the risk of displacement in either form. This is important, as a situation may shift from shrinkage to growth in just a few years, and possibly cause green spaces to affect social issues in a completely different way.

Green gentrification is important to discuss, as it puts a clear emphasis on equity and fairness. These aspects must be given more attention in the debate and practice of green space and urban planning, and regeneration. Incentives aimed at enhancing the quantity and quality of green spaces must still be pursued, while the exclusion of people from the resulting benefits must be avoided. These issues deserve more attention, particularly in the large number of European cities experiencing new growth after shrinkage, because post-shrinkage is often considered to be a context where pressure on urban land and densification are supposed to be minor challenges, and where brownfield site redevelopment is considered to represent a way of improving neighborhoods without larger social costs. The example of Leipzig shows that this is not the case. With this example we may add an atypical case to the debate on green gentrification through brownfield redevelopment, which, up to now, is based majorly on cases from continuously growing cities with contested housing markets [49–51].

Cities with large amounts of unused land and vacancies may, in only a few years, turn into places where urban land and housing markets become more contested. If this is the case, regulations are needed in order to not undermine the social dimension of sustainability and to ensure that all residents equally benefit from urban greening.

Rather than focusing solely on the concept of green gentrification, our results suggest it is worth considering green spaces as a parameter within the gentrification framework, which consists of different impacting factors, without putting too much weight on this single factor. While environmental improvements can serve as legitimation for developments such as rising rents, other amenities do that as well (for example, gentrification also occurs in areas with many good cafés, cf. Int.4). Looking at the LVP case, it is possible to clearly see how different factors must be considered when trying to explain gentrification processes, as well as urban transformation processes in general. It is insufficient to merely focus on the impact of green spaces, because green gentrification is a multifaceted, dynamic and symbiotic process [96]. But it is also true that "the urban natural environment plays an important and understudied role in shaping gentrification processes" [46] (p. 578). Or to put it differently: These debates have not been sufficiently related to one another. In the case study presented, the park has contributed to and accelerated upgrading and residential change, and has increased the likelihood of direct and indirect displacement—but it did not cause those processes. When discussing issues of regeneration and undesired consequences such as displacement and segregation, the debate should, therefore, consider greening and green spaces as an additional factor and investigate the way in which they interact with real estate and housing market forces.

6. Conclusions

Based on the results of the LVP case study, our discussion and the relevant literature, we have come up with three main conclusions.

Firstly, there are no clearly determined or predictable effects of green space development in urban development processes that allow for any 'one size fits all' solutions. Such solutions would not do justice to the importance of the respective urban or neighborhood development context, which can vary widely. As such, green space development can be a stabilizing factor in times of shrinkage, yet contributes to upgrading in times of (re)growth.

Secondly, although green space development can contribute to or act as a trigger for gentrification, the concept of green gentrification and its applicability in specific cases should be handled with care—as it might imply that urban green itself may lead to or trigger gentrification and displacement. In our study, we instead observed an interlinkage of different factors. First and foremost, it was market forces in urban real estate that sped up gentrification dynamics, proving that urban green is not a causal but a catalyst or accelerating factor.

Thirdly, the case of Leipzig shows that it can take only a few years for an urban context to fundamentally change. Cities experiencing a shift from shrinkage to regrowth can face situations whereby green spaces have completely different—positive or negative—socio-spatial and socio-economic effects. This is probably the main lesson urban policymakers and (green space, housing) planners can learn from our case study.

Last but not least, we have to emphasize that our conclusions are based on one case study and cannot be simply applied to other cases, even if we assume that Leipzig is representative of a large number of regrowing European cities and our study might, therefore, be helpful for understanding the dynamics of green space and urban development in cities with a similar trajectory to Leipzig or regrowing cities in general. Despite a growing body of green gentrification literature, urban green is still understudied in gentrification theory. The debates need to be more closely interlinked, and further research and practical experience is needed in order to effectively evaluate the potential effects of greening in a specific housing market, and to avoid unintended negative social side-effects of urban green space development.

Author Contributions: Conceptualization, L.A., A.H. and S.H.; data curation, L.A.; formal analysis, L.A.; investigation, L.A. and A.H.; methodology, L.A., A.H. and S.H.; project administration, L.A.; supervision, A.H. and S.H.; visualization, L.A.; writing—original draft, L.A. and A.H.; writing—review and editing, L.A., A.H. and S.H. All authors have read and agreed to the published version of the manuscript.

Funding: This research received no external funding.

Acknowledgments: The authors would like to thank the interviewees for their contribution, which was essential for this study. We acknowledge support by the German Research Foundation and the Open Access Publication Funds of TU Berlin for publishing this paper.

Conflicts of Interest: The authors declare no conflicts of interest.

References

1. Haase, A.; Wolff, M.; Rink, D. From shrinkage to regrowth. The nexus between urban dynamics, land use change and ecosystem service provision. In *Urban Transformations—Sustainable Urban Development towards Resource Efficiency, Quality of Life and Resilience*; Kabisch, S., Koch, F., Gawel, E., Haase, A., Knapp, S., Krellenberg, K., Zehnsdorf, A., Eds.; Future City Series; Springer: Cham, Germany, 2018; pp. 197–219.

2. BMVBS (Bundesministerium für Verkehr, Bau und Stadtentwicklung); BBR (Bundesamt für Bauwesen und Raumordnung). *Perspektiven für die Innenstadt im Stadtumbau—3. Statusbericht der Bundestransferstelle Stadtumbau Ost*; BMVBS: Berlin, Germany, 2008.

3. BMUB (Bundesministerium für Umwelt, Naturschutz, Bau und Reaktorsicherheit). *Grün Stadt. Eine Lebenswerte Zukunft. Grünbuch Stadtgrün*; BMUB: Berlin, Germany, 2015.

4. Stadt Leipzig. Leipzig 2020. Integriertes Stadtentwicklungskonzept (SEKo). In *Beiträge zur Stadtentwicklung 50*; Stadt Leipzig: Leipzig, Germany, 2009; Available online: http://www.leipzig.de/fileadmin/mediendatenbank/leipzig-de/Stadt/02.6_Dez6_Stadtentwicklung_Bau/61_Stadtplanungsamt/Stadtentwicklung/Stadtentwicklungskonzept/SEKo_Pdfs/SEKo_BlaueReihe_50_Web.pdf (accessed on 19 October 2019).

5. Stadt Leipzig. Integriertes Stadtentwicklungskonzept Leipzig 2030 (INSEK). In *Zielbild und Stadtentwicklungsstrategie*; Stadt Leipzig: Leipzig, Germany, 2018.

6. Friedrichs, J. Gentrification. Forschungsstand und methodologische Probleme. In *Gentrification. Theorie und Forschungsergebnisse*; Friedrichs, J., Kecskes, R., Eds.; Leske + Budrich: Opladen, Germany, 1996; pp. 13–40.

7. Dooling, S. Ecological Gentrification: A Research Agenda Exploring Justice in the City. *Int. J. Urban Reg. Res.* **2009**, *33*, 621–639. [CrossRef]

8. Hannemann, C. Schrumpfende Städte in Ostdeutschland—Ursache und Folgen einer Stadtentwicklung ohne Wirtschaftswachstum. *Aus Polit. Zeitgesch.* **2003**, *B28*, 16–23.

9. Stadt Chemnitz. Fortschreibung des Räumlichen Handlungskonzepts Wohnen. In *Gesamtstädtische Betrachtungen*; Stadtumbau—Werkstatt Chemnitz: Chemnitz, Germany, 2005.

10. Stadt Leipzig. Neue Freiräume im Leipziger Osten. In *Stadterneuerung—Das Neue Leipzig*; Stadt Leipzig: Leipzig, Germany, 2005.

11. Stadt Görlitz. *Integriertes Stadtentwicklungskonzept INSEK. Demographie, Fachkonzepte Städtebau und Denkmalschutz, Wohnen*; Fortschreibung 2009/2010; Stadt Görlitz: Görlitz, Germany, 2009; Available online: https://www.goerlitz.de/uploads/02-Buerger-Dokumente/INSEK.pdf (accessed on 28 October 2019).

12. BBSR (Bundesinstitut für Bau-, Stadt- und Raumforschung). Wachsen und Schrumpfen von Städten und Gemeinden. Kartenanwendung: Laufende Raumbeobachtung des BBSR. 2017. Available online: https://gis.uba.de/mapapps/resources/apps/bbsr/index.html?lang=de (accessed on 28 August 2017).

13. Haase, A.; Herfert, G.; Kabisch, S.; Steinführer, A. Reurbanizing Leipzig (Germany): Context Conditions and Residential Actors (2000–2007). *Eur. Plan. Stud.* **2012**, *20*, 1173–1196. [CrossRef]

14. Heinig, S.; Herfert, G. Leipzig—intraregionale und innerstädtische Reurbanisierungspfade. In *Reurbanisierung. Materialität und Diskurs in Deutschland*; Brake, K., Herfert, G., Eds.; VS Verlag für Sozialwissenschaften: Wiesbaden, Germany, 2012; pp. 323–342.

15. Stadt Dresden. *Zukunft Dresden 2025+. Integriertes Stadtentwicklungskonzept Dresden (INSEK)*; Fortschreibung; Stadt Dresden: Dresden, Germany, 2014; Available online: http://www.dresdner-debatte.de/sites/default/files/discussion-info/downloads/2014-04-08_insek_abcd_dbob.pdf (accessed on 28 October 2019).

16. Stadt Leipzig. Stadterneuerung und Stadtumbau in Leipzig: Gestern-heute-morgen. In *Beiträge zur Stadtentwicklung 43*; Stadt Leipzig: Leipzig, Germany, 2005.

17. Stadt Chemnitz. *Städtebauliches Entwicklungskonzept—Chemnitz 2020*; Beschlussvorlage; Stadt Chemnitz: Chemnitz, Germany, 2009; Available online: https://www.chemnitz.de/chemnitz/media/unsere-stadt/stadtentwicklung/seko/seko_2020.pdf (accessed on 28 October 2019).

18. Naturkapital Deutschland—TEEB DE. *Ökosystemleistungen in der Stadt—Gesundheit Schützen und Lebensqualität Erhöhen*; Kowarik, I., Bartz, R., Brenck, M., Eds.; TU Berlin, UFZ Leipzig: Berlin/Leipzig, Germany, 2016.

19. BfN (Bundesamt für Naturschutz) (Ed.) Urbane Grüne Infrastruktur. Grundlage für attraktive und zukunftsfähige Städte; Hinweise für die kommunale Praxis. 2017. Available online: https://www.bfn.de/fileadmin/BfN/planung/siedlung/Dokumente/UGI_Broschuere.pdf (accessed on 29 June 2019).

20. UBA (Umweltbundesamt); BFS (Bundesamt für Strahlenschutz); BSR (Bundesinstitut für Risikobewertung); RKI (Robert Koch Institut) (Eds.) Themenheft Umweltgerechtigkeit. In *UMID: Umwelt und Mensch—Informationsdienst 2/2011*; UBA: Berlin, Germany, 2011; Available online: https://www.umweltbundesamt.de/sites/default/files/medien/515/publikationen/umid0211_0.pdf (accessed on 18 October 2019).

21. Amoly, E.; Dadvand, P.; Forns, J.; López-Vicente, M.; Basagaña, X.; Julvez, J.; Alvarez-Pedrerol, M.; Nieuwenhuijsen, M.J.; Sunyer, J. Green and blue spaces and behavioral development in Barcelona schoolchildren: The BREATHE project. *Environ. Health Perspect.* **2014**, *122*, 1351–1358. [CrossRef] [PubMed]

22. Krekel, C.; Kolbe, J.; Wüstemann, H. The greener, the happier? The effect of urban land use on residential well-being. *Environ. Econ.* **2016**, *121*, 117–127. [CrossRef]

23. Friedrichs, J. Gentrification. In *Großstadt. Soziologische Stichworte*; Häußermann, H., Ed.; Leske + Budrich: Opladen, Germany, 1998; pp. 57–66.

24. Krajewski, C. *Urbane Transformationsprozesse in zentrumsnahen Stadtquartieren: Gentrifizierung und Innere Differenzierung am Beispiel der Spandauer Vorstadt und der Rosenthaler Vorstadt in Berlin*; Institut für Geographie der Westfälischen Wilhelms-Universität Münster: Münster, Germany, 2006.

25. Bernt, M.; Rink, D.; Holm, A. Gentrificationforschung in Ostdeutschland: Konzeptionelle Probleme und Forschungslücken. *Ber. Dtsch. Landeskd.* **2010**, *84*, 185–203.

26. Glatter, J.; Wiest, K. Gentrifizierungstendenzen unter den Bedingungen des Mietermarktes? Zum Wandel innenstadtnaher Quartiere in ostdeutschen Städten seit der Wiedervereinigung. In *Jahrbuch StadtRegion 2007/08. Schwerpunkt: Arme Reiche Stadt*; Gestring, N., Glasauer, H., Hannemann, C., Petrowsky, W., Pohlan, J., Eds.; Barbara Budrich: Opladen, Germany, 2008; pp. 55–72.

27. Rink, D.; Schneider, A.; Haase, A.; Wolff, M. Vom Leerstand zur Knappheit. *Kreuzer* **2017**, *2*, 28.

28. Rink, D. Leipzig—Stadt der Extreme. *Leipz. Blätter Sonderh.* **2015**, *1000*, 4–7.

29. Holm, A.; Marcinczak, S.; Ogrodowczyk, A. New-build gentrification in the post-socialist city: Łódź and Leipzig two decades after socialism. *Geografie* **2015**, *120*, 164–187.

30. Wiest, K.; Hill, A. Gentrification in ostdeutschen Cityrandgebieten? Theoretische Überlegungen zum empirischen Forschungsstand. *Ber. Dtsch. Landeskd.* **2004**, *78*, 25–39.

31. Marcuse, P. Gentrification, Abandonment, and Displacement: Connections, Causes, and Policy Responses in New York City. *J. Urban Contemp. Law* **1985**, *28*, 195–240.

32. Bolte, G.; Bunge, C.; Hornberg, C.; Köckler, H.; Mielck, A. (Eds.) *Umweltgerechtigkeit. Chancengleichheit bei Umwelt und Gesundheit: Konzepte, Datenlage und Handlungsperspektiven*; Hans Huber: Bern, Germany, 2012.

33. BMUB (Bundesministerium für Umwelt, Naturschutz, Bau und Reaktorsicherheit) (Ed.) *Quartiersmanagement Soziale Stadt. Eine Arbeitshilfe für die Umsetzung vor Ort*; BMUB: Berlin, Germany, 2016.

34. Hoffmann, A.; Gruehn, D. *Bedeutung von Freiräumen und Grünflächen in Deutschen Groß- und Mittelstädten für den Wert von Grundstücken und Immobilien*; LLP-Report 010; TU/LLP: Dortmund, Germany, 2010.

35. Liebelt, V.; Bartke, S.; Schwarz, N. Hedonic pricing analysis of the influence of urban green spaces onto residential prices: The case of Leipzig, Germany. *Eur. Plan. Stud.* **2018**, *26*, 133–157. [CrossRef]

36. Kabisch, N.; Frantzeskaki, N.; Pauleit, S.; Naumann, S.; Davis, M.; Artmann, M.; Haase, D.; Knapp, S.; Korn, H.; Stadler, J.; et al. Nature-based solutions to climate change mitigation and adaptation in urban areas: Perspectives on indicators, knowledge gaps, barriers, and opportunities for action. *Ecol. Soc.* **2016**, *21*, 39. [CrossRef]

37. Haase, D.; Kabisch, S.; Haase, A.; Andersson, E.; Banzhaf, E.; Baró, F.; Brenck, M.; Fischer, L.K.; Frantzeskaki, N.; Kabisch, N.; et al. Greening cities—To be socially inclusive? About the alleged paradox of society and ecology in cities. *Habitat Int.* **2017**, *64*, 41–48. [CrossRef]

38. Keeler, B.L.; Hamel, P.; McPhearson, T.; Hamann, M.H.; Donahue, M.L.; Prado, K.A.M.; Arkema, K.K.; Bratman, G.N.; Brauman, K.A.; Finlay, J.C.; et al. Social-ecological and technological factors moderate the value of urban nature. *Nat. Sustain.* **2019**, *2*, 29–38. [CrossRef]

39. Rigolon, A.; Németh, J. Green gentrification or 'just green enough': Do park location, size and function affect whether a place gentrifies or not? *Urban Stud.* **2019**. [CrossRef]

40. Gould, K.; Lewis, T. *Green Gentrification—Urban Sustainability and the Struggle for Environmental Justice*; Routledge: London, UK; New York, NY, USA, 2017.

41. Quastel, N. Political Ecologies of Gentrification. *Urban Geogr.* **2009**, *30*, 694–725. [CrossRef]

42. Goodling, E.; Green, J.; McClintock, N. Uneven development of the sustainable city: Shifting capital in Portland, Oregon. *Urban Geogr.* **2015**, *36*, 504–527. [CrossRef]

43. Checker, M. Wiped Out by the 'Greenwave': Environmental Gentrification and the Paradoxical Politics of Urban Sustainability. *City Soc.* **2011**, *23*, 210–229. [CrossRef]

44. Pearsall, H. From brown to green? Assessing social vulnerability to environmental gentrification in New York City. *Environ. Plan. C* **2010**, *28*, 872–886. [CrossRef]

45. Curran, W.; Hamilton, T. Just green enough: Contesting environmental gentrification in Greenpoint, Brooklyn. *Local Environ.* **2012**, *17*, 1027–1042. [CrossRef]

46. Bryson, J. The nature of gentrification. *Geogr. Compass* **2013**, *7*, 578–587. [CrossRef]

47. While, A.; Jonas, A.; Gibbs, D. The Environment and the Entrepreneurial City. Searching for the Urban ,Sustainability Fix' in Manchester and Leeds. *Int. J. Urban Reg. Res.* **2004**, *28*, 549–569. [CrossRef]

48. Essoka, J. The gentrifying effects of brownfields redevelopment. *West. J. Black Stud.* **2010**, *34*, 299–315.

49. Millington, N. From urban scar to 'park in the sky': Terrain vague, urban design, and the remaking of New York City's High Line Park. *Environ. Plan. A* **2015**, *47*, 2324–2338. [CrossRef]

50. Eckerd, A. Cleaning up without clearing out? A spatial assessment of environmental gentrification. *Urban Aff. Rev.* **2011**, *47*, 31–59. [CrossRef]

51. Kern, L. From toxic wreck to crunchy chic: Environmental gentrification through the body. *Environ. Plan. D* **2015**, *33*, 67–83. [CrossRef]

52. Wolch, J.R.; Byrne, J.; Newell, J.P. Urban green space, public health, and environmental justice: The challenge of making cities 'just green enough'. *Landsc. Urban Plan.* **2014**, *125*, 234–244. [CrossRef]

53. Holm, A. Ein ökosoziales Paradoxon: Stadtumbau und Gentrifizierung. *Politische Ökologie* **2011**, *124*, 45–53.

54. Banzhaf, H.; McCormick, E. *Moving beyond Cleanup: Identifying the Crucibles of Environmental Gentrification*; Andrew Young School of Policy Studies Research Paper Series: Working Paper 07-29.05; Georgia State University Atlanta: Atlanta, GA, USA, 2007.

55. Pearsall, H. Moving out or moving in? Resilience to environmental gentrification in New York City. *Local Environ.* **2012**, *17*, 1013–1026. [CrossRef]

56. Dale, A.; Newman, L. Sustainable development for some: Green urban development and affordability. *Local Environ.* **2009**, *14*, 669–681. [CrossRef]

57. Jaffé, E. How Parks Gentrify Neighborhoods, and How To Stop It. 2014. Available online: http://www.fastcodesign.com/3037135/evidence/how-parks-gentrify-neighborhoods-and-how-to-stop-it (accessed on 24 November 2018).

58. Pearsall, H. The contested future of Philadelphia's Reading Viaduct: Blight, neighborhood amenity or global attraction? In *Just Green Enough: Urban Development and Environmental Gentrification*; Curran, W., Hamilton, T., Eds.; Routledge: London, UK, 2017; pp. 197–208.

59. Anguelovski, I. From Toxic Sites to Parks as (Green) LULUs? New challenges of inequity, Privilege, Gentrification, and Exclusion for Urban Environmental Justice. *J. Plan. Lit.* **2016**, *31*, 23–36. [CrossRef]

60. Pearsall, H.; Anguelovski, I. Contesting and Resisting Environmental Gentrification: Responses to New Paradoxes and Challenges for Urban Environmental Justice. *Sociol. Res. Online* **2016**, *21*, 6. [CrossRef]

61. Haase, A.; Rink, D. Inner-city transformation between reurbanization and gentrification: Leipzig, eastern Germany. *Geografie* **2015**, *120*, 226–250.

62. Rink, D.; Haase, A.; Bernt, M.; Arndt, T.; Ludwig, J. *Urban Shrinkage in Leipzig, Germany*; Research Report, EU 7 FP Project Shrink Smart (Contract No. 225193), WP2. UFZ Report 1/2011; Helmholtz Centre for Environmental Research—UFZ: Leipzig, Germany, 2011.
63. Rink, D.; Behne, S. Grüne Zwischennutzungen in der wachsenden Stadt: Die Gestattungsvereinbarung in Leipzig. *Stat. Quart.* **2017**, *1*, 39–44.
64. Grossmann, K.; Arndt, T.; Haase, A.; Rink, D.; Steinführer, A. The influence of housing oversupply on residential segregation. Exploring the post-socialist city of Leipzig. *Urban Geogr.* **2015**, *36*, 550–577. [CrossRef]
65. Rink, D. Housing market and housing policy in Leipzig, presentation. 2019; unpublished.
66. Stadt Leipzig. *Wohnungspolitisches Konzept der Stadt Leipzig*; Fortschreibung 2015; Stadt Leipzig: Leipzig, Germany, 2015; Available online: http://www.leipzig.de/fileadmin/mediendatenbank/leipzig-de/Stadt/02.6_Dez6_Stadtentwicklung_Bau/61_Stadtplanungsamt/Stadtentwicklung/Leipzig_weiter_denken/Wohnen/Entwurf_Wohnungspolitisches_Konzept.pdf (accessed on 10 December 2019).
67. Stadt Leipzig. Monitoringbericht Wohnen 2018. 2019. Available online: https://static.leipzig.de/fileadmin/mediendatenbank/leipzig-de/Stadt/02.6_Dez6_Stadtentwicklung_Bau/61_StaStadtplanungs/Stadtentwicklung/Monitoring/Monitoting_Wohnen/Monitoringbericht_Wohnen_2018.pdf (accessed on 10 December 2019).
68. Stadt Leipzig. *Lebendig Grüne Stadt am Wasser (Green Space Strategy)*; Stadt Leipzig: Leipzig, Germany, 2017.
69. Stadt Leipzig. Konzeptioneller Stadtteilplan Leipziger Osten (KSP Ost). Stadt umbauen! In *Beiträge zur Stadtentwicklung 38*; Stadt Leipzig: Leipzig, Germany, 2001; Available online: http://www.leipziger-osten.de/fileadmin/UserFileMounts/Redakteure/Inhaltsbilder/Stadtumbau/KSP_LeipzigerOsten.pdf (accessed on 19 September 2019).
70. Stadt Leipzig. Integriertes Stadtentwicklungskonzept Leipziger Osten (STEK LeO). 2013. Available online: https://www.leipzig.de/fileadmin/mediendatenbank/leipzig-de/Stadt/02.6_Dez6_Stadtentwicklung_Bau/64_Amt_fuer_Stadterneuerung_und_Wohnungsbaufoerderung/Leipziger_Osten_Fotos/Publikationen/Broschuere_Integriertes_Stadtentwicklungskonzept_Leipzig.pdf (accessed on 19 October 2019).
71. Stadt Leipzig. Teilaufhebung der Sanierungssatzung, Leipzig-Reudnitz' Vorgesehen. 2015. Available online: http://www.leipzig.de/news/news/teilaufhebung-dersanierungssatzung-leipzig-reudnitz-vorgesehen/ (accessed on 30 March 2017).
72. Stadtarchiv Leipzig. *'Eilenburger Bahnhof' around 1947*; Photo Collection Signature: BA 1983/16747; Stadtarchiv Leipzig: Leipzig, Germany, 2019.
73. Stadt Leipzig. Masterplan Parkbogen Ost. 2017. Available online: http://www.leipzig.de/fileadmin/mediendatenbank/leipzig-de/Stadt/02.6_Dez6_Stadtentwicklung_Bau/64_Amt_fuer_Stadterneuerung_und_Wohnungsbaufoerderung/Projekte/Flyer_Broschueren_PDFs/Parkbogen_Ost/Parkbogen_Ost_Masterplan.pdf (accessed on 19 October 2019).
74. BBSR (Bundesinstitut für Bau-, Stadt- und Raumforschung). Neuer Freiraum auf innerstädtischer Bahnbrache. Leipzig-Reudnitz: 'Stadtteilpark Reudnitz'. 2008. Available online: https://www.nationale-stadtentwicklungspolitik.de/NSP/SharedDocs/Projekte/WSProjekte_DE/Leipzig_Reudnitz_Stadtteilpark_Reudnitz.html (accessed on 12 October 2019).
75. Stadt Leipzig. *Statistischer Quartalsbericht III/2016 [09/16]*; Leipziger Statistik und Stadtforschung; Stadt Leipzig: Leipzig, Germany, 2016.
76. Richter, B.; Grunewald, K.; Meinel, G. Analyse von Wegedistanzen in Städten zur Verifizierung des Ökosystemleistungsindikators, Erreichbarkeit städtischer Grünflächen'. *AGIT J. Angew. Geoinformatik* **2016**, *2*, 472–480.
77. WHO Europe (World Health Organization). *Urban Green Spaces and Health. A Review of Evidence*; WHO Regional Office for Europe: Copenhagen, Denmark, 2016.
78. Amt für Statistik und Wahlen Leipzig. Ortsteilkatalog 2018. Strukturdaten der Ortsteile und Stadtbezirke. 2019. Available online: https://static.leipzig.de/fileadmin/mediendatenbank/leipzig-de/Stadt/02.1_Dez1_Allgemeine_Verwaltung/12_Statistik_und_Wahlen/Raumbezug/Ortsteilkatalog/Ortsteilkatalog_2018.pdf (accessed on 13 October 2019).
79. SystemImmo. Immobilien als Kapitalanlage. Leipzig. 2016. Available online: https://www.systemimmo.de/leipzig-iq-apartments (accessed on 5 June 2016).

80. Capital. Immobilienpreise und Mietspiegel: Leipzig-Reudnitz-Thonberg. Immobilien-Kompass-Karte. 2016. Available online: http://www.capital.de/immobilien-kompass/leipzig/suedost/reudnitz-thonberg.html (accessed on 28 December 2016).

81. Rometsch, J. Hohe Mieten im Neubau. Leipzig Wächst Jedes Jahr um 3000 Wohnungen—Aber Das Reicht Noch Nicht. *Leipziger Volkszeitung.* 22 March 2017. Available online: http://www.lvz.de/Leipzig/Lokales/Leipzig-waechstjedes-Jahr-um-3000-Wohnungen-aber-das-reicht-noch-nicht (accessed on 22 March 2017).

82. Immaxi Immobilien. Leipzig Reudnitz: Pulsierender Stadtteil Östlich der City. 2017. Available online: http://www.immaxi.de/reudnitz (accessed on 2 February 2017).

83. Immaxi Immobilien. Immobilien Entwicklung 2019 in Leipzig. 2019. Available online: https://www.immaxi.de/immobilien-trend-2019 (accessed on 27 September 2019).

84. PISA Immobilienmanagement (Ed.) *Marktbericht 2016. Der Wohnungsmarkt in Leipzig. Miete-Wohneigentum-Investment;* W&R Media KG: Leipzig, Germany, 2016.

85. Amt für Statistik und Wahlen Leipzig. Kommunale Bürgerumfrage 2017. Ergebnisbericht. 2018. Available online: https://static.leipzig.de/fileadmin/mediendatenbank/leipzig-de/Stadt/02.1_Dez1_Allgemeine_Verwaltung/12_Statistik_und_Wahlen/Stadtforschung/Buergerumfrage2017.pdf (accessed on 20 September 2019).

86. Amt für Statistik und Wahlen Leipzig. Leipzig-Informationssystem LIS: Kommunale Bürgerumfragen. 2019. Available online: http://statistik.leipzig.de/statpubl/index.aspx?cat=2&rub=1&obj=0 (accessed on 23 September 2019).

87. Holl, J. Leipzig Instead of Berlin. *New York Magazine*, 25 October 2013. Available online: http://nymag.com/nymag/travel/leipzig/(accessed on 12 January 2017).

88. Raymunt, M. From Leipzig to Hypezig—Hipsters eye new playground. *Chicago Tribune*, 21 February 2014.

89. Immobilien Scout. Immobiliensuche. Reudnitz-Thonberg. 2016. Available online: https://www.immobilienscout24.de/Suche/S-T/Wohnung-Miete/Sachsen/Leipzig/Reudnitz-Thonberg?enteredFrom=one_step_search (accessed on 16 November 2016).

90. PWIB Wohnungs-Infobörse. Entwicklung der Immobilienpreise für Wohnungen in Leipzig. 2016. Available online: http://www.wohnungsboerse.net/mietspiegel-Leipzig/7390 (accessed on 1 November 2016).

91. Immobilien Scout. Mietpreisentwicklung für Wohnungen. 2019. Available online: https://atlas.immobilienscout24.de/orte/deutschland/sachsen/leipzig/reudnitz-thonberg?marketingFocus=APARTMENT_RENT#/preisentwicklung (accessed on 2 October 2019).

92. Bunce, S. *Sustainability Policy, Planning and Gentrification in Cities;* Routledge: London, UK, 2017.

93. Stadt Leipzig. *Gebietsbezogenes Integriertes Handlungskonzept EFRE. Nachhaltige Stadtentwicklung Leipziger Osten;* Stadt Leipzig: Leipzig, Germany, 2015.

94. Pearsall, H. New Directions in Urban Environmental/Green Gentrification Research. In *The Handbook of Gentrification Studies;* Lees, L., Phillips, M., Eds.; Edward Elgar: Cheltenham/Northampton, UK, 2018; pp. 329–345.

95. Amt für Statistik und Wahlen Leipzig. Statistischer Quartalsbericht IV/2016 [4/17]. Leipziger Statistik und Stadtforschung. 2017. Available online: https://static.leipzig.de/fileadmin/mediendatenbank/leipzig-de/Stadt/02.1_Dez1_Allgemeine_Verwaltung/12_Statistik_und_Wahlen/Statistik/Statistischer_Quartalsbericht_Leipzig_2016_4.pdf (accessed on 2 October 2019).

96. Goossens, C. 'They Have Ruined Everything': Green Gentrification in Ghent, Belgium. *Sociol. Urbana e Rural.* **2019**, *119*, 62–81. [CrossRef]

land

Article

How Are Urban Green Spaces and Residential Development Related? A Synopsis of Multi-Perspective Analyses for Leipzig, Germany

Nina Schwarz [1,2,*], Annegret Haase [3], Dagmar Haase [2,4], Nadja Kabisch [3,4], Sigrun Kabisch [3], Veronika Liebelt [2,5,6], Dieter Rink [3], Michael W. Strohbach [7], Juliane Welz [8] and Manuel Wolff [3,4]

[1] Department Urban and Regional Planning and Geo-Information Management (PGM), Faculty of Geo-Information Science and Earth Observation (ITC), University of Twente, 7500 AE Enschede, The Netherlands

[2] Department Computational Landscape Ecology, Helmholtz Centre for Environmental Research—UFZ, 04318 Leipzig, Germany; dagmar.haase@ufz.de (D.H.); veronika.liebelt@idiv.de (V.L.)

[3] Department of Urban and Environmental Sociology, Helmholtz Centre for Environmental Research—UFZ, 04318 Leipzig, Germany; annegret.haase@ufz.de (A.H.); nadja.kabisch@geo.hu-berlin.de (N.K.); sigrun.kabisch@ufz.de (S.K.); dieter.rink@ufz.de (D.R.); manuel.wolff@hu-berlin.de (M.W.)

[4] Department of Geography, Humboldt-Universität zu Berlin, 10099 Berlin, Germany

[5] German Centre for Integrative Biodiversity Research (iDiv), Biodiversity Economics, Halle-Jena-Leipzig, 04103 Leipzig, Germany

[6] Department of Economics, Leipzig University, 04109 Leipzig, Germany

[7] Landscape Ecology and Environmental Systems Analysis, Institute of Geoecology, Technische Universität Braunschweig, 38106 Braunschweig, Germany; m.strohbach@tu-braunschweig.de

[8] Fraunhofer Center for International Management and Knowledge Economy IMW, 04109 Leipzig, Germany; juliane.welz@imw.fraunhofer.de

* Correspondence: n.schwarz@utwente.nl; Tel.: +31-534-891-455

check for updates

Citation: Schwarz, N.; Haase, A.; Haase, D.; Kabisch, N.; Kabisch, S.; Liebelt, V.; Rink, D.; Strohbach, M.W.; Welz, J.; Wolff, M. How Are Urban Green Spaces and Residential Development Related? A Synopsis of Multi-Perspective Analyses for Leipzig, Germany. *Land* **2021**, *10*, 630. https://doi.org/10.3390/land10060630

Academic Editor: Thomas Panagopoulos

Received: 24 March 2021
Accepted: 9 June 2021
Published: 12 June 2021

Publisher's Note: MDPI stays neutral with regard to jurisdictional claims in published maps and institutional affiliations.

Abstract: The relationship between urban green spaces (UGS) and residential development is complex: UGS have positive and negative immediate impacts on residents' well-being, residential location choice, housing, and land markets. Property owners and real estate agents might consider how prospective clients perceive UGS and act accordingly, while urban planners influence UGS location and management as well as aim at steering the built environment. Typically, studies focus on one of these perspectives at a time. Here, we provide a synopsis of results from studies, taking different perspectives for a single case study: Leipzig, Germany. We summarise and discuss the findings of eight studies on UGS and residential development. In detail, these studies focus on spatial pattern analysis, hedonic pricing analysis, mixed-methods studies on experts' perspectives, surveys, and choice experiments exploring residents' perceptions of UGS. We reflect on the feasibility of deriving a synthesis out of these independent studies and to what extent context matters. We conclude that both triangulating of data and methods, as well as long-term and context-sensitive studies are needed to explain the interlinkages between UGS and residential development and their context dependency.

Keywords: multi-method approach; residential development; urban development; urban green

1. Introduction

Green spaces are a vital part of cities. Many types of urban green spaces (UGS) exist, i.e., green locations in cities which provide opportunities for recreation or relaxation, or for just being there or passing through. This encompasses diverse UGS such as parks, cemeteries, urban forests, gardens, street trees, allotments, or agricultural land [1]. The importance of UGS for urban residents is reflected in many policies at different international, national, and local levels, most prominently in the Sustainable Development Goals. For example, SDG 11 says "Make cities inclusive, safe, resilient and sustainable", and target

31

11.7 states that "by 2030 general access to safe, inclusive and accessible green spaces and public spaces will be guaranteed" [2].

The effects of UGS are manifold and clearly context-dependent, with significant differences between Global North and Global South contexts. In this paper, we focus on Leipzig as a second-tier European city for investigating the relation between UGS and residential development. Second-tier cities in Europe represent the backbone of the large urban system. They are not capital cities, have more than 200,000 inhabitants, and operate as hubs for the economy, education, culture, and mobility [3]. Typically, second-tier cities developed in the period of industrialisation. UGS consist of a number of larger spaces in central locations of city-wide importance and a network of UGS in the different neighbourhoods of the city. In Leipzig, one of the large UGS is the floodplain forest that runs from northwest to southeast through the city, passing its central parts as well.

The effects of UGS can be observed on different levels, from immediate effects on residents to residential location choice and decisions taken by planners and developers. First, UGS can directly or indirectly impact health, social integration, and well-being of residents. A number of effects of UGS on individuals' well-being were documented, for instance, on physical and psychological health, using a variety of indicators such as body mass index, stress level, birth outcomes, or depression, amongst others [4,5]. Furthermore, UGS deliver additional urban ecosystem services beneficial for residents, including local climate regulation moderating extreme events such as heat stress or food provision [6–8]. However, UGS can also have both direct and indirect negative effects on residents, such as causing allergic reactions due to pollen [9], being habitats for disease vectors, or reducing perceived or actual safety [10], for so-called wild spaces: [11]. Regarding the social value of UGS, several authors confirm their potential in bringing together members of different social backgrounds, even only fleetingly, and therefore promoting community integration [12,13].

Second, these positive and negative effects of UGS can influence residents' location choices as well as residential duration [14]. Indeed, residents are often willing to pay more for living closer to UGS (e.g., [15]), and effects of UGS on housing prices have been extensively studied in hedonic pricing studies. These studies reveal statistically significant positive effects of various UGS characteristics on selling or renting prices (review by [16–18]) as well as urban land prices [19]. Such market effects of UGS may force low-income residents to move away to areas with lower environmental quality, a process discussed as eco-gentrification [5,20,21].

Third, the potential effects of UGS are increasingly considered by urban planners and developers. For instance, urban renewal schemes sometimes employed in cities of the Global North, involving the creation of new UGS, assume that UGS increase the attractiveness of neighbourhoods for residents with higher incomes (e.g., [22–24]).

The relationship between UGS and residential development is context-dependent, i.e., related to demographic, social, economic, or political characteristics and their interrelations in the respective cities or neighbourhoods. In particular, there are different impacts on residential development in shrinking and in growing cities. In shrinking cities, a decline in population and an increase in housing vacancy and empty commercial spaces is often met with the demolition of buildings. This demolition, in turn, is an opportunity to create new UGS but—due to a lack of funding—often as unmanaged UGS with spontaneous vegetation termed "urban wilderness" [25]. However, such "urban wilderness" has been acknowledged poorly by the local population and rather causes a decrease in investments and fosters the decline of the neighbourhood [25]. Notwithstanding, greening and establishing new UGS in shrinking neighbourhoods have been a common strategy to increase life quality and make people stay or attract new residents [6,26,27]. On the contrary, in growing cities, UGS are a scarce and valued resource, sometimes at the centre of land-use conflicts between economic interests of further housing and commercial/industrial construction versus nature conservation [22,28]. What is more, historical factors such as designing urban parks for specific purposes [29] and traditions of gardening [30] also frame the relationship between UGS and residential development.

Clearly, we need to consider different actors and their perspectives to better understand the relationship of UGS and residential development. Residents perceive and make use (or decide not to make use) of UGS and choose where to live in a city depending on their preferences, their budget, and other constraints (e.g., [31,32]). Urban planners consider UGS in their planning, including the creation or renouncement of UGS, and use planning instruments to influence the built environment, such as zoning [33]. Property owners decide upon when, where, and how to invest and what price to ask for, while real estate agents consider how to advertise for houses, flats, etc.

Empirical studies on UGS often have a strong background in one of the disciplines involved, for instance, demography, economy, planning, and geography [34]. Thus, they typically focus on one piece of the jigsaw puzzle at a time, following their individual research question, for instance, by investigating residents' decision-making. A more comprehensive picture about the relationship between UGS and residential development is also limited by the heterogeneity on how UGS are conceptualised in such studies (e.g., [6]) and the heterogeneity in viewpoints among different socio-economic groups (e.g., [35]). Finally, neither the perceptions of urban residents nor the composition of UGS are static: Urban populations and the urban fabrics are constantly changing, and so is the structure of UGS, including their visual appearance [36].

Leipzig, Germany, is a city where many of the aspects mentioned above have been studied in recent decades. This paper attempts to condense findings of the relationship between UGS and residential development based on eight empirical studies (detailed in Section 2.3). Specifically, we try to answer the following two research questions: 1. What are the relationships between UGS and residential developments in Leipzig beyond different disciplinary boundaries? 2. How can a synopsis of several studies for a given city (here, Leipzig) contribute to a better understanding of those relationships? With the aim of knowledge integration, we provide a synoptic view of several independent studies and use the triangulation of methods and data as an approach. According to Thurmond ([37], p. 254), the benefits of a synthesis of different methodological approaches in a kind of "a posteriori" mixed-method triangulation include "increasing confidence in research data, creating innovative ways of understanding a phenomenon, revealing unique findings, challenging or integrating theories, and providing a clearer understanding of the problem". By the conscious combination of different theoretical concepts, qualitative and quantitative methods, empirical and statistical as well as GIS data sets, and results [38,39], we intend to obtain a broader, more diverse, and deeper insight into the relationship between UGS and residential development and to look for consistencies or discrepancies. Using a synopsis of different studies carried out in the same study area, we aim to gain new insights from previous research and provide a more balanced picture compared to what a single study could achieve.

We conceptualize "residential development" as the dynamic shaping a city's built-environment in physical and structural terms such as changes in the number, size, or quality of housing; changes in costs/prices; changes in occupation by different groups of residents; and the choice of people for housing and the related residential mobility. In this understanding, the term encompasses different perspectives, including structures, processes, and decisions related to housing and the residential built environment. Similarly, we also consider different types of UGS here, be they privately owned (e.g., courtyards) or public green spaces (e.g., parks), maintained UGS, or not (vegetated brownfields). This broad understanding of residential development and UGS allows us to investigate various ways in which UGS and residential developments interact.

After a brief introduction of the case study (Section 2.1), we summarise our approach (Section 2.2) and systematically describe and structure the main findings of eight studies as well as the methods used (Section 2.3). Then, we combine the individual studies by triangulating their findings in a causal-loop diagram, thus answering our first research question (Section 3). Based on this diagram, we discuss our second research question, specifically the added value of the synopsis and, furthermore, to what extent we can move from a synopsis (i.e., a brief summary of the studies) to a synthesis (i.e., consolidating findings into a deeper understanding) (Section 4). Finally, we draw conclusions for both further research and urban planning strategies (Section 5).

2. Methodological Design

2.1. Leipzig—The Case Study

As already mentioned, Leipzig is a second-tier European city. The city of Leipzig encompasses 297 sq.km in total, with about 12 sq.km of forests and 24 sq.km of other green spaces within its administrative boundary (estimated from [40], Figure 1), together representing 12% of the city area in 2012. A floodplain forest stretches from the south to the northwest of the city. As an additional value, Leipzig is embedded in a lake district with large green spaces where former open-pit mines have been converted into an attractive recreation area. Leipzig's population number has changed profoundly in recent decades (Figure 2). Before the German reunification in 1990, the population number declined from 611,000 in the year 1982 to 511,079 in 1990. The post-socialist transformation was then, accompanied by large-scale deindustrialisation, leading to a high unemployment. In consequence, many people moved out to other prosperous regions in Germany. Population loss continued in the course of post-socialist transformation, with the lowest population number in the year 1998 (437,000) before it stabilized in the 2000s and dynamically increased in the 2010s (2020: 605,400). Measures like converting inner-urban brownfields into urban forests [41] and other UGS in order to increase residential quality during population decline have given way to questions of re-densification [42] as well as reurbanisation and to some extent gentrification [43].

As mentioned above, Leipzig's development is being characterized by long-term shrinkage from the 1960s to the end of the 1990s. Shrinkage was the most massive during the 1990s, when the city lost 100,000 inhabitants in 10 years. After a short stabilization period, Leipzig experienced dynamic regrowth since 2010, with yearly growth rates of >2%. While the housing market in the time of shrinkage suffered from abandonment and high vacancies, it turned into a contested market with rising housing costs since 2010. New construction takes place today mainly in the upmarket segment, and the availability of modestly priced housing is decreasing. Since 2000, Leipzig's inner city has seen reurbanization and an exchange of the residential population [43]. Being extreme in the scope of both shrinkage and regrowth, in its basic development, Leipzig stands for a larger group of second-rank cities across Europe [28].

2.2. Analysis of Existing Studies

We have investigated Leipzig's highly dynamic development in terms of population growth and decline (Figure 2) and accompanying green space development in various studies. Here, we selected those eight studies that mainly address the relation between residential and urban green space development (compilation in Figure 3). In our in-depth analysis, we excluded studies that solely addressed one of the two components (for instance, describing land use or cover classifications) or that did not use empirical data (for instance, modelling studies).

First, we compiled the primary outcomes and characteristics of the methodologies for all eight studies (compilation step in Figure 3). In an iterative process, we then addressed the complementarity of the studies (instead of only comparing them) along three poles (synopsis step in Figure 3): synthesising the main results of each study and reflecting on the potential matches in time and space of the chosen analysis (see Box 1) resulted in key relationships of the elements (UGS and socio-demographic attributes). We discovered a complex web of links between elements that the studies used to characterise UGS, such as size, proximity, or accessibility, as well as residential development, including residential location choice, residential quality, and the real estate market. We visualised these links in a causal loop diagram as interactions between different variables around the relationship between residential development and urban green spaces (Figure 4). Therefore, Figure 4 is an outcome of the communication process in our author group. This was possible as, in the compilation phase, we accounted for the diversity of UGS types, socio-demographic groups, and spatio-temporal scales, which have been studied in each of the studies. At least one co-author of each study was involved in writing this synopsis and answered questions related to the individual studies. Following a double bottleneck approach, we finally derived three viewpoints that synthesise our findings (synthesis step in Figure 3).

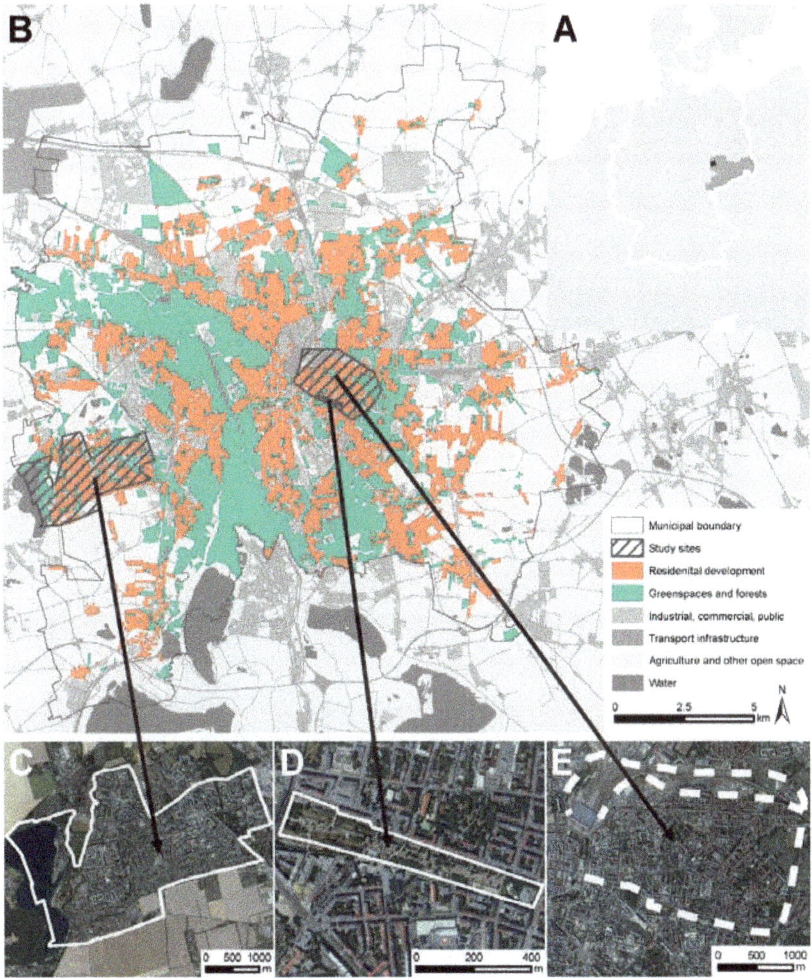

Figure 1. (**A**) The location of Leipzig in the state of Saxony (dark gray) in Germany (white line). (**B**) The urban fabric and green areas of the city of Leipzig in 2012. (**C**) The location of Study 1. (**D**) The location of study 4. (**E**) The location of study 5. (Source basemap A: ESRI Basemap Europe; B: Urban Atlas LCLU 2012 [40]; C: GeoSN).

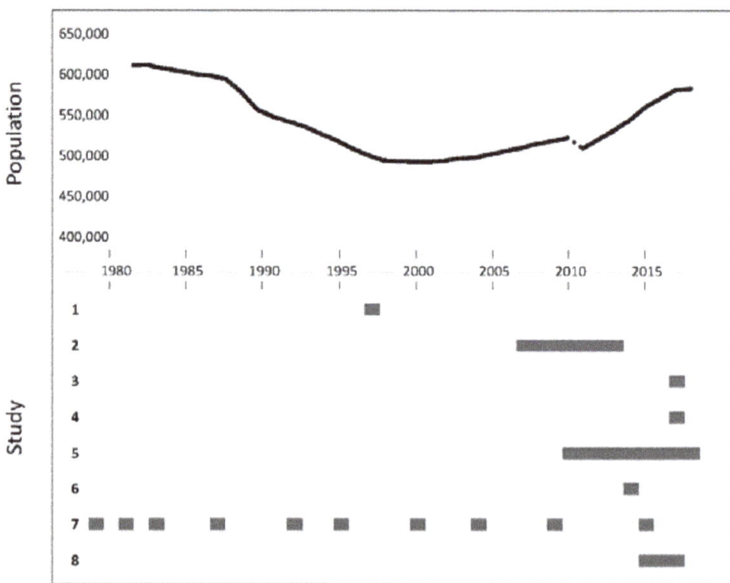

Figure 2. Population development (upper part) in Leipzig and timing of the eight studies (lower part). Source of population numbers: Regional Database Germany and Statistical Office for the Free State of Saxony; corrected to fit the current extent of the municipality. The drop in population after 2011 is due to corrections after the 2011 census.

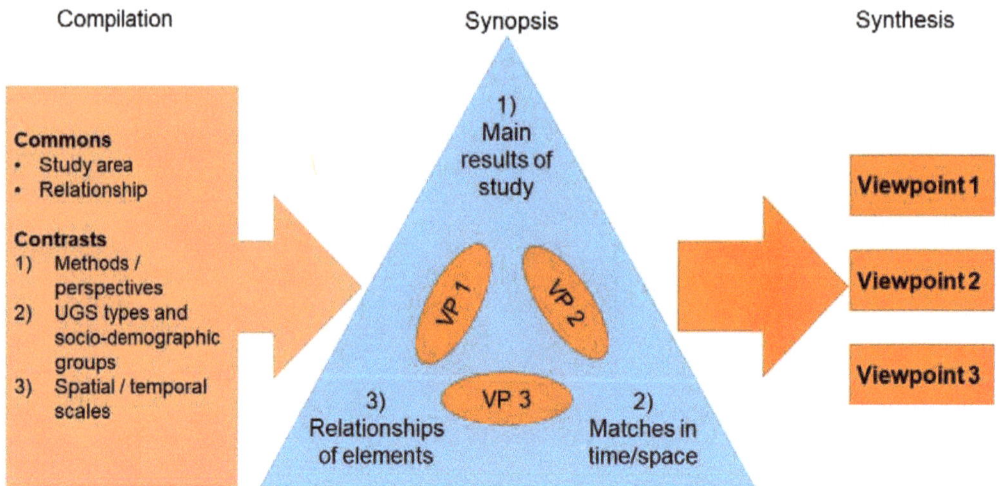

Figure 3. Approach of the study via compilation and synopsis, leading to three viewpoints (VPs) on synthesis.

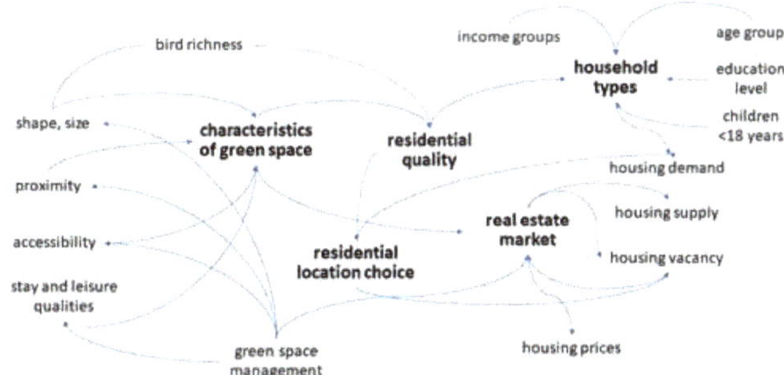

Figure 4. Interacting variables around the relationship of UGS and residential development according to the eight case studies. Dotted arrow: link cannot yet be precisely defined as causality or correlation. Core terms used in the eight studies (see Box 1) are marked bold.

2.3. Overview of the Studies

The eight studies discuss various aspects of UGS in Leipzig between 1979 to 2018. Based on the key methods and data sets employed, we grouped these eight studies into three categories as a base for triangulation of methods: (1) analysing spatial patterns, (2) investigating experts' perspectives, (3) accessing residents' perceptions (see Box 1). Two of the eight studies referred to spatial patterns: Strohbach et al. [44] analysed bird species richness and how it related to the socio-economic status of residents at the scale of all Leipzig's districts, with UGS being the mediator. Liebelt et al. [45,46] employed hedonic pricing analysis to quantify the relationship between UGS and housing costs. Three studies are based on experts' perspectives. One study deals with the broader design concept of specific UGS sites: Lene-Voigt park, which shall become part of the larger Parkbogen Ost greenway [47]. The other study refers only to Lene-Voigt Park [26]. Finally, one expert-focused study investigates the acceptance of urban forests created in the 2010s in different neighbourhoods across the city [41,48,49]. Three of the eight empirical studies analyse residents' perspectives. Two of them refer to the municipality, as such [50,51], whereas another study applied a long-term research design to the district Grünau [52,53]. By experts, we refer to a person who is very knowledgeable about the particular area of residential-green development. Residents live in Leipzig and the particular areas/districts we studied.

These studies have been carried out by researchers with different disciplinary backgrounds such as ecology (study 1), economics (study 2), sociology (studies 3 to 5), and geography (studies 6 to 8) (Box 1).

Box 1. Structured overview of the eight studies.

Study 1. Strohbach et al. 2009 on spatial patterns
time frame. 1997
Spatial scale. Citywide
Focus of content. Correlation between bird diversity near homes and socioeconomic and demographic characteristics of residents.
Methods. Spatial and statistical analysis based on bird atlas data, land use maps, and municipal statistics.
Types of UGS and residents covered and differentiated. UGS: land uses differentiated (parks, allotments, cemeteries, grassland, forest). Residents: differentiated by income level, age and household size based on statistical data from the municipality on district level.
Main results. High species richness along floodplains with higher income and higher population density
High-density and less-well-off districts are species-poor. Prefabricated large housing estates have high shares of UGS but are not particularly species-rich.

Box 1. *Cont.*

Study 2. Liebelt et al. 2018a; 2018b on spatial patterns
Time frame. 2007–2013
Spatial scale. Citywide
Focus of content. (2018a): Influence of urban green spaces on the rental and sale prices of residential property. (2018b): District-level preferences as revealed from property prices depending on district characteristics.
Methods. Hedonic pricing analysis on the city (2018a) and district level (2018b).
Types of UGS and residents covered and differentiated. UGS: not differentiated. Residents: 2018a not differentiated, 2018b: indirectly differentiated at the district level based on various district properties.
Main results. Impact on housing prices mainly by type of housing; housing size, distance to city centre, and balcony; UGS less important but statistically significant at the city level. District level: the direction of effects (increase or decrease of prices in relation to UGS variables) varies. Example: Higher distance to UGS within a district: prices of housing close to UGS increase.
Study 3. Konzack 2017 on experts' perspectives
Time frame. 2017
Spatial scale. 5 km length
Focus of content. Importance of "Parkbogen Ost" project on housing market and local investors, housing situation around Lene-Voigt Park.
Methods. Mapping and spatial analysis, expert, and investor interviews.
Types of UGS and residents covered and differentiated. UGS: not differentiated, Residents: indirectly differentiated via housing conditions.
Main results. Residential buildings close to park in top condition; buildings from Wilhelminian time; often renovated, incl. balconies. Experts and civil society stakeholders see numerous benefits of new park. First conflicts visible with residents demanding apartments instead of new urban forest.
Study 4. Ali et al. 2020 on experts' perspectives
Time frame. 2017
Spatial scale. 11 hectares
Focus of content. Impact of Lene-Voigt park on residential change in inner-city neighbourhood.
Methods. Statistical analysis, mapping, in situ observations, analysis of real estate announcements, expert interviews.
Types of UGS and residents covered and differentiated. UGS: one park. Residents: indirectly differentiated via population, migration, and housing market data.
Main results. Housing rents close to the park increased and now slightly higher than further away. Park is seen as attractor for the area; arrival of younger middle-class households. Concerns about future high-end renovations.
Study 5. Rink/Arndt 2016, Mathey et al. 2018, Schmidt et. al. 2018 on experts' perspectives
Time frame. 2010–2018
Spatial scale. 13.6 hectares
Focus of content. Acceptance of newly created urban forests and impact on residential change, particularly housing and commercial vacancies.
Methods. Statistical analysis, mapping, in situ observations, analysis of real estate announcements, expert interviews.
Types of UGS and residents covered and differentiated. UGS: newly created urban forests. Residents: differentiated via gender, age, income, qualification, household type.
Main results. New urban forests mostly accepted. Despite less biodiversity, residents prefer new urban forests over brownfields due to recreation options. Vacancies reduced; price effects difficult to detect.
Study 6. Welz et al. 2017 on residents' perspectives
Time frame. 2014
Spatial scale. City-wide
Focus of content. Residential mobility in and to Leipzig; focus on housing preferences of urban immigrants.
Methods. Quantitative household survey; statistical analysis focussing on residential profiles.
Types of UGS and residents covered and differentiated. UGS: not differentiated. Residents: age groups, household types etc.
Main results. UGS not among factors triggering moving out of current home nor for residential choice
Proximity to UGS important for families with child(ren), single-parent families, and pensioner couples.
Private green: balcony preferred over garden or courtyard.

Box 1. *Cont.*

> **Study 7. Kabisch et al. 2013, and 2018 on residents' perspectives**
> *Time frame.* 1979–2020
> *Spatial scale.* 1000 hectares
> *Focus of content.* Social, built and ecological development of a large housing estate.
> *Methods.* Long-term observation of the estate since its erection: 11 comparable surveys, quantitative analysis, observations, mapping.
> *Types of UGS and residents covered and differentiated.* All UGS in the area (old parks, meadows, alley, playgrounds, green yards, pocket parks, pocket gardens, allotment gardens). Residents: age and socioeconomic groups, household types etc.
> *Main results.* UGS appreciated in evaluation of whole estate, specifically by households with children,
> UGS not decisive for selecting an apartment, but important in decision to stay or to move, in some neighbourhood's access to close-by UGS was restricted in favour of local residents (security issues).
> **Study 8. Scheuer et al. 2018 on residents' perspectives**
> *Time frame.* 2015–2017
> *Spatial scale.* City-wide
> *Focus of content.* Housing preferences.
> *Methods.* Choice experiments leading to decision trees.
> *Types of UGS and residents covered and differentiated.* UGS: parks and UGS in low/further distance, house green/house garden. Residents: differentiated by age groups, household types etc.; bias of young households
> *Main results.* Rent, location, and type of housing highest impact on accepting or declining a flat UGS as neighbourhood amenities of minor importance.

3. Synopsis of the Eight Studies

In Section 3, we summarise the key similarities of the eight studies and answer our first research question: What are the relationships between UGS and residential developments in Leipzig beyond different disciplinary boundaries? Section 4 later discusses different perspectives on how to interpret these findings. The findings of all eight studies are summarized in Box 1. The links between essential elements used in the studies are visualized in Figure 3. Three studies (studies 6, 7, 8) show that UGS belong to the criteria influencing residential location choice apart from flat or house characteristics and locational issues. For instance, being close to parks was of higher importance for families with children, single parents, and pensioner couples compared to other household types (study 6). However, variables describing the flat or house (e.g., flat size, type of housing) are the most important factors influencing which flat or house to decide for (next to rental prices) (studies 6 and 8).

UGS are also a factor influencing residential quality, i.e., the quality of life in a specific neighbourhood (through the shaping of new green areas (study 5)), and satisfaction of residents with their neighbourhood (study 7). Again, UGS are particularly important for specific household types, e.g., families and single parents (study 7).

Figure 4 shows the close and variegated bundle of relations between UGS and housing/real estate market processes. Housing prices are, under the circumstances, related to UGS (studies 2, 3, 4) as well. A statistical relationship between UGS size, distance to the next UGS and their shape (in the sense of rugged versus compact), and housing prices can be observed (study 2). Both the socio-economic characteristics of the residential population and the availability of UGS in the districts play a role in explaining the price effects of UGS at the district level. For instance, in those districts with high mean distances of housing units to UGS, people pay more for living closer to UGS (study 2). Knowing or at least anticipating such relations, developers use UGS to advertise for housing, especially in contexts where housing markets get more contested and benefits of real estate businesses increase (study 4).

Consequently, residential buildings close to newly created UGS are renovated and include elements that increase living standards, such as balconies (studies 3 and 4). Developing UGS and subsequent rise in the neighbourhoods' attractiveness put pressure on surrounding lots for upgrading and even turning UGS into residential buildings (study 3).

Green space planning also anticipates such developments and uses the creation or renovation of UGS as a tool to steer the future development of specific neighbourhoods (studies 3, 4, 5) and to reduce housing vacancies (study 5). Likewise, developers take UGS management into consideration (studies 3 and 4).

Study 1 investigated a city-wide relationship of UGS and residential development, namely that bird species richness is correlated with the income of residents. The connecting factor is UGS: High bird species richness can be found in districts with a large amount of UGS, and these areas also tend to be of higher income. On the contrary, higher building density leaves little room for green space and thus supports fewer species. As a result, the opportunity to experience high species richness around the home is unevenly distributed in Leipzig. Whether species-rich neighbourhoods attract more well-off residents or vice versa cannot be distinguished.

4. Discussion: From Synopsis to Synthesis?

Here, we discuss our second research question: How can a synopsis of several studies for a given city (City of Leipzig) contribute to a better understanding of those relationships? When working on the synopsis of the studies by triangulating methods and data, we realised that we had different viewpoints on how to interpret their findings: Viewpoint 1: We may condense some of the results of the combined studies into general insights into the relationship of UGS and residential development in Leipzig. This is despite the different temporal and spatial contexts of the studies and methods employed. Viewpoint 2: We may refrain from attempting a synthesis and rather reflect on the context- and methods-dependency of the individual studies. Viewpoint 3: In addition to viewpoints 1 and 2, we identified remaining challenges.

4.1. Viewpoint 1: A Synthesis

The first viewpoint states that triangulating the findings and methods of the eight studies concerned permits drawing some valid overarching consistencies that could not have been identified by the individual studies. First, a majority of studies dealing with the whole city identified common patterns of UGS and residential quality. Large and high-quality green spaces (with higher bird biodiversity)—especially parks—are close to high-quality and high-price residential estates ([44], study 1; [23], study 4). Conversely, several studies were able to point out low-income residential areas that are—majorly—underserved with attractive large green spaces (studies 3 and 4). Both patterns echo findings already described in the literature (general review: [5]; for postsocialist cities: [54]; for Global South cities: [55]). Residential vacancy contributes to the unattractiveness of these residential areas. Thus, through combining different methods (i.e., from qualitative and qualitative analyses) and perspectives (i.e., experts perspective, residents perspective, spatial patterns-perspective), we can confirm and verify that UGS does matter for residential quality.

Second, we corroborate that UGS and residential location choices are interrelated with each other. Proximity to UGS impacts rent prices and is an important feature of the positive assessment of the residential environment and satisfaction. Recent other hedonic studies also found that UGS proximity increases selling prices [56,57] but counter-intuitively decreases renting prices [58]. Furthermore, upgrading residential areas and increasing immigration to residential areas surrounded by renewed and maintained UGS has been observed, indicating eco-gentrification [5,20,21].

Third, concluding from spatial relations to causal relations is challenging: Several of the studies found that—at the household level—the vicinity of UGS is not one of the primary decision variables for renting or buying a flat (studies 6, 7, 8). The studies showed that the interaction of UGS with residential choice, preference, or relocation decision is second-ranked; what is much more important is other locational factors, costs, and type of flat/building stock (e.g., summarized by [59]). Nevertheless, the decision to stay in a neighbourhood is strongly related to UGS (study 7). Furthermore, study 8 shows that UGS becomes important once other housing-related variables are matching.

The synthesis perspective can include the different temporal and spatial scales. We found that the spatial scale (see Box 1) at which a study was undertaken matters; household, neighbourhood, and city-scale are crucial to coming to the aforementioned three conclusions. Combining these scales permits drawing a more comprehensive picture of the relationship between UGS and residential development than the individual studies. In a similar vein, the different scientific disciplines and their specific methods enrich the overall synthesis as patterns found in one study can be verified using the results of another, and gaps, as well as conflicting results, help to get a deeper understanding.

4.2. Viewpoint 2: Context Matters

The second viewpoint focuses on the importance of the spatio-temporal context. Residential development in Leipzig has changed substantially within the time frame of the eight studies (Figure 2). Consequently, the fact that the studies were conducted at different points in time could very well be one reason for diverging results. For instance, the study on bird richness is based on data from 1997, a time when Leipzig was still losing population. On the contrary, the hedonic pricing analysis was conducted for a later period in which the population number was already increasing. The two studies tackling the perspective of experts took place when population numbers were growing rapidly. Moving from a shrinking to a growing population has changed densities [28] and completely altered the opportunities of creating UGS. Instead, it puts pressure on planned or interim UGS, such as temporal green spaces [60–62]. The temporal context might also have effects on individual studies and the interpretation of their results. For example, the hedonic pricing approach assumes that prices can clearly distinguish between goods. However, during the time period covered by the hedonic pricing analysis, housing prices just started to rise. Therefore, the price differences between a highly valued housing unit and a non-attractive housing unit might have been rather small compared to other housing markets. This notwithstanding, these price differences might still be too large for low-income households (as indicated by study 4). Obviously, conducting these different studies at the same time, or—even better—as long-term research would be more suitable to deal with such dynamics over time. This temporal context also comes into play when choosing an appropriate method. It is crucial to be aware of whether all factors potentially influencing a decision (e.g., location choice) are assessed independently from each other (as, for instance, in typical survey questions) or combined (e.g., choice experiments).

Spatial scales have been used in a twofold function. First, they are used for defining the spatial extent of the study, such as a neighbourhood, UGS, household, or the whole city. Second, the resolution—such as trees or specific land use categories—is used to define the analytical unit that is studied. The way the two functions of scale—as a container and analytical lens—are interrelated can be expressed in the degree of spatial heterogeneity [63]. Of course, studies with a small spatial extent such as a single park can go into much more detail, for instance, about the perspectives of different actors (studies 3 and 4, 7), but need to neglect other areas that follow a completely different logic. When focussing on the city scale and a specific spatial resolution, distinctive patterns and dynamics, which are observable only at more (dis)aggregated scales, such as the region or a park, are hidden. The hedonic pricing analysis comparing different spatial scales explicitly tackled this aspect ([46], here: study 2). This scale-dependency can lead to a fallacy, i.e., the false perceptions of the spatial organization when macro-level data are used to draw conclusions at a more disaggregated level [64].

4.3. Viewpoint 3: Heterogeneity Challenges

Although we may align ourselves on different points on the "synthesis" versus "context matters"—scale, we agree on what the important open questions are that need further investigation. These are related to the heterogeneity of residents and UGS, the context-dependency of factors, and the potential of empirical material to re-question existing theories.

Comparing the outcomes of the studies made us more aware of the potentials and pitfalls of the employed methods. We discussed the extent to which the methods can account for the heterogeneity of residents and UGS. Residents are heterogeneous in many respects, for instance, forming household types according to their income, age group, education level, or presence of children. Such socio-demographic characteristics may influence the importance of UGS for residential quality, location choice, and the real estate market.

UGS are also clearly heterogeneous, for instance, in terms of their functions: cemeteries, forests, or parks are used for different purposes, and residents might have different attitudes towards them [65]. What is more, differentiating UGS into types based on functionality (e.g., a cemetery is often differentiated from a park, but there are cemeteries that are used for recreation) might not be enough, since biodiversity within UGS can again follow a completely different pattern. Research has shown that while people value species-rich UGS [66], perceived and real species richness are not necessarily correlated ([67] but see also [68]). In many of the studies, different UGS were pragmatically grouped in one summary category, i.e., "urban green".

The studies presented here provide some indication of heterogeneity in the relationship between UGS and residential development—something that is rarely studied (see also the reviews by [69,70]). Some of the methods seem more suited for this task than others. Surveys and choice experiments can unveil heterogeneity among residents if respondents are stratified and enough data points are available. However, a classic hedonic pricing study cannot cope with heterogeneity among residents at all, since it investigates housing units' prices and typically cannot relate these findings to who has bought or rented a specific housing unit.

We believe that a combination of different methods is required to arrive at a comprehensive knowledge integration about the relationship between UGS and residential development. Addressing many different variables describing heterogeneous UGS in surveys or choice experiments is in principle possible but would lead to lengthy questionnaires. Likewise, the number of independent variables in a hedonic pricing analysis would increase dramatically if one would want to investigate the distance to playgrounds, parks, forests, or other types of UGS separately. Other methods employed to study the quality of life in cities could also be helpful here [71].

Future research also needs to refer more explicitly to the question of which residents actually can afford to consider UGS in their location choice. Only focusing on the attractiveness of neighbourhoods and preferences for location choice neglects financial and other constraints.

Our attempt to synthesize the results of our various studies might also represent the starting point of a theorising-back process in which local and context-specific findings are synthesised in a way that they may challenge existing theories and eventually lead to their advancement or adaptation [72]. To fully allow for such a process, developing robust conceptual frames is essential.

5. Concluding Remarks

With our study, we contribute towards a more integrated perspective on UGS and residential development, as, for instance, asked for by [34]. We brought together eight empirical studies to investigate the relationship between UGS and residential development for the city of Leipzig, Germany, to answer two research questions: What are the relationships between UGS and residential developments in Leipzig beyond different disciplinary boundaries? How can a synopsis of several studies for a given city (here, Leipzig) contribute to a better understanding of those relationships? The eight studies were grouped into three broad categories: analysing spatial patterns with correlations and regression approaches, experts' perspective (interviews, mapping, and observation), and residents' perspectives (mainly surveys and choice experiments). Various methods were used in the different studies embedded in different spatio-temporal contexts. When discussing the findings on the relationship between UGS and residential development, we condensed common patterns and stressed the context-dependency of individual studies. We decided to take the above-mentioned differences as an advantage, and instead of offering one answer at the end of this paper, we offer a more nuanced answer in the form of viewpoints when answering our second research question. We decided on this solution to show the added value of multiple interpretations of cases. In terms of future research needed, we argue for a stronger focus on the heterogeneity of residents and UGS as well as factors constraining residential location choice. Finally, we highlight that a structured comparison of thematic studies might produce very different types of knowledge: We found arguments for digging deeper into the content of residential and UGS development in the form of synthesis but also to deal with the reasons for different interpretations due to context-sensitivity. Comparisons can produce cross-cutting insights, also for variegated targets (e.g., searching for synthetic knowledge or seeking to explain differences).

Very few studies so far combine several methods; for instance, [18] combine a hedonic pricing analysis with a survey. Depending on the specific context and purpose of the study, multi-method approaches could together provide a broader picture with multiple access roads on the relationship between UGS and residential development. Furthermore, establishing more long-term social-ecological research and monitoring sites would help to better understand processes, their respective impacts, and their context-dependency.

Here, we have focussed on one single case study city and the complexity therein. Comparisons of case studies are still rare but needed to elicit general patterns from them in meta-studies, as, for instance, called for in the land-use science in general [73]. With our approach, we have focused on the role of human actors and decision-making within the causal chain [73] and highlighted how complex such an endeavour is even within a single case study.

Investigating the relationship of UGS and residential development in parallel in the Global North and Global South might again challenge our understanding completely: UGS likely are under much higher pressure to provide housing for a rapidly growing urban population, and their effects on residential development could be quite different. It has been argued that cities in the Global South constitute a type of city distinct from cities in the Global North [74], with very diverse narratives about, for instance, urban resilience towards environmental risks [75]. Thus, the relations we investigated for one specific case study could change drastically for other contexts, indicating great potential for future research.

Author Contributions: Conceptualisation, N.S., A.H., D.H., N.K., S.K., V.L., D.R., M.W.S., J.W. and M.W.; Writing—original draft, N.S., A.H. and D.H.; Writing—review and editing, N.K., S.K., V.L., D.R., M.W.S., J.W. and M.W. Visualisation, D.H., N.S., M.W.S. and M.W. All authors have read and agreed to the published version of the manuscript.

Funding: M.W.S. acknowledges funding by the program "Science for Sustainable Development" of the Volkswagen Foundation and the Ministry for Science and Culture of Lower Saxony (METAPOLIS, grant no. ZN3121). D.H., N.K., M.W.S. and N.S. acknowledge funding by the European Commission's Sixth Framework Programme for research (EC FP6 Contract No. 036921). V.L. gratefully acknowledges the support of iDiv funded by the German Research Foundation (DFG–FZT 118, 202548816), Federal Ministry of Education and Research (marEEShift, grant no. 01LC1826A), ESCALATE. This research was partly supported by the project ENABLE, funded through the 2015–2016 BiodivERsA COFUND call for research proposals, with the national funders The Swedish Research Council for Environment, Agricultural Sciences, and Spatial Planning, Swedish Environmental Protection Agency, German Aeronautics and Space Research Centre, National Science Centre (Poland), The Research Council of Norway, and the Spanish Ministry of Economy and Competitiveness. J.W., A.H., and S.K. acknowledge funding by the City of Leipzig. N.K.'s work was supported by the German Federal Ministry of Education and Research, Grant/Award Number: 01LN1705A; Environmental-Health Interactions in Cities (GreenEquityHEALTH)—Challenges for Human Well-Being under Global Changes.

Institutional Review Board Statement: Not applicable.

Informed Consent Statement: Not applicable.

Data Availability Statement: Data sharing not applicable.

Acknowledgments: We wish to thank all residents and experts for their participation in our studies and providing data, specifically the Federal Agency for Nature Conservation as well as the department for urban planning of the city of Leipzig. V.L. thanks Immobilien Scout GmbH for providing data. Furthermore, we are thankful to four anonymous reviewers for their comments on a previous version of this manuscript and to Lina Zech for her help in editing the manuscript.

Conflicts of Interest: The authors declare no conflict of interest.

References

1. Boulton, C.; Dedekorkut-Howes, A.; Byrne, J. Factors shaping urban greenspace provision: A systematic review of the literature. *Landsc. Urban Plan.* **2018**, *178*, 82–101. [CrossRef]
2. UN General Assembly. Transforming Our World: The 2030 Agenda for Sustainable Development, 21 October 2015, A/RES/70/1. Available online: https://www.refworld.org/docid/57b6e3e44.html (accessed on 11 June 2021).
3. Camagni, R.; Capello, R. *Second Rank Cities in Europe. Structural Dynamics and Growth Potential*; Routledge: London, UK, 2016.
4. Nesbitt, L.; Hotte, N.; Barron, S.; Cowan, J.; Sheppard, S.R. The social and economic value of cultural ecosystem services provided by urban forests in North America: A review and suggestions for future research. *Urban For. Urban Green.* **2017**, *25*, 103–111. [CrossRef]
5. Wolch, J.R.; Byrne, J.; Newell, J.P. Urban green space, public health, and environmental justice: The challenge of making cities 'just green enough'. *Landsc. Urban Plan.* **2014**, *125*, 234–244. [CrossRef]
6. Haase, D.; Larondelle, N.; Andersson, E.; Artmann, M.; Borgström, S.; Breuste, J.; Gomez-Baggethun, E.; Gren, Å.; Hamstead, Z.; Hansen, R.; et al. A Quantitative Review of Urban Ecosystem Service Assessments: Concepts, Models, and Implementation. *Ambio* **2014**, *43*, 413–433. [CrossRef]
7. Schwarz, N.; Moretti, M.; Bugalho, M.N.; Davies, Z.G.; Haase, D.; Hack, J.; Hof, A.; Melero, Y.; Pett, T.; Knapp, S. Understanding biodiversity-ecosystem service relationships in urban areas: A comprehensive literature review. *Ecosyst. Serv.* **2017**, *27*, 161–171. [CrossRef]
8. du Toit, M.J.; Cilliers, S.S.; Dallimer, M.; Goddard, M.; Guenat, S.; Cornelius, S.F. Urban green infrastructure and ecosystem services in sub-Saharan Africa. *Landsc. Urban Plan.* **2018**, *180*, 249–261. [CrossRef]
9. Cariñanos, P.; Grilo, F.; Pinho, P.; Casares-Porcel, M.; Branquinho, C.; Acil, N.; Andreucci, M.B.; Anjos, A.; Bianco, P.M.; Brini, S.; et al. Estimation of the Allergenic Potential of Urban Trees and Urban Parks: Towards the Healthy Design of Urban Green Spaces of the Future. *Int. J. Environ. Res. Public Health* **2019**, *16*, 1357. [CrossRef] [PubMed]
10. Lyytimäki, J.; Petersen, L.K.; Normander, B.; Bezák, P. Nature as a nuisance? Ecosystem services and disservices to urban lifestyle. *Environ. Sci.* **2008**, *5*, 161–172. [CrossRef]
11. Threlfall, C.G.; Kendal, D. The distinct ecological and social roles that wild spaces play in urban ecosystems. *Urban For. Urban Green.* **2018**, *29*, 348–356. [CrossRef]
12. Barbosa, O.; Tratalos, J.A.; Armsworth, P.R.; Davies, R.G.; Fuller, R.A.; Johnson, P.; Gaston, K.J. Who benefits from access to green space? A case study from Sheffield, UK. *Landsc. Urban Plan.* **2007**, *83*, 187–195. [CrossRef]
13. Krellenberg, K.; Welz, J.; Reyes-Päcke, S. Urban green areas and their potential for social interaction—A case study of a socio-economically mixed neighbourhood in Santiago de Chile. *Habitat Int.* **2014**, *44*, 11–21. [CrossRef]
14. Łaszkiewicz, E.; Kronenberg, J.; Marcińczak, S. Attached to or bound to a place? The impact of green space availability on residential duration: The environmental justice perspective. *Ecosyst. Serv.* **2018**, *30*, 309–317. [CrossRef]
15. Tu, G.; Abildtrup, J.; Garcia, S. Preferences for urban green spaces and peri-urban forests: An analysis of stated residential choices. *Landsc. Urban Plan.* **2016**, *148*, 120–131. [CrossRef]
16. Brander, L.; Koetse, M.J. The value of urban open space: Meta-analyses of contingent valuation and hedonic pricing results. *J. Environ. Manag.* **2011**, *92*, 2763–2773. [CrossRef]
17. Czembrowski, P.; Kronenberg, J. Hedonic pricing and different urban green space types and sizes: Insights into the discussion on valuing ecosystem services. *Landsc. Urban Plan.* **2016**, *146*, 11–19. [CrossRef]
18. Ardeshiri, A.; Ardeshiri, M.; Radfar, M.; Shormasty, O.H. The values and benefits of environmental elements on housing rents. *Habitat Int.* **2016**, *55*, 67–78. [CrossRef]
19. Gruehn, D. Economic Valuation of Urban Open Spaces and their Contribution to Life Quality in European Cities. In *Geografia Konferencja*; 2008; pp. 59–66. Available online: Cejsh.icm.edu.pl/cejsh/element/bwmeta1.element.hdl_11089_1797/c/gruehn_Economic_Valuation_of_Urban_Open_Spaces.pdf (accessed on 11 June 2021).
20. Dooling, S. Ecological Gentrification: A Research Agenda Exploring Justice in the City. *Int. J. Urban Reg. Res.* **2009**, *33*, 621–639. [CrossRef]
21. Haase, D.; Kabisch, S.; Haase, A.; Andersson, E.; Banzhaf, E.; Baró, F.; Brenck, M.; Fischer, L.K.; Frantzeskaki, N.; Kabisch, N.; et al. Greening cities—To be socially inclusive? About the alleged paradox of society and ecology in cities. *Habitat Int.* **2017**, *64*, 41–48. [CrossRef]

22. Swanwick, C.; Dunnett, N.; Woolley, H. Nature, Role and Value of Green Space in Towns and Cities: An Overview. *Built Environ.* **2003**, *29*, 94–106. [CrossRef]

23. Bryson, J. The Nature of Gentrification. *Geogr. Compass* **2013**, *7*, 578–587. [CrossRef]

24. Checker, M. Wiped Out by the "Greenwave": Environmental Gentrification and the Paradoxical Politics of Urban Sustainability. *City Soc.* **2011**, *23*, 210–229. [CrossRef]

25. Rink, D. Wilderness: The Nature of Urban Shrinkage? The Debate on Urban Restructuring and Restoration in Eastern Germany. *Nat. Cult.* **2009**, *4*, 275–292. [CrossRef]

26. Ali, L.; Haase, A.; Heiland, S. Gentrification through Green Regeneration? Analyzing the Interaction between Inner-City Green Space Development and Neighborhood Change in the Context of Regrowth: The Case of Lene-Voigt-Park in Leipzig, Eastern Germany. *Land* **2020**, *9*, 24. [CrossRef]

27. Danford, R.S.; Strohbach, M.W.; Warren, P.S.; Ryan, R.L. Active Greening or Rewilding the city: How does the intention behind small pockets of urban green affect use? *Urban For. Urban Green.* **2018**, *29*, 377–383. [CrossRef]

28. Haase, A.; Wolff, M.; Rink, D. From shrinkage to regrowth. The nexus between urban dynamics, land use change and ecosystem service provision. In *Urban Transformations—Sustainable Urban Development towards Resource Efficiency, Quality of Life and Resilience, Future City Series*; Kabisch, S., Koch, F., Gawel, E., Haase, A., Knapp, S., Krellenberg, K., Zehnsdorf, A., Eds.; Springer: Berlin/Heidelberg, Germany, 2018; pp. 197–219.

29. Cranz, G. *The Politics of Park Design: A History of Urban Parks in America*; MIT Press: Cambridge, UK, 1982; Available online: https://direct.mit.edu/books/book/5052/The-Politics-of-Park-DesignA-History-of-Urban (accessed on 11 June 2021).

30. Ignatieva, M. Plant Material for Urban Landscapes in the Era of Globalization: Roots, Challenges and Innovative Solutions. In *Applied Urban Ecology*; Wiley: Hoboken, NJ, USA, 2011; pp. 139–151.

31. Bouzarovski, S.; Haase, A.; Hall, R.; Steinführer, A.; Kabisch, S.; Ogden, P.E. Household Structure, Migration Trends, and Residential Preferences in Inner-city León, Spain: Unpacking the Demographies of Reurbanization. *Urban Geogr.* **2010**, *31*, 211–235. [CrossRef]

32. Storper, M.; Manville, M. Behaviour, Preferences and Cities: Urban Theory and Urban Resurgence. *Urban Stud.* **2006**, *43*, 1247–1274. [CrossRef]

33. Nuissl, H.; Couch, C. Lines of Defence: Policies for the Control of Urban Sprawl. In *Urban Sprawl in Europe*; Wiley: New York, NY, USA, 2008; pp. 217–241.

34. Clark, W.A. Residential mobility in context: Interpreting behavior in the housing market. *Pap. Rev. Sociol.* **2017**, *102*, 575. [CrossRef]

35. Borth, K.; Summers, R. Segmentation of Homebuyers by Location Choice Preferences. *Hous. Policy Debate* **2017**, *28*, 428–442. [CrossRef]

36. Pickett, S.; Cadenasso, M.; Grove, J.; Boone, C.G.; Groffman, P.M.; Irwin, E.; Kaushal, S.S.; Marshall, V.; McGrath, B.P.; Nilon, C.; et al. Urban ecological systems: Scientific foundations and a decade of progress. *J. Environ. Manag.* **2011**, *92*, 331–362. [CrossRef] [PubMed]

37. Thurmond, V.A. The Point of Triangulation. *J. Nurs. Sch.* **2001**, *33*, 253–258. [CrossRef] [PubMed]

38. Altrichter, H.; Feldman, A.; Posch, P.; Somekh, B. *Teachers Investigate their Work*; An Introduction to Action Research Across the Professions; Routledge: London, UK, 2006.

39. O'Donoghue, T.; Punch, K. *Qualitative Educational Research in Action: Doing and Reflecting*; Routledge: London, UK, 2003.

40. EEA European Environment Agency. Urban Atlas. 2015. Available online: https://www.eea.europa.eu/data-and-maps/data/copernicus-land-monitoring-service-urban-atlas (accessed on 11 June 2021).

41. Rink, D.; Arndt, T. Investigating perception of green structure configuration for afforestation in urban brownfield development by visual methods—A case study in Leipzig, Germany. *Urban For. Urban Green.* **2016**, *15*, 65–74. [CrossRef]

42. Wolff, M.; Haase, A.; Haase, D.; Kabisch, N. The impact of urban regrowth on the built environment. *Urban Stud.* **2016**, *54*, 2683–2700. [CrossRef]

43. Haase, A.; Rink, D. Inner-city transformation between reurbanization and gentrification: Leipzig, eastern Germany. *Geografie.* **2015**, *120*, 226–250. [CrossRef]

44. Strohbach, M.W.; Haase, D.; Kabisch, N. Birds and the City: Urban Biodiversity, Land Use, and Socioeconomics. *Ecol. Soc.* **2009**, *14*. [CrossRef]

45. Liebelt, V.; Bartke, S.; Schwarz, N. Hedonic pricing analysis of the influence of urban green spaces onto residential prices: The case of Leipzig, Germany. *Eur. Plan. Stud.* **2018**, *26*, 133–157. [CrossRef]

46. Liebelt, V.; Bartke, S.; Schwarz, N. Revealing Preferences for Urban Green Spaces: A Scale-sensitive Hedonic Pricing Analysis for the City of Leipzig. *Ecol. Econ.* **2018**, *146*, 536–548. [CrossRef]

47. Konzack, A. On the Role of Greening Projects in Urban Development—An Analysis of the "Parkbogen Ost" in Leipzig. Master's Thesis, Institute for Geography, Humboldt University, Berlin, Germany, 2017. Available online: https://www.geographie.hu-berlin.de/de/abteilungen/landschaftsoekologie/projekte/abgeschlossen/greensurge (accessed on 11 June 2021).

48. Mathey, J.; Arndt, T.; Banse, J.; Rink, D. Public perception of spontaneous vegetation on brownfields in urban areas—Results from surveys in Dresden and Leipzig (Germany). *Urban For. Urban Green.* **2018**, *29*, 384–392. [CrossRef]

49. Schmidt, C.; Böttner, S.; Meier, M.; Schmidt, U. Urbane Wälder, Modul Stadtumbau (Urban Forests, Module Urban Restructuring), TU Dresden, Report. 2018. Available online: http://urbane-waelder.de/index_htm_files/Modul_Stadtumbau.pdf (accessed on 11 June 2021).

50. Scheuer, S.; Haase, D.; Haase, A.; Kabisch, N.; Wolff, M.; Schwarz, N.; Großmann, K. Combining tacit knowledge elicitation with the SilverKnETs tool and random forests—The example of residential housing choices in Leipzig. *Environ. Plan. B Urban Anal. City Sci.* **2018**, *47*, 400–416. [CrossRef]

51. Welz, J.; Haase, A.; Kabisch, S. Zuzugsmagnet Grossstadt—Profile aktueller Zuwanderer. *disP Plan. Rev.* **2017**, *53*, 18–32. [CrossRef]

52. Kabisch, S.; Grossmann, K. Challenges for large housing estates in light of population decline and ageing: Results of a long-term survey in East Germany. *Habitat Int.* **2013**, *39*, 232–239. [CrossRef]

53. Kabisch, S.; Ueberham, M.; Schlink, U.; Hertel, D.; Mohamdeen, A.M.S. Local residential quality from an inter-disciplinary perspective: Combining individual perception and micrometeorological factors. In *Urban Transformations—Sustainable Urban Development through Resource Efficiency, Quality of Life and Resilience*; Future City 10; Kabisch, S., Koch, F., Gawel, E., Haase, A., Knapp, S., Krellenberg, K., Nivala, J., Zehnsdorf, A., Eds.; Springer: Berlin/Heidelberg, Germany, 2018; pp. 235–255.

54. Kronenberg, J.; Haase, A.; Łaszkiewicz, E.; Antal, A.; Baravikova, A.; Biernacka, M.; Dushkova, D.; Filčak, R.; Haase, D.; Ignatieva, M.; et al. Environmental justice in the context of urban green space availability, accessibility, and attractiveness in postsocialist cities. *Cities* **2020**, *106*, 102862. [CrossRef]

55. Rigolon, A.; Browning, M.H.E.M.; Lee, K.; Shin, S. Access to Urban Green Space in Cities of the Global South: A Systematic Literature Review. *Urban Sci.* **2018**, *2*, 67. [CrossRef]

56. Franco, S.F.; Macdonald, J.L. Measurement and valuation of urban greenness: Remote sensing and hedonic applications to Lisbon, Portugal. *Reg. Sci. Urban Econ.* **2018**, *72*, 156–180. [CrossRef]

57. Plant, L.; Rambaldi, A.; Sipe, N. Evaluating Revealed Preferences for Street Tree Cover Targets: A Business Case for Collaborative Investment in Leafier Streetscapes in Brisbane, Australia. *Ecol. Econ.* **2017**, *134*, 238–249. [CrossRef]

58. Donovan, G.; Butry, D.T. The effect of urban trees on the rental price of single-family homes in Portland, Oregon. *Urban For. Urban Green.* **2011**, *10*, 163–168. [CrossRef]

59. Schirmer, P.M.; Van Eggermond, M.A.; Axhausen, K.W. The role of location in residential location choice models: A review of literature. *J. Transp. Land Use* **2014**, *7*, 3–21. [CrossRef]

60. Rall, E.L.; Haase, D. Creative intervention in a dynamic city: A sustainability assessment of an interim use strategy for brownfields in Leipzig, Germany. *Landsc. Urban Plan.* **2011**, *100*, 189–201. [CrossRef]

61. Rink, D.; Behne, S. Grüne Zwischennutzungen in der wachsenden Stadt: Die Gestattungsvereinbarung in Leipzig, (Green interim uses in the growing city: The permission agreement in Leipzig). *Statistischer Quartalsbericht* **2017**, *1*, 39–44.

62. Haase, D.; Kabisch, N.; Haase, A. Endless Urban Growth? On the Mismatch of Population, Household and Urban Land Area Growth and Its Effects on the Urban Debate. *PLoS ONE* **2013**, *8*, e66531. [CrossRef]

63. Andersson, E.; McPhearson, T.; Kremer, P.; Gomez-Baggethun, E.; Haase, D.; Tuvendal, M.; Wurster, D. Scale and context dependence of ecosystem service providing units. *Ecosyst. Serv.* **2015**, *12*, 157–164. [CrossRef]

64. Resende, G.M. Spatial Dimensions of Economic Growth in Brazil. *ISRN Econ.* **2013**, *2013*, 1–19. [CrossRef]

65. Schindler, M.; Le Texier, M.; Caruso, G. Spatial sorting, attitudes and the use of green space in Brussels. *Urban For. Urban Green.* **2018**, *31*, 169–184. [CrossRef]

66. Fuller, A.R.; Irvine, K.N.; Devine-Wright, P.; Warren, P.H.; Gaston, K.J. Psychological benefits of greenspace increase with biodiversity. *Biol. Lett.* **2007**, *3*, 390–394. [CrossRef]

67. Dallimer, M.; Irvine, K.N.; Skinner, A.M.J.; Davies, Z.G.; Rouquette, J.R.; Maltby, L.; Warren, P.H.; Armsworth, P.R.; Gaston, K.J. Biodiversity and the Feel-Good Factor: Understanding Associations between Self-Reported Human Well-being and Species Richness. *Bioscience* **2012**, *62*, 47–55. [CrossRef]

68. Fischer, L.; Honold, J.; Botzat, A.; Brinkmeyer, D.; Cvejić, R.; Delshammar, T.; Elands, B.; Haase, D.; Kabisch, N.; Karle, S.; et al. Recreational ecosystem services in European cities: Sociocultural and geographical contexts matter for park use. *Ecosyst. Serv.* **2018**, *31*, 455–467. [CrossRef]

69. Kloek, M.E.; Buijs, A.E.; Boersema, J.J.; Schouten, M.G. Crossing Borders: Review of Concepts and Approaches in Research on Greenspace, Immigration and Society in Northwest European Countries. *Landsc. Res.* **2013**, *38*, 117–140. [CrossRef]

70. Roy, N.; Dubé, R.; Després, C.; Freitas, A.; Légaré, F. Choosing between staying at home or moving: A systematic review of factors influencing housing decisions among frail older adults. *PLoS ONE* **2018**, *13*, e0189266. [CrossRef] [PubMed]

71. D'Acci, L. Monetary, Subjective and Quantitative Approaches to Assess Urban Quality of Life and Pleasantness in Cities (Hedonic Price, Willingness-to-Pay, Positional Value, Life Satisfaction, Isobenefit Lines). *Soc. Indic. Res.* **2014**, *115*, 531–559. [CrossRef]

72. Wolff, M.; Haase, A. Viewpoint: Dealing with trade-offs in comparative urban studies. *Cities* **2020**, *96*, 102417. [CrossRef]

73. Van Vliet, J.; Magliocca, N.R.; Büchner, B.; Cook, E.; Benayas, J.M.R.; Ellis, E.C.; Heinimann, A.; Keys, E.; Lee, T.M.; Liu, J.; et al. Meta-studies in land use science: Current coverage and prospects. *Ambio* **2016**, *45*, 15–28. [CrossRef]

74. Schindler, S. Towards a paradigm of Southern urbanism. *City* **2017**, *21*, 47–64. [CrossRef]

75. Borie, M.; Pelling, M.; Ziervogel, G.; Hyams, K. Mapping narratives of urban resilience in the global south. *Glob. Environ. Chang.* **2019**, *54*, 203–213. [CrossRef]

Article

Landscape-Based Visions as Powerful Boundary Objects in Spatial Planning: Lessons from Three Dutch Projects

Sabine van Rooij [1], Wim Timmermans [1], Onno Roosenschoon [1], Saskia Keesstra [1,2], Marjolein Sterk [1] and Bas Pedroli [3,*]

[1] Wageningen Environmental Research, P.O. Box 47 6700 AA Wageningen, The Netherlands; sabine.vanrooij@wur.nl (S.v.R.); wim.timmermans@wur.nl (W.T.); onno.roosenschoon@wur.nl (O.R.); saskia.keesstra@wur.nl (S.K.); marjolein.sterk@wur.nl (M.S.)

[2] Civil, Surveying and Environmental Engineering, The University of Newcastle, Callaghan 2308, Australia

[3] Land Use Planning Group, Wageningen University, P.O. Box 47 6700 AA Wageningen, The Netherlands

[*] Correspondence: bas.pedroli@wur.nl; Tel.: +31-317-485-396

Abstract: In a context of a rapidly changing livability of towns and countryside, climate change and biodiversity decrease, this paper introduces a landscape-based planning approach to regional spatial policy challenges allowing a regime shift towards a future land system resilient to external pressures. The concept of nature-based solutions and transition theory are combined in this approach, in which co-created normative future visions serve as boundary concepts. Rather than as an object in itself, the landscape is considered as a comprehensive principle, to which all spatial processes are inherently related. We illustrate this approach with three projects in the Netherlands in which landscape-based visions were used to guide the land transition, going beyond the traditional nature-based solutions. The projects studied show that a shared long-term future landscape vision is a powerful boundary concept and a crucial source of inspiration for a coherent design approach to solve today's spatial planning problems. Further, they show that cherishing abiotic differences in the landscape enhances sustainable and resilient landscapes, that co-creation in the social network is a prerequisite for shared solutions, and that a landscape-based approach enhances future-proof land-use transitions to adaptive, circular, and biodiverse landscapes.

Keywords: nature-based solutions; transition; regional planning; landscape management; future vision; circularity; resource management; biodiversity

Citation: van Rooij, S.; Timmermans, W.; Roosenschoon, O.; Keesstra, S.; Sterk, M.; Pedroli, B. Landscape-Based Visions as Powerful Boundary Objects in Spatial Planning: Lessons from Three Dutch Projects. *Land* **2021**, *10*, 16. https://dx.doi.org/10.3390/land10010016

Received: 19 November 2020
Accepted: 23 December 2020
Published: 28 December 2020

Publisher's Note: MDPI stays neutral with regard to jurisdictional claims in published maps and institutional affiliations.

1. Introduction

The livability of the city and the countryside is under great pressure all over the world. Cities face major challenges, such as expanding housing development, densification, flood prevention, and biodiversity decline [1]. In rural areas, waterlogging and drought are increasingly uncertain factors for agriculture. Drought is also a problem for nature areas, with nitrogen deposition recently labeled an acute additional threat to biodiversity [2,3]. At the same time, there is a growing awareness among governments, citizens and the business community that the planning tasks facing these challenges cannot be realized without a coherent vision of the future of our landscape [4]. In addition to emergency measures for the short term, well-thought-out long-term adjustments to changing conditions are necessary. Reference is made here to transition theory [5]: once a real transition is required, a regime shift should take place. In a spatial context, the transition should be based on the landscape as a vehicle for spatial planning since the landscape provides both the physical and the perceived baseline for spatial development [6]. Given the complex character of the current landscape planning issues, climate-robust biodiversity and circularity are key principles for responsible landscape adaptation in such spatial processes [7]. In this context, it is crucial that all actors take part in the spatial planning process and that they

47

can contribute their local knowledge and opinions so that the emerging strategies are not only tailored to the biophysical, but also to the social landscape [8].

Finding solutions for this ill-defined Gordian knot of challenges spatial planning is facing means that actions of different sectors and actors need to be aligned. When each sector or actor approaches the challenges as a well-structured sectoral problem, there is little chance that visions and actions will come together in an adequate strategy for a region [4,6]. Another approach is to acknowledge that the solution for the knot of challenges is a boundary concept. A solution for these challenges is not part of our present-day infrastructure, our conventional social arrangements and technologies [9]. A future vision for a region, based on a set of conceptual principles such as "landscape-based" can serve as a boundary concept, offering common ground to the scientists and practitioners with different backgrounds, values and interests that were involved [6]. A boundary concept is flexible enough to adapt to local needs and to different perspectives, but also robust enough to maintain conceptual coherence across scientific disciplines and across the science-practice boundary [9,10]. Planning literature [11] reflects this shift from single fixed quantitative targets, via multiple, qualitative concepts to the guidance of interactions in a multi-stakeholder perspective and finally even fuzzy planning approaches.

Nature-based solutions are spatial interventions that use natural materials and especially natural principles. Nature-based solutions can be considered an umbrella term for all related applications of ecosystem services, natural capital and "lessons from nature" [12]. For governments and companies (construction, infrastructure), these types of solutions are an encouraging reference because multiple goals can be combined in one measure. When applying nature-based solutions, the emphasis is often on solving problems in urban and rural planning by making use of the processes and patterns of nature [13]. So far, solutions are often sought within the scale level of the project area. For adaptation to changing circumstances in urban and rural areas, it is important to search for solutions on a broader scale level, the landscape level, because the city and the countryside are interconnected systems [6]. Nature-based solutions are ideally suited to link those scales and to use a systems approach that uses the natural processes instead of working against them.

In order to strive for an integral solution in a spatial context and incorporating the social environment [14], this paper introduces a landscape-based planning approach aiming at regional spatial policies allowing for a community-based transition in a relatively urbanized countryside resilient to various external pressures. Three examples from the Netherlands serve as an illustration of three topical transition issues in a metropolitan context: climate adaptation, biodiversity enhancement and flood risk. Questions to be answered are how a landscape-based approach adds to nature-based solutions, how the abiotic landscape can be considered in terms of opportunities, how future visions can support sound transitions and how local and regional planning can reinforce each other.

2. Landscape: A Concept rather than an Object

2.1. Multiplicity of Landscape

Landscape according to the European Landscape Convention is an area, as perceived by people, whose character is the result of the action and interaction of natural and/or human factors [15]. In this paper, we consider landscape as a vehicle for spatial planning, rather than as an object for planning itself. Starting from the basic abiotic differentiation underlying all landscape processes, a landscape-based approach to spatial planning should make use of the opportunities offered by the landscapes further differentiated by societal expectations and cultural norms, instead of designing the landscape according to the economic ambitions of today's users only as, at the end of the day, is still often the dominant practice [16–18].

2.2. Urban and Rural Relationships

Soil, topography, water and historical patterns define the opportunities of spatial planning to a large extent. Neglecting these patterns and the associated processes through intensification and scale-enlargement of land use has led to numerous examples of tragic degradation [19]. Although the countryside is often considered as the rural opposite of the city, both have always been tightly connected [20,21]. The first cities were directly related to the provision of sufficient food, water and energy in the surrounding land and their defense depended on their situation in the landscape. Industrialization favored the growth of cities connected to the availability of resources such as coal and metal and to transport networks of rivers, over seas and over land. Brenner [22] describes the next step of planetary urbanization with cities worldwide better connected to each other than to their surrounding landscape. Timmermans, Woestenburg, Annema, Jonkhof, Shlakku and Yano [21] consider European capitals as cities that, through centuries, have successfully managed to profit from the different chances offered by their surrounding landscapes, while other cities faced degeneration when the landscape did not offer enough of what they needed for further growth.

2.3. Land use Transitions

Landscape as a concept is a crucial element in land-use transitions [23]. Transition science addresses the interplay between humans and the systems around them in which they operate. The key characteristic is that it is oriented on system and policy innovation. This is essential, as business as usual, or even innovations that optimize the current situation, are insufficient to resolve the issues of our time [24]. As a key contribution to transition science, Rotmans describes a new world view that can help to define actions to facilitate the now urgently needed transitions in our society [5,25]. In this world view, problems are solved in a cooperation model (as opposed to an exploitation model), in which business models are not focused on economic return, but on societal return and value creation instead of value extraction. Key to this framework is the transition cross (X-curve) of Visser, Keesstra, Maas and de Cleen [8] in which they explain the process from the old system towards a new system. In the X-curve, transition is described as a process of construction and demolition, which usually has a long pre-development phase (decades) and the real transition phase is relatively short (years), and characterized by chaotic and disruptive events (compare the adaptive cycle in ecosystems of Holling [26]). In Figure 1, this conceptual model is further elaborated for transitions in landscape adaptation processes. When we follow the green line, the lower left part represents the start of innovative new approaches. They start as a niche product that is developed in an "Experiment" phase. Once the approach or product has proven to be useful, it gains popularity in the "Acceleration" and "Emergence" phases, but still is seen as a niche product. To get to the other side of the chaotic transformation phase, enablers are needed to shift to the "Institutionalization" and "Stabilization" phases in which the approach or product becomes the new normal. Apart from this positive transition towards more sustainable approaches and products, it is also important to have attention for the phasing out of unsustainable approaches and products, which is depicted by the grey line. Many innovations are not truly transformative, and only optimize business as usual. The phasing out needs to incorporate the dependencies and lock-ins that form barriers for change. This process has several steps: "corrective barriers", "reduce dependencies" and "reduce relevance" to reach the phasing out.

This framework aims to enable changes and actions that should be taken to support sustainability in the short- and long-term and give direction to necessary actions in land restoration, sustainable land use and management and land and soil policy. The framework can provide the required intensive guidance to (i) analyze the impact of incentives, (ii) identify new reference points in the transition and (iii) stimulate transition catalysts, and (iv) innovate by testing cutting edge policy instruments in close cooperation with society [8].

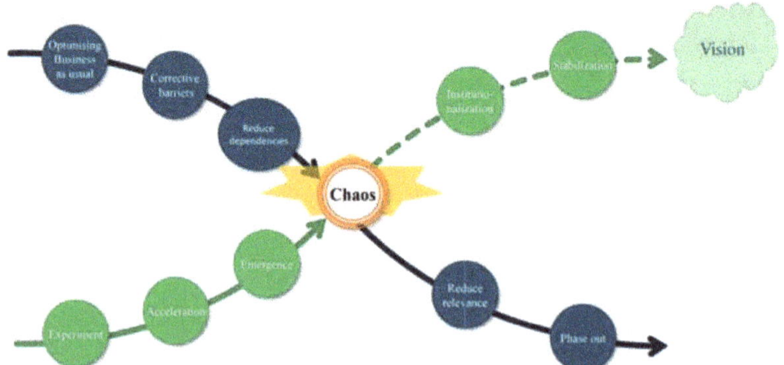

Figure 1. The X-curve transition cross (adapted after Visser, Keesstra, Maas and de Cleen [8]).

3. Characteristics of Landscape-Based Planning

Much has been written about nature-based solutions. In this paper, we do consider nature-based solutions a guiding principle for solving the knot of challenges mentioned in the introduction that society is facing today. Still, we prefer to slightly reframe the concept:

- using the term "landscape-based planning" rather than nature-based solutions, as it combines the social and environmental systems, instead of focussing on ecosystem processes. The concept will identify from the start of the planning process the opportunities that the landscape dimension (the spatial-temporal dimension including biophysical, social and cultural aspects) can offer [27].
- using "landscape-based" rather than "nature-based": Herewith we follow the reasoning of Termorshuizen and Opdam [28] that the concept of landscape services in metropolitan Europe is more appropriate than the concept of ecosystem services. They chose the concept of landscape services over the concept of ecosystem services as it better associates with pattern–process relationships, it better unifies scientific disciplines and it is more relevant and legitimate to local practitioners. People live in landscapes, not in ecosystems. It informs the actors with sound knowledge about how to best reconcile their needs to the landscape structure and processes.
- using "landscape-based planning" rather than "landscape-based solutions". In northwest Europe, one can observe a strong tradition of strict land-use planning on our scarcely available land. We have a history of dominating strong natural processes like flooding instead of letting land-use be the consequence of these processes. Further, a rich knowledge basis has been developed about the conditions under which landscape services can emerge [29]. This knowledge is a vital key to restore the landscape services that are needed as a solution for spatial planning challenges. This requires well-thought-out land-use planning, principally fostering a co-creating practice to fully account for the societal aspects of future developments. Focus is not so much on solutions for individual problems in the spatial development, but rather on a comprehensive planning approach encompassing as much as possible the potentially emerging problems in the future.

The next section describes some examples of a landscape-based approach, highlighting the guidance steps as defined in the previous section.

4. Examples of Landscape-Based Approaches

4.1. Example A: Regional Adaptation Strategy for the Region Vallei and Veluwe

4.1.1. Reflection on Incentives

Climate change is happening, and it is forecasted that we will face more extreme weather events, more often. Therefore, in the Netherlands, a National Adaptation Policy was put into place. As a follow-up to that, in the Vallei and Veluwe region (2456 km^2), a regional coalition of 28 municipalities, the water board, two provinces, the drinking water company and the regional security board co-created a Regional Adaptation Strategy for this strategy, a landscape-based approach was adopted. This approach is analogous to a co-production principle for landscape governance and transformation [29,30] and underpins a new governance style of local urban climate adaptation [31].

The Veluwe is a hilly mosaic landscape with dry sandy areas where the villages were traditionally located and low-lying areas with shallow groundwater tables with new neighborhoods where heavy rainfall causes floods and droughts are frequent in dry periods. For the area, a "climate effect atlas" was developed, including information about the long-term effects of floods and droughts [32].

4.1.2. Definition of New Reference Points for Transition

To help the process of transition to a situation where water management is adapted to the changing climate, a so-called "climate effect atlas" was produced, containing important reference points for adaptation. The climate effect atlas was produced in three steps: (1) data from the current climate and the expected climate in 2050 were retrieved from the Dutch national weather service (Royal Netherlands Meteorological Institute, KNMI); (2) a hydrological model was made, using the climate data and current landscape data, with which maps with the frequency of flood events now and in the future could be generated; (3) a map was produced that reflects the hydrological functioning of the natural system of the region, based on the combination of soil and geomorphological data.

In the Regional Adaptation Strategy, the current system and climate effect atlas were compared, which showed that the current natural and water management systems are unable to adequately regulate the water volumes associated with these excessive rains and prolonged droughts.

The climate effect atlas gives regional actors more insight into the functioning and the limits of the natural system regarding run-off, the impact of climate change on this system and on different types of land use. These insights were used to set up a Regional Adaptation Strategy and will further find their way in a Regional Adaptation Plan, that will be developed as a follow-up.

4.1.3. Identification of Transition Catalysts

The main catalyst of the Vallei and Veluwe approach was the awareness of upcoming climate change. This resulted in a National Adaptation Policy that is now regionally implemented. The catalyzing factor that led to choosing an innovative attitude, where technology and engineering follow natural processes in the system instead of working against them was a visionary group of policymakers and politicians who succeeded in convincing the stakeholders in the region to adopt a systems-, or landscape-based approach. Interestingly, this might be considered the refurbishing of the Design with Nature concept of McHarg [33], in this case emphasizing a participative stakeholder approach.

4.1.4. Innovation by Testing Policy Instruments

In the next steps, the science-base will be introduced to the Regional Adaptation Plan, defining a new policy for dealing with drought and flood events and new engineering standards. Administrative support is now in place for a landscape-based approach, integrating traditional man-made infrastructures such as technical sewage systems and water bodies, with landscape-based measures and upcoming nature-based techniques such as community-based initiatives of rainwater filtration and green roofs. A reflective participation approach [34] will be used to test the policy instruments defined.

4.2. Example B: Regional Case of Bee Landscapes: A Socio-Ecological Network for Pollinators

4.2.1. Reflection on Incentives

Wild pollinators have drastically declined in occurrence and diversity at local and regional scales in North-West Europe during the last decades [17,35]. Land-use change, and the current land management intensity are the most important causes [17]. Greater landscape-scale habitat diversity often results in more diverse pollinator communities and more effective crop and wild plant pollination. Semi-natural habitats, habitat corridors, landscape heterogeneity and landscape configuration are propagated as incentives to mitigate the negative effects of intensive land use on wild pollinators [35,36].

4.2.2. Definition of New Reference Points for Transition

As a response to these incentives, a regional authority (Province of South Holland), a company (Heineken) and a research institute (Wageningen University and Research) joined forces in 2013 in the "Green Circles" program, and decided to create a "Bee-Landscape" in the region around Leiden, Zoetermeer and Alphen aan de Rijn, as much as possible involving local stakeholders [37,38]. The aim was to initiate and stimulate a transition towards a more sustainable region in the province of South-Holland. The intensive, hands-on and parity-based cooperation between these different parties provides a very new reference point for transition. Using the concept of socio-ecological networks, a social network was set up to stimulate and enable coordinated action to realize an ecologically functional Bee Landscape.

In the initial phase of this initiative, the knowledge institute was asked to give scientifically sound substance to the term "Bee Landscape". The scientific knowledge was communicated in an attractive and easy to understand manner to local stakeholders. In 2016, the growing group of stakeholders that were involved in the Bee Landscape drew up and signed a covenant, in which measurable ambitions were described for the Bee Landscape.

4.2.3. Identification of Transition Catalysts

Green Circles launched the ambition to create a "Bee Landscape" and involved local and regional partners to do so, without specifying what a bee landscape exactly would be. The term "Bee Landscape" served as a boundary concept. The network of actors could lean on the sound knowledge of research institutes involved such as the preconditions for a high diversity of wild pollinators in the landscape and the strategic areas for the improvement of the habitat network. A so-called "Bee Landscape helpdesk" provided the opportunity for actors to invite pollinator specialists to their property for free advice. Further, a monitoring scheme for wild pollinators was put into place, to monitor progress, which was actively shared with the actors in the network.

The network grew from a dozen organizations in 2014 to over 30 in 2020. In 2016, a shared vision on a sustainable Bee Landscape and an associated covenant was drawn up and signed by 20 organizations. In this covenant, the time horizon taken was vague, but far away ("for the future"). This was in line with the spirit of the network: organizations working together and inspiring and helping each other, rather than cooperation based on control and accountability. Also, the use of the term Bee Landscape turned out to be helpful as it served as a strong boundary concept. It helped to move both the social and the bee network forward, even though not all actors had the same image of a comprehensive bee landscape. Last but not least, the participation of leading companies such as the Heineken brewery and AKZO, helped to motivate other parties to join the network.

4.2.4. Innovation by Testing Policy Instruments

The province chose to invest in a network coordinator to set up the network, support the exchange of knowledge and experiences so that the network became a true learning network, supporting the tailoring of scientific knowledge to the needs of the network and in providing a monitoring scheme for pollinators. In addition, they co-financed measures for wild pollinators in the field. This approach of investing in the emergence of self-managing networks that are committed to solving socio-ecological problems is quite new in the Netherlands and now several other initiatives have used the same approach.

4.3. Example C: Regional Case—The Plan Ooievaar (Plan "Stork")

4.3.1. Reflection on Incentives

The Dutch riverine area with the Rhine and Meuse rivers is centrally located in the Netherlands. Due to normalization and embankments, the riverbed between the dikes is now higher in the landscape caused by sedimentation, while the surrounding area has become lower, due to soil subsidence. The resulting river landscape was used almost only for intensive cattle grazing. This all has led to a situation where the high river discharges became a threat for the surrounding area, the biodiversity decreased, and the spatial quality of the area (i.e., the characteristic functional coherence of patterns and processes in the landscape, after [39]) was low.

Therefore, in 1985, a competition was held to find the best design for the landscape development of the riverine area at a regional level. The Plan "Ooievaar" [40], which won the competition, contained a number of new, appealing principles in terms of managing rivers, nature, agriculture and extraction of minerals (clay, sand or gravel). The plan advocated new interactions between the natural dynamics of a river system, the resulting visual expression and spatial quality, and land use. As a result, in the river landscape, agriculture and nature development would go hand in hand by making full use of the agricultural system and ecosystem potentials.

Plan "Ooievaar" promoted interweaving river management, nature development and landscape architecture, it was followed up by several experiments. When in 2001 the Deputy Minister separately commissioned Rijkswaterstaat and Regional governments to develop the National Strategy Room for the River, both parties decided to develop it together, based on positive previous experiences of collaboration [41].

4.3.2. Definition of New Reference Points for Transition

In this plan, the entire Dutch river area was an object of design. This scale level was innovative at that time. Another essential element in this plan is the combination of what is constructed and what unfolds naturally. The man-made part is drawn, described and calculated. The part that develops naturally is a matter of speculation, which does not however mean it happens by chance. The very opposite is true: it is fed by expertise. However, this self-same expertise teaches that the process is a game in which uncertainty and surprise are influencing the outcome.

4.3.3. Identification of Transition Catalysts

"Ooievaar" is the Dutch word for "Stork", and the makers of this plan, relying on their ecological expertise, expected the interventions suggested in the plan to lead to new nature values and an increase in biodiversity. This newly created natural environment would appeal to the black stork, a species characteristic of highly varied river ecosystems, which had left the Netherlands a long time ago. The label "Stork" can be considered as a boundary concept, enabling different sectors to work together. The main transition catalysts were the Non-Governmental Organizations: (i) growing societal resistance to dike reinforcement and (ii) a growing belief that these measures alone could not deliver future flood safety [42]. Then, in 1993 and 1995, two large flood events occurred which showed the vulnerability of our riverine area. The confrontation with the acute flood risk, combined with the first awareness of the effects of climate change on river discharges, contributed to a paradigm shift in flood management towards accommodating floods in a co-creative process.

4.3.4. Innovation by Testing Policy Instruments

In the early 2000s, the National Room for the River Program was launched to increase flood safety by giving the rivers literally more room, combined with increased spatial quality of landscape, nature and culture [43,44]. This program changed the topography and the water regime of the Dutch rivers profoundly, giving the river dynamics free rein, provoking a chain of transformations that increases biodiversity. Since a new landscape arises—with the black stork as a bonus. The policy instruments used were setting civil engineering boundary conditions for flood safety and navigation and co-creation of nature rehabilitation and landscape plans within this framework.

5. Discussion

5.1. Lessons Learned from the Examples

Comparing the three examples, common characteristics of the transition pathway become clear, as illustrated in the transition model of Figure 1. Incentives in the field of water management, climate change and biodiversity put pressure on the business as usual. In our highly organized and specialized society, these incentives can affect a wide range of actors. An adequate reaction to these incentives therefore involves a multitude of sectors that are closely interrelated. This requires a transformation of the business as usual. In the examples, a landscape-based approach is used as a guideline for this transition (Figure 2). Starting from the undefined need for change ("chaos"), the abiotic landscape system defines the safe operating space for sustainable development in which a future vision should fit. Visions that are unsustainable in the long term will appear to be unfeasible in an early stage. Shared visions are comprised of a multidimensional target space. In the first phase, pathways should be defined to arrive within the limits of this space, varying under the influence of external factors, such as climate change and societal changes. In time, the boundary concept takes shape and increasingly inspires stakeholders and policymakers, constituting a second phase. At the end of the day, a normative design represents a more narrowly defined point on the planning horizon, that is, a reasonably explicit description of the future landscape. In fact, this visioning of a normative design implies a back-casting approach, implying that the safe operating space is not the only norm. This process will need to be repeated each time new pressing issues appear to become incentives (cfr. adaptive planning [45]).

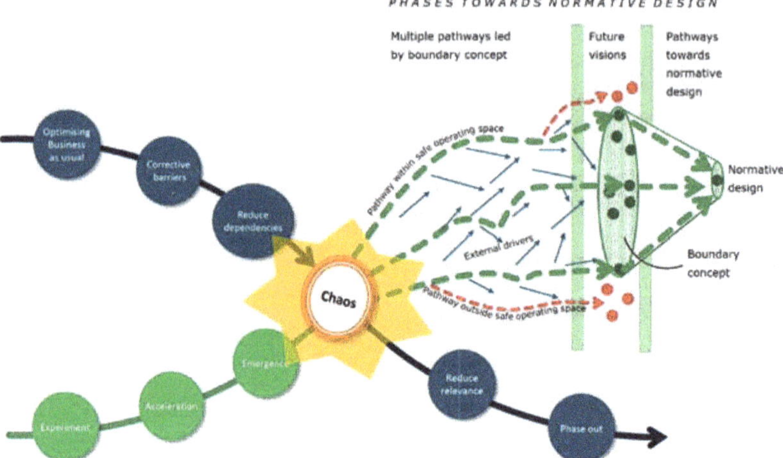

Figure 2. The role of visions as boundary concepts. Several sustainable visions (dark green spots) are comprised of multidimensional target space, whereas unsustainable visions (red dots) are unfeasible. Pathways should be defined—and followed—to arrive within the limits of this space, varying under the influence of external factors, such as climate change and societal changes.

The examples can be compared on the basis of the characteristics described in Section 3 (see Table 1).

Table 1. Comparison of the examples.

Characteristics of Land-Use Plan		Example A: Vallei and Veluwe	Example B: Bee Landscape	Example C: Plan Ooievaar
Goal:		Climate adaptation preventing flooding and droughts	Stimulate population development of pollinating insects	Decrease regional flood risk while stimulating biodiversity and spatial landscape quality
Boundary concept adopted:		Natural landscape system	Bee landscape	Black stork
Landscape dimension	Spatial dimensions defined by:	Water dynamics (watershed level)	Population dynamics (landscape level)	Water dynamics (regional/national level)
	Time horizon:	Medium to long term	Long term	Medium to long term
Landscape services	Biodiversity:	Positive effects of nature-based solutions	More diverse pollinator communities	Newly created natural environment increases biodiversity
	Circularity:	Circular practices in wastewater and solid waste management	Pollination and honey production as a win-win situation	Extraction of river sediment for construction industry while alleviating flood risk
	Adaptation to climate change:	Flood prevention by many measures; several water-retention solutions; counteracting urban heat island effect	Climate robust habitats for high diversity of pollinators	Adaptation to changing river discharges and extreme flood and drought events
Comprehensive planning	Incentives:	Threat of extreme weather events	Decline of pollinators	Increasing flood risk along the rivers
	New Reference points for transition:	2050 climate scenario in map	Parity based hands-on transition management	DNA of the River (basic landscape principles to be taken into consideration)
	Transition catalysts:	Waterboard as responsible for climate regional adaptation plan	Diverse network of actors; capable network coordinator	Non-governmental organizations and later sectoral stakeholders
	Innovative policy instruments:	Adopting a systems approach to overcome local–regional dichotomy	Financial support for building and running a learning network	Clear boundary conditions for co-creation of nature and landscape rehabilitation
Landscape-based approach		Build on the characteristics of the physical landscape to adapt to changing precipitation patterns	Build on the motivation and willingness in the social landscape system to reverse the decline of pollinators in their landscape	Build on the original natural characteristics of river sections on a national and regional level, to accommodate higher discharges

The steps and requirements that we distill from the example cases are the following:

- Boundary concept. The introduction of a boundary concept is often summarized in a catchy term and can help actors to agree on goals on an abstract level. This enhances constructive conversations and cooperation. The boundary concept in our examples is the underlying landscape vision, where a catchy term is often the label to which the vision is attached.
- Landscape dimension. Actors need to realize that the landscape as an underlying system is crucial in solving the challenges that they face both in space and in time.

- Landscape services, especially biodiversity, circularity and climate adaptation. Understanding the natural and societal system is a first step in the planning process so that it becomes clear how the landscape can provide the landscape services that people need and to assess the physical and societal capacity of the system to offer opportunities for new arrangements of functions. This gives actors insight into the direction that they need to move, and the principles that need to be at the basis of a vision for the future.
- Comprehensive planning. Once the four steps describing the examples allow translation into change directions in a land-use map or landscape visual, this may very well result in a phase of chaos (Figure 1). This phase of the transition process is crucial: actors get a reality check on the guiding principles that they agreed on. This may result in a landscape that is not meeting all of their needs. An interplay between actors will then take place, resulting in the shift of needs or in the shift of guiding principles, that will lead to a landscape plan that is better fulfilling the needs, reflecting the power balance between the actors involved within the context of the landscape system. Finally, when this is settled, the landscape-based plan can be translated to action perspectives of the different types of stakeholders, resulting in a clearly defined pathway towards a new stabilized situation.

5.2. Cherishing Abiotic Differences Enhances Sustainable and Resilient Landscapes

The cases on water management and flood protection, in the Vallei and Veluwe and Plan Ooievaar respectively, show the importance of understanding natural processes in landscapes and knowledge of the potential that different zones in the landscape offer to enhance natural processes that deliver essential landscape services [46]. Especially regarding water management challenges, such as adaptation to climate change, the natural system offers diversity in opportunities on how measures can contribute to water management. Tailoring the measures to the opportunities the landscape offers, will make the measures more effective and efficient in delivering landscape services [47]. Also, it will result in measures that will be more sustainable, as they are in accordance with or compatible with the local natural circumstances. This will lead to a landscape where differences in natural characteristics due to hydrologic, topographic or soil differences are emphasized and considered to be an asset instead of a threat. This results in a landscape in which the natural characteristics can be highlighted instead of smoothed out, which will add value to the identity of an area within its regional context. This adds to the connection people have with the soil and landscape around them, enhancing human health and wellbeing [16,48].

5.3. Strategies for Landscapes are More Efficient than Those for Administrative Units

From the examples, we learned that "working with nature" needs to be done at the proper scale level, the scale level of the natural processes that deliver the desired landscape services [49–51]. In the example of the regional adaptation strategy, solutions for local floods and droughts are sought in the regional natural system, on which hydrological processes take place, connecting rural and urban areas, which is similar to other cases in different parts of the world [52–54]. The interaction between local and regional planning dimensions is crucial here [10]. In the case of the Bee Landscape, expert knowledge is used to design the required type, amount and coherence of habitat for viable populations of pollinating insects. In the case of Plan Ooievaar, the whole Dutch riverine area situated in more than four provinces and in numerous municipalities was considered as one natural system. This led to a transitional change in river management, where natural river dynamics are embraced instead of combatted [54,55]. The highly adjusted riverbed system was adapted to a more biodiverse and flood resilient system [40]. In all three cases, administrative borders were overruled, and cooperation took place on the landscape level: the level on which key natural processes take place for landscape services that were needed. Therefore, for an efficient implementation of land management strategies, it is important to take the natural limits of the system into account [56,57]. For the biosphere part of the system, the landscape as a unit may serve as the natural planning limit, thus reducing the complexity of the whole socio-economic/bio-physical system.

5.4. Landscape Visions are Powerful Boundary Concepts to Define Pathways towards A Desired Future

Thinking about a desired future is always an inspiring activity. Having envisaged a specific future, pathways towards such a future can be defined [58]. A desired future may not be a strongly delineated image, also a co-created vision can be a basis for normative scenario design [30,59,60]. Landscape visions can very well serve as the vehicle for discussing future land use. As such they can be called boundary concepts in collaborative landscape governance, in analogy to the landscape services used as such by Westerink, Opdam, Van Rooij and Steingröver [29]. The image of the Bee Landscape was clearly a very inspiring local boundary concept, to bring a substantial number of parties together for a shared future of the regional landscape. Also, the idea of natural rejuvenation of the meandering river system appeared to be a powerful boundary concept to inspire many municipalities and other institutions to join forces in nature rehabilitation in the floodplains. In the Vallei and Veluwe, the "basic natural system" at a regional scale was put forward as an inspiring principle, that will enable the scientific and governance community to identify adequate local pathways towards a desired future.

5.5. Co-Creation Is Essential to Safeguard Adaptive, Circular and Biodiverse Landscapes

Successful landscape-based planning is characterized by working on the required scale level, choosing the required dimension of the planning area, always embedded in the next higher level. The use of landscape services emphasizes the potential of the landscape as a means to realize a desired future. All sectors present in an area are involved in the planning, which is a basic character of co-creation, where public and knowledge institutions collaborate not only with private bodies but also with civil society to innovate services and products [61,62]. Therefore, as all sectors have their own perspective on issues, language and context, it appears that a boundary concept helps to enable the conversation and overcome differences between stakeholders. In addition, the cases show that a time horizon in the far future helps to overcome discussions on current problems and enables focus on possibilities. Recently, a vision on a natural future for the Netherlands in 2120 appeared, showing how the Netherlands would look like when taking the landscape as a guiding principle for adaptation to climate change in a biodiverse and circular environment. Stakeholders and authorities of all sorts found this very interesting and mind-shifting, giving way to further, out of the box discussions on the issues that are ahead [18,63].

5.6. Landscape-Based Approaches Enhance Future-Proof Land-Use Transitions

Transitions in the context of spatial planning can only be recognized as such after completion. The evaluation of the cases shows that the landscape-based planning approach stands for land-use transitions based on a landscape-based spatial development, that is, an approach to spatial processes that takes into account all the relationships in the landscape: both the physical landscape with its layers and functions and the socio-ecological landscape with its different scales and actors. This provides a basis for a transition framework towards sustainable land use in the long term, which in turn gives direction to necessary short-term actions in land restoration, nature rehabilitation and innovative forms of land use. Although change is often difficult to bring about, the landscape itself shows the traces of constant change in a positive way—its particular character does not need to suffer from change. Normative futures as boundary concepts can help—instead of building our present on our past—to learn and build our present on our future, on what is possible, instead of merely on what has gone before [4,8,30].

6. Conclusions

As illustrated with the three examples, the landscape-based planning approach enhances a development towards a future land system resilient to external pressures—at least the foreseeable ones. The concepts of nature-based solutions and transition theory are fundamentally combined in this approach, where co-created normative future visions serve as boundary concepts in the regional spatial planning debate. A shared long-term vision of what our future landscape should look like is a crucial source of inspiration for a coherent design approach to solve today's spatial planning problems, such as climate adaptation, biodiversity enhancement and circular resource management. The landscape-based approach principally uses the natural characteristics of the landscape system as an opportunity instead of a limitation. It gives direction to the technical-economic preconditions for sustainable landscape development, such as drainage standards and environmental quality. Rather than as an object in itself, the landscape is considered as a comprehensive principle, to which all spatial processes are inherently related. Local planning can only be adequate when logically embedded in a regional perspective. The main recommendation for future research is therefore that solutions to regional planning problems should be studied that go beyond the traditional nature-based solutions, by emphasizing the spatial dimension, the specific time horizon considered and the interaction of all sectoral considerations of the urban and rural landscape. Special attention should be paid to the adequateness of social involvement and participation, which is to be defined for each case differently, and which could easily play a disturbing role. Also, the dominance of strong short-term economic functions such as transport, housing, energy provision is currently often triggering trade-offs, especially when the shared long-term vision is not accompanied by (inter)national instruments to guide sustainable developments at lower spatial scales [64,65]. At the end of the day, however, the landscape-based planning approach should allow professionals, researchers, stakeholders and citizens alike, to participate in the transition to a forward-looking normative design. Working towards such a future, pathways can be defined towards a shared vision, observing the boundaries of a safe operating space.

Author Contributions: Conceptualization, S.v.R., W.T., O.R. and B.P.; methodology, S.v.R., W.T., O.R. and S.K.; writing—original draft preparation, S.v.R., B.P. and M.S.; writing—review and editing, B.P. and M.S.; visualization, S.K., W.T. and S.v.R.; supervision, S.v.R.; project administration, O.R.; funding acquisition, S.v.R. and O.R. All authors have read and agreed to the published version of the manuscript.

Funding: This research was funded by the Netherlands Ministry of Agriculture, Nature and Food Quality, grant number KB-34A-007-008 1-2C-6 (Circular & Climate Neutral), and KB-36-005-006/008 (Nature-inclusive Transitions).

Conflicts of Interest: The authors declare no conflict of interest. The funders had no role in the design of the study; in the collection, analyses or interpretation of data; in the writing of the manuscript or in the decision to publish the results.

References

1. Adam, M.; Ab Ghafar, N.; Ahmed, A.; Nila, K. A systematic review on city liveability global research in the built environment: Publication and citation matrix. *J. Des. Built Environ.* **2017**, 62–72. [CrossRef]
2. Decina, S.M.; Hutyra, L.R.; Templer, P.H. Hotspots of nitrogen deposition in the world's urban areas: A global data synthesis. *Front. Ecol. Environ.* **2020**, *18*, 92–100. [CrossRef]
3. Roth, T.; Kohli, L.; Bühler, C.; Rihm, B.; Meuli, R.G.; Meier, R.; Amrhein, V. Species turnover reveals hidden effects of decreasing nitrogen deposition in mountain hay meadows. *PeerJ* **2019**, *7*, e6347. [CrossRef] [PubMed]
4. Senge, P.; Scharmer, C.O.; Jaworski, J.; Flowers, B.S. *Presence: Human Purpose, and the Field of the Future*; Society for Organizational Learning: Cambridge, MA, USA, 2004.
5. Rotmans, J.; Loorbach, D. Towards a better understanding of transitions and their governance. A systemic and reflexive ap-proach. In *Transitions to Sustainable Development—New Directions in the Study of Long Term Transformation Change*; Routledge: New York, NY, USA, 2010; pp. 105–220.
6. Nassauer, J.I. Landscape as medium and method for synthesis in urban ecological design. *Landsc. Urban Plan.* **2012**, *106*, 221–229. [CrossRef]
7. Amenta, L.; Van Timmeren, A. Beyond wastescapes: Towards circular landscapes. Addressing the spatial dimension of circularity through the regeneration of wastescapes. *Sustainability* **2018**, *10*, 4740. [CrossRef]
8. Visser, S.; Keesstra, S.; Maas, G.; de Cleen, M. Soil as a Basis to Create Enabling Conditions for Transitions Towards Sustainable Land Management as a Key to Achieve the SDGs by 2030. *Sustainability* **2019**, *11*, 6792. [CrossRef]

9. Star, S.L. This is not a boundary object: Reflections on the origin of a concept. *Sci. Technol. Hum. Values* **2010**, *35*, 601–617. [CrossRef]

10. Opdam, P.; Nassauer, J.I.; Wang, Z.; Albert, C.; Bentrup, G.; Castella, J.-C.; McAlpine, C.; Liu, J.; Sheppard, S.; Swaffield, S. Science for action at the local landscape scale. *Landsc. Ecol.* **2013**, *28*, 1439–1445. [CrossRef]

11. De Roo, G.; Silva, E.A. *A Planner's Encounter with Complexity*; Ashgate Publishing, Ltd.: London, UK, 2010.

12. Potschin, M.; Kretsch, C.; Haines-Young, R.; Furman, E.; Berry, P.; Baró, F. Nature-based solutions. In *OpenNESS Ecosystem Services Reference Book. EC FP7 Grant Agreement No. 308428*; Potschin, M., Jax, K., Eds.; Available online: www.openness-project.eu/library/reference-book (accessed on 11 December 2020).

13. Scott, M.; Lennon, M.; Haase, D.; Kazmierczak, A.; Clabby, G.; Beatley, T. Nature-based solutions for the contemporary city/Re-naturing the city/Reflections on urban landscapes, ecosystems services and nature-based solutions in cities/Multifunctional green infrastructure and climate change adaptation: Brownfield greening as an adaptation strategy for vulnerable communities?/Delivering green infrastructure through planning: Insights from practice in Fingal, Ireland/Planning for biophilic cities: From theory to practice. *Plan. Theory Pract.* **2016**, *17*, 267–300. [CrossRef]

14. Opdam, P.; Steingröver, E. How could companies engage in sustainable landscape management? An exploratory perspective. *Sustainability* **2018**, *10*, 220. [CrossRef]

15. COE. *European Landscape Convention (Florence Convention), Treaty Series Nr. 176*; Council of Europe: Strasbourg, France, 2000.

16. Albert, C.; Schröter, B.; Haase, D.; Brillinger, M.; Henze, J.; Herrmann, S.; Gottwald, S.; Guerrero, P.; Nicolas, C.; Matzdorf, B. Addressing societal challenges through nature-based solutions: How can landscape planning and governance research contribute? *Landsc. Urban Plan.* **2019**, *182*, 12–21. [CrossRef]

17. IPBES. *The Assessment Report of the Intergovernmental Science-Policy Platform on Biodiversity and Ecosystem Services on Pollinators, Pollination and Food Production*; IPBES: Bonn, Germany, 2016; p. 552.

18. Kalantari, Z.; Ferreira, C.S.S.; Deal, B.; Destouni, G. Nature-based solutions for meeting environmental and socio-economic challenges in land management and development. *Land Degrad. Dev.* **2019**, *31*, 1867–1870. [CrossRef]

19. Fabos, J. *Planning the Total Landscape: A Guide to Intelligent Land Use*; Routledge: Abingdon, UK, 2019.

20. Soja, E.W. Beyond postmetropolis. *Urban Geogr.* **2011**, *32*, 451–469. [CrossRef]

21. Timmermans, W.; Woestenburg, M.; Annema, H.; Jonkhof, J.; Shlakku, M.; Yano, S. *The Rooted City: European Capitals and Their Connection with the Landscape*; Blauwdruk: Wageningen, The Netherland, 2015.

22. Brenner, N. Open City or the Right to the City? Demand for a democratisation of urban space. *Topos Eur. Landsc. Mag.* **2013**, *85*, 42–45.

23. Pinto Correia, T.; Primdahl, J.; Pedroli, B. *European Landscapes in Transition. Implications for Policy and Practice*; Cambridge University Press: Cambridge, UK, 2018.

24. Doppelt, B. *Leading Change toward Sustainability: A Change-Management Guide for Business, Government and Civil Society*; Routledge: New York, NY, USA, 2017.

25. Rotmans, J. *Change of an Era; Our World in Transition*; Boom Uitgevers: Amsterdam, The Netherlands, 2017.

26. Holling, C.S. Understanding the complexity of economic, ecological, and social systems. *Ecosystems* **2001**, *4*, 390–405. [CrossRef]

27. Arts, B.; Buizer, M.; Horlings, L.; Ingram, V.; Van Oosten, C.; Opdam, P. Landscape approaches: A state-of-the-art review. *Annu. Rev. Environ. Resour.* **2017**, *42*, 439–463. [CrossRef]

28. Termorshuizen, J.; Opdam, P. Landscape services as a bridge between landscape ecology and sustainable development. *Landsc. Ecol.* **2009**, *24*, 1037–1052. [CrossRef]

29. Westerink, J.; Opdam, P.; Van Rooij, S.; Steingröver, E. Landscape services as boundary concept in landscape governance: Building social capital in collaboration and adapting the landscape. *Land Use Policy* **2017**, *60*, 408–418. [CrossRef]

30. Turnhout, E.; Metze, T.; Wyborn, C.; Klenk, N.; Louder, E. The politics of co-production: Participation, power, and transformation. *Curr. Opin. Environ. Sustain.* **2020**, *42*, 15–21. [CrossRef]

31. Trell, E.; van Geet, M. The governance of local urban climate adaptation: Towards participation, collaboration and shared responsibilities. *Plan. Theory Pract.* **2019**, *20*, 376–394. [CrossRef]

32. Wageningen Environmental Research; Climate Adaptation Services; SWECO. Klimaateffectatlas Vallei en Veluwe. Available online: https://klimaatvalleienveluwe.nl/atlas/ (accessed on 14 December 2020).

33. McHarg, I.L. *Design with Nature*; American Museum of Natural History: New York, NY, USA, 1969.

34. Flyvbjerg, B. *Making Social Science Matter: Why Social Inquiry Fails and How It Can Succeed Again*; Cambridge University Press: Cambridge, UK, 2001.

35. Habel, J.C.; Samways, M.J.; Schmitt, T. Mitigating the precipitous decline of terrestrial European insects: Requirements for a new strategy. *Biodivers. Conserv.* **2019**, *28*, 1343–1360. [CrossRef]

36. Senapathi, D.; Goddard, M.A.; Kunin, W.E.; Baldock, K.C. Landscape impacts on pollinator communities in temperate systems: Evidence and knowledge gaps. *Funct. Ecol.* **2017**, *31*, 26–37. [CrossRef]

37. Green Circles on the Bee Landscape. Available online: https://www.greencircles.nl/bee-landscape (accessed on 14 December 2020).

38. Van Der Zon, A.P.M. Audit Report for the Enrollment of the Green Circles Bee Landscape as a Verified Conservation Area. Available online: https://www.earthmind.org/sites/default/files/2019-03-GreenCircleBee-Plan-Audit.pdf (accessed on 14 December 2020).

39. Busscher, T.; Van Den Brink, M.; Verweij, S. Strategies for integrating water management and spatial planning: Organising for spatial quality in the Dutch "Room for the River" program. *J. Flood Risk Manag.* **2019**, *12*, e12448. [CrossRef]

40. De Bruin, D.; Hamhuis, D.; Nieuwenhuijzen, L.; Overmars, W.; Sijmons, D.; Vera, F. *Plan Ooievaar, de Toekomst van het Rivierengebied*; Stichting Gelderse Milieufederatie: Arnhem, The Netherland, 1987.

41. Peters, B.; Overmars, W.; Kurstjens, G.; Rademakers, J.G.M. Van Plan Ooievaar tot Smart Rivers; 25 jaar ecologisch herstel van het rivierengebied tegen veranderende achtergronden (in Dutch). *De Levende Nat.* **2014**, *115*, 78–83.

42. Van Der Brugge, R.; Rotmans, J.; Loorbach, D. The transition in Dutch water management. *Reg. Environ. Chang.* **2005**, *5*, 164–176. [CrossRef]

43. Rijke, J.; Van Herk, S.; Zevenbergen, C.; Ashley, R. Room for the River: Delivering integrated river basin management in the Netherlands. *Int. J. River Basin Manag.* **2012**, *10*, 369–382. [CrossRef]

44. Schut, M.; Leeuwis, C.; van Paassen, A. Room for the River: Room for research? The case of depoldering De Noordwaard, the Netherlands. *Sci. Public Policy* **2010**, *37*, 611–627. [CrossRef]

45. Hersperger, A.M.; Bürgi, M.; Wende, W.; Bacău, S.; Grădinaru, S.R. Does landscape play a role in strategic spatial planning of European urban regions? *Landsc. Urban Plan.* **2020**, *194*, 103702. [CrossRef]

46. Maes, J.; Jacobs, S. Nature-based solutions for Europe's sustainable development. *Conserv. Lett.* **2017**, *10*, 121–124. [CrossRef]

47. Cerdà, A.; Rodrigo-Comino, J. Is the hillslope position relevant for runoff and soil loss activation under high rainfall conditions in vineyards? *Ecohydrol. Hydrobiol.* **2020**, *20*, 59–72. [CrossRef]

48. Brevik, E.C.; Steffan, J.J.; Rodrigo-Comino, J.; Neubert, D.; Burgess, L.C.; Cerdà, A. Connecting the public with soil to improve human health. *Eur. J. Soil Sci.* **2019**, *70*, 898–910. [CrossRef]

49. Guerrero, P.; Haase, D.; Albert, C. Locating spatial opportunities for nature-based solutions: A river landscape application. *Water* **2018**, *10*, 1869. [CrossRef]

50. Hallett, P.D.; Lichner, L.u.; Cerdà, A. Biohydrology: Coupling biology and soil hydrology from pores to landscapes. *Ecohydrology* **2010**, *3*, 379–381. [CrossRef]

51. Keshavarzi, A.; Kumar, V.; Bottega, E.L.; Rodrigo-Comino, J. Determining land management zones using pedo-geomorphological factors in potential degraded regions to achieve land degradation neutrality. *Land* **2019**, *8*, 92. [CrossRef]

52. Abbas, F.; Hammad, H.M.; Fahad, S.; Cerdà, A.; Rizwan, M.; Farhad, W.; Ehsan, S.; Bakhat, H.F. Agroforestry: A sustainable environmental practice for carbon sequestration under the climate change scenarios—A review. *Environ. Sci. Pollut. Res.* **2017**, *24*, 11177–11191. [CrossRef] [PubMed]

53. Hălbac-Cotoară-Zamfir, R.; Keesstra, S.; Kalantari, Z. The impact of political, socio-economic and cultural factors on implementing environment friendly techniques for sustainable land management and climate change mitigation in Romania. *Sci. Total Environ.* **2019**, *654*, 418–429. [CrossRef]

54. Thorslund, J.; Jarsjo, J.; Jaramillo, F.; Jawitz, J.W.; Manzoni, S.; Basu, N.B.; Chalov, S.R.; Cohen, M.J.; Creed, I.F.; Goldenberg, R. Wetlands as large-scale nature-based solutions: Status and challenges for research, engineering and management. *Ecol. Eng.* **2017**, *108*, 489–497. [CrossRef]

55. Addy, S.; Cooksley, S.; Dodd, N.; Waylen, K.; Stockan, J.; Byg, A.; Holstead, K. *River Restoration and Biodiversity*; IUCN: Gland, Switzerland, 2016.

56. Keesstra, S.; Nunes, J.P.; Saco, P.; Parsons, T.; Poeppl, R.; Masselink, R.; Cerdà, A. The way forward: Can connectivity be useful to design better measuring and modelling schemes for water and sediment dynamics? *Sci. Total Environ.* **2018**, *644*, 1557–1572. [CrossRef]

57. Saco, P.M.; Rodríguez, J.F.; Moreno-de las Heras, M.; Keesstra, S.; Azadi, S.; Sandi, S.; Baartman, J.; Rodrigo-Comino, J.; Rossi, M.J. Using hydrological connectivity to detect transitions and degradation thresholds: Applications to dryland systems. *Catena* **2020**, *186*, 104354. [CrossRef]

58. Metzger, M.J.; Lindner, M.; Pedroli, B. Towards a roadmap for sustainable land use in Europe. *Reg. Environ. Chang.* **2018**, *18*, 707–713. [CrossRef]

59. McDowall, W.; Eames, M. Forecasts, scenarios, visions, backcasts and roadmaps to the hydrogen economy: A review of the hydrogen futures literature. *Energy Policy* **2006**, *34*, 1236–1250. [CrossRef]

60. Pérez-Soba, M.; Paterson, J.; Metzger, M.J.; Gramberger, M.; Houtkamp, J.; Jensen, A.; Murray-Rust, D.; Verkerk, P.J. Sketching sustainable land use in Europe by 2040: A multi-stakeholder participatory approach to elicit cross-sectoral visions. *Reg. Environ. Chang.* **2018**, *18*, 775–787. [CrossRef]

61. Puerari, E.; De Koning, J.I.; Von Wirth, T.; Karré, P.M.; Mulder, I.J.; Loorbach, D.A. Co-Creation dynamics in urban living labs. *Sustainability* **2018**, *10*, 1893. [CrossRef]

62. Arnkil, R.; Järvensivu, A.; Koski, P.; Piirainen, T. *Exploring Quadruple Helix Outlining User-Oriented Innovation Models*; Tampereen Yliopisto: University of Tampere: Tampere, Finland, 2010.

63. Baptist, M.; van Hattum, T.; Reinhard, S.; van Buuren, M.l.; de Rooij, B.; Hu, X.; van Rooij, S.; Polman, N.; van den Burg, S.; Piet, G. *A Nature-Based Future for the Netherlands in 2120*; Wageningen University & Research: Wageningen, The Netherlands, 2019. [CrossRef]

64. Girard, C.; Pulido-Velazquez, M.; Rinaudo, J.-D.; Pagé, C.; Caballero, Y. Integrating top–down and bottom–up approaches to design global change adaptation at the river basin scale. *Glob. Environ. Chang.* **2015**, *34*, 132–146. [CrossRef]

65. Gupta, J.; Vegelin, C. Sustainable development goals and inclusive development. *Int. Environ. Agreem. Politics Law Econ.* **2016**, *16*, 433–448. [CrossRef]

 *land* 

Article

Not Simply Green: Nature-Based Solutions as a Concept and Practical Approach for Sustainability Studies and Planning Agendas in Cities

Diana Dushkova [1] and Dagmar Haase [1,2,*]

[1]   Department of Geography, Humboldt University Berlin, Unter den Linden 6, 10099 Berlin, Germany; diana.dushkova@geo.hu-berlin.de

[2]   Helmholtz Centre for Environmental Science—UFZ, Department of Comp. Landscape Ecology, Permoserstr. 15, 04318 Leipzig, Germany

*   Correspondence: dagmar.haase@geo.hu-berlin.de; Tel.: +49-030-2093-9445

Received: 21 December 2019; Accepted: 4 January 2020; Published: 11 January 2020

Abstract: The concept of a nature-based solution (NBS) has been developed in order to operationalize an ecosystem services approach within spatial planning policies and practices, to fully integrate the ecological dimension, and, at the same time, to address current societal challenges in cities. It exceeds the bounds of traditional approaches that aim 'to protect and preserve' by considering enhancing, restoring, co-creating, and co-designing urban green networks with nature that are characterized by multifunctionality and connectivity. NBSs include the main ideas of green and blue infrastructure, ecosystem services, and biomimicry concepts, and they are considered to be urban design and planning tools for ecologically sensitive urban development. Nowadays, NBSs are on their way to the mainstream as part of both national and international policies. The successful implementation of NBSs in Europe and worldwide, which is becoming increasingly common, highlights the importance and relevance of NBS for sustainable and livable cities. This paper discusses the roles, development processes, and functions of NBSs in cities by taking Leipzig as a case study. Using data from interviews conducted from 2017 to 2019, we study the past and current challenges that the city faces, including the whole process of NBS implementation and successful realization. We discuss the main drivers, governance actors, and design options of NBSs. We highlight the ecosystem services provided by each NBS. We discuss these drivers and governance strategies by applying the framework for assessing the co-benefits of NBSs in urban areas in order to assess the opportunities and challenges that NBSs may have. This way, we are able to identify steps and procedures that help to increase the evidence base for the effectiveness of NBS by providing examples of best practice that demonstrate the multiple co-benefits provided by NBSs.

Keywords: urban nature-based solutions; societal challenges; sustainability; ecosystem services; green infrastructure; Leipzig

---

1. Introduction

The concept of nature-based solutions (NBS) has been developed in order to operationalize an ecosystem services approach within spatial planning policies and practices, to fully integrate the ecological dimension, and, at the same time, to address current societal challenges [1,2]. This concept exceeds the bounds of traditional approaches that aim 'to protect and preserve' by considering the enhancing, restoring, co-creating, and co-designing new green networks with nature that are characterized by multifunctionality and connectivity [2,3].

In this context, an NBS includes the main ideas of green and blue infrastructure, ecosystem services, and biomimicry concepts, and it is considered to be an urban design and planning tool for ecologically sensitive urban development [3–5]. Defined by the European Commission [4] as "actions [...] and solutions to societal challenges [...] which are inspired by, supported by, or copied from nature", NBSs provide multiple environmental, social, and economic co-benefits at the same time, such as the improvement of place attractiveness, of health and quality of life, the creation of green jobs, etc. An increasing number of scientific publications understand NBSs as a new concept for climate change adaptation and mitigation [6–9].

Thus, NBSs incorporate four interrelated goals. Firstly, they enhance sustainable urbanization by ensuring essential ecosystem functions and by promoting urban regeneration. Secondly, they restore the functionality of degraded ecosystems and their services. Thirdly, they develop aspects of climate change adaptation and mitigation, including the redesign of human-made infrastructure and the integration of gray with green and blue infrastructure. Fourthly, they improve risk management and resilience by utilizing a nature-based design that combines multiple functions and benefits such as pollution reduction, carbon storage, biodiversity conservation, reducing heat stress, and enhanced water retention [1,3].

There has been a growing awareness of NBSs that involve elements of ecosystems and seek to use natural elements to improve the adaptive capacity of human and natural systems by providing ecosystem services to cope with the adverse effects of climate change [6–10]. Nowadays, NBSs are on their way to the mainstream in national and international policies. The successful implementation of NBSs in Europe and worldwide, which are becoming increasingly common, highlights the importance and relevance of NBSs for sustainable and livable cities [5]. A great number of the ongoing EU research framework programme HORIZON 2020 projects endeavor to explore how NBSs work in different urban contexts with regard to the political, social, cultural, institutional, environmental, and economic background.

This paper presents the results of the research within one such project: CONNECTING Nature (COproductioN with NaturE for City Transitioning, INnovation and Governance). CONNECTING Nature is a research and innovation project that aims to accelerate the scaling of NBSs in European cities in which the main idea is to innovate with nature for the development and implementation of NBSs for urban sustainability issues. The CONNECTING Nature project aims to 'co-produce' with scientific partners and different stakeholders—i.e., municipalities, SMEs (small and medium enterprises), NGOs, and citizens—a design framework for the development and implementation of the concrete steps, activities, and tools that are used in the co-production of NBSs. In order to identify the gaps in the existing research and actions on NBS (e.g., EU projects, reference frameworks, and scientific literature on the impact of NBSs), a comprehensive scoping exercise of existing nature-based and grey solutions projects in Europe was carried out within the several working packages of the CONNECTING Nature Project. Thus, the project developed an NBS database that reports a full range of NBS interventions from European cities and creates a basic profile of NBS implementation across Europe. In doing so, it generates and provides knowledge beyond the analysis of various individual NBS projects that have been implemented in cities or urban regions by conducting the first systematic survey and review of NBS interventions in urban environments in Europe. The structural analysis of such a database helps not only to present an overview of urban NBS interventions in Europe but also to identify the limitation, success, and failure factors for NBSs. One of the tasks was to recognize knowledge gaps for research and development in the field of NBSs. The most important ones are:

- lack of knowledge about the potential of NBSs to address the challenges and which to best implement;
- lack of knowledge related to the potential co-benefits that result from NBSs;
- lack of knowledge about the functions nature provides to cities;
- lack of technical knowledge on how to plan, build, and maintain NBSs;
- gaps in the knowledge regarding the different stages of NBS implementation;
- lack of monitoring on the impact of NBS.

To fill these gaps and allocate resources for producing knowledge for the recognized gaps of the performance of various NBSs, this paper presents the results of the study, focusing in particular on the role, development process (co-creation, co-design, and co-development), and functions of NBS in cities by taking Leipzig as a case study. We aim to explore how the NBS concept at the local level, considering a city that experiences a high dynamic: Leipzig could contribute to bridge the gaps between research and innovation, focusing on nature-based solutions. The specific objectives are to explore how the nature-based solution concept could help link research and innovation in the area of biodiversity and ecosystem services, and what actions are needed to further support the knowledge base for nature-based solutions by presenting key recommendations for overcoming barriers and bridging gaps in order to more effectively uptake and promote NBSs.

Leipzig was selected as a case study of particular interest and exemplary character because the city managed to transform a former industrial area, which had to cope with increasing unemployment rates after industry collapsed, into a green and more liveable place. There are several exemplary NBS cases that are worth considering in the city and that have been addressed for the issues of urban regeneration, coping with former industrial and neglected areas, e.g., brownfields, and the demand for green space for several urban districts, which is currently referred to as the biggest challenge for several cities in Europe and beyond. Firstly, examples of greening initiatives include the creation of wilderness patches in the areas that surround the former industrial area, which are known as New Lake Land (Leipziger Neuseenland). Secondly, green areas also include the central green parts of the city, the floodplains (Leipziger Auenwald), and the renaturation of rivers—which are especially highly polluted during the time of industrialization. Thirdly, a novel network of green–blue interconnected cycling pathways has been implemented, including green roads, green walls, etc., in order to connect the green and blue spaces of the city center with those in the suburban areas.

By considering past and current challenges that Leipzig addresses, we analyze the whole process from the first conception of NBS creation to its implementation and successful realization. The overall aim of this paper is to show and discover more evidence for the effectiveness of NBSs by providing examples of best practices that demonstrate the multiple co-benefits provided by NBSs. In order to achieve this, this paper first presents an analysis of a broad spectrum of NBS cases, explores success factors in the governance of NBSs, and provides a methodology for evaluating NBS case studies. Secondly, this paper compiles case studies that demonstrate the cost-effectiveness and efficiency of the delivery of ecosystem services that have provided an equitable distribution of multiple benefits. Thirdly, we assess the indicators of success and failure, which can be useful when comparing to related issues in other cities.

By providing this analysis, we examine how NBSs relate to existing concepts and sustainability in general as well as what implications can be drawn for NBS research, its applications, and policies. By applying the framework for assessing the co-benefits of NBSs in urban areas developed by [6,9], while also reflecting [1,10] and actualizing within the EU funded H2020 CONNECTING Nature Project impact assessment of NBSs [11], we assess the opportunities and challenges that face NBSs while also addressing all the related issues of urban society and sustainable development.

2. Study Area

The city of Leipzig, which is situated in the eastern part of Germany, formerly the socialist GDR, is a "city of extremes" [12]. The city went through a long period of contraction, especially after Germany's reunification in the 1990s when the city lost more than 20% of its population (about 100,00 inhabitants of lost population). This was due to industrial decline, for example lignite coal industry contraction, job losses, massive out-migration, and a decline in birth rate as well as residential suburbanization into the city's periphery [13]. In the beginning of the 2000s, Leipzig became the capital of "vacant housing", where about 20% of buildings were vacant [12,14]. Due to industrial decline after the reunification of Germany, Leipzig has become famous for thousands of brownfield sites. At the same time, population decline led to a sudden improvement of the environmental situation (decontamination and a decrease in air pollution).

From 2000 onwards, shrinkage has become less visible, and since 2010, Leipzig has seen a dynamic regrowth of its population at a rate of about 2% per year, reaching 600,000 inhabitants in 2019 [7,15]. The extreme population growth of the city has proceeded from about 2010 onward and puts existing green infrastructure, open spaces, and brownfield sites under extreme pressure to be converted into sealed or built space. In this way, it has become necessary to strengthen the ecosystem services in the city and implement new urban NBS initiatives with the goal of improving the quality of the environment and creating a greener, more livable city (Figure 1).

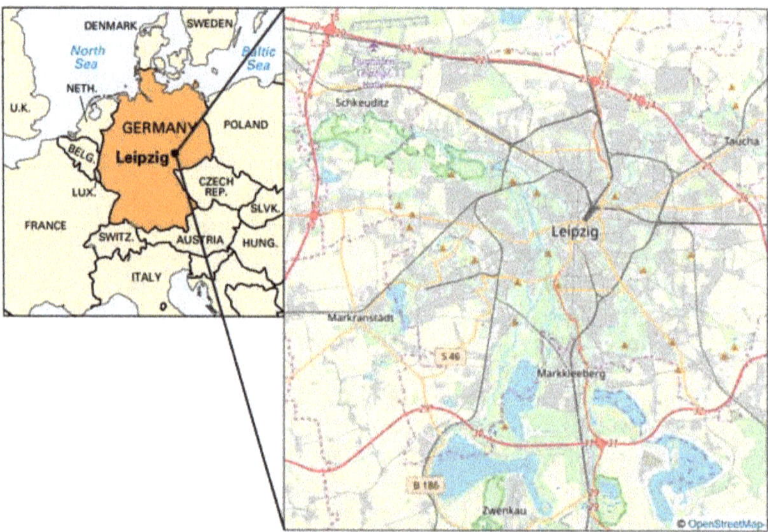

Figure 1. Map of Leipzig (data sources: OpenStreetMap; ArcGIS).

The city currently faces major challenges (Figure 2), including large-scale social, economic, political, and ecological changes over the last 150 years that have been caused by industrialization, the city's socialist past, post-industrial development with active out-migration, and the side effects of rapid development (from a shrinking city to becoming one of Germany's most rapidly growing cities in only a few decades). Other large challenges include changes in Leipzig's natural, semi-natural, and urban landscape, which underline the need to redevelop large open-pit lignite mines on the outskirts of the city, former military training grounds (e.g., to the northeast), and other derelict industrial infrastructures. Environmentally, these challenges also included dealing with the consequences of past industrial uses and landscape changes, including water and air pollution as well as changes to the water system. On the one hand, Leipzig is characterized by extensive and unique urban woodlands, alluvial forests, municipal parks, and garden colonies within easy reach of the city center, as well as by a varied system of rivers, brooks, and recently re-opened canals. On the other hand, its environmental richness came under serious duress during more than a century of heavy industrialization, including large-scale open-pit lignite mines, power stations, and dirty chemical industries.

The second group of challenges includes those caused by environmental risks connected to climate change—i.e., flooding and heat waves. The last big floods in Germany (stemming from the Elbe, Oder, Mulde, and Rhine) drew attention to the fact that technical or manufactured water protection is not 100% effective, particularly when taking climate change and longer heavy precipitation events into consideration [16,17]. Moreover, with its hazardous bursting of large dams, the last Elbe flood in Germany in 2013 made it very clear that the failure of technological solutions can result in enormous damage and casualties. This is because the capacities of technological solutions are enormous in terms of how much water they can hold back, which is much more than any NBS. In this regard, when restoring and maintaining the functionality of wetlands and floodplains or when revitalizing large and smaller rivers NBSs manifest a continuous and more natural inflow and distribution of rainfall water in a larger area and have been shown to improve protection against catastrophic flood events along rivers and coasts [17].

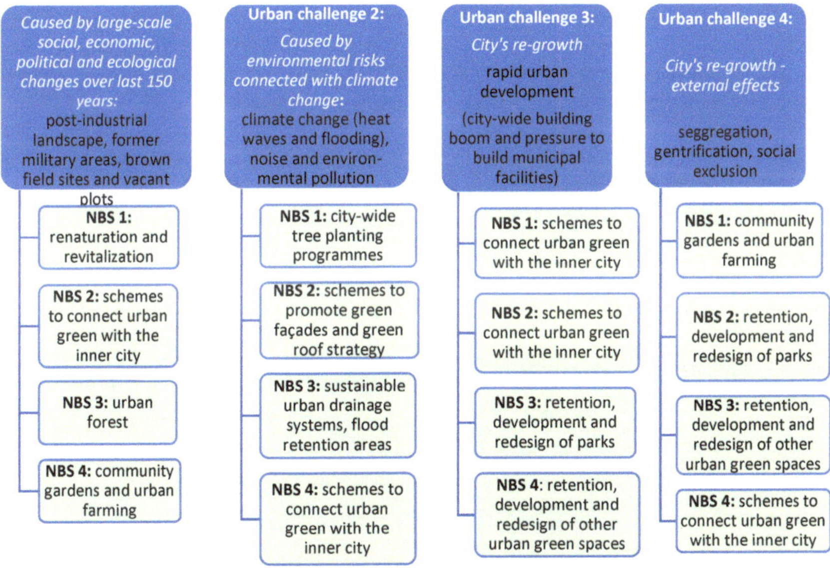

Figure 2. Current major sustainability challenges systematized for the city of Leipzig and related to the nature-based solutions (NBSs) that are the focus of this study (source: authors).

Other sustainability challenges for Leipzig are connected to the stimulation of economic growth, which provides employment opportunities and counteracts population decline and social exclusion alongside emergent socio-spatial differentiation—partly population segregation and what is called ecogentrification—between urban districts. Figure 2 provides an overview of how these challenges are addressed by using the concept of NBS in the case of Leipzig. These challenges will be analyzed in detail in the next sections of the paper.

3. Material and Methods

In order to research the selected NBS cases, we analyzed the multiple benefits provided by NBSs. These include benefits related to climate change adaptation and mitigation, the conservation of biodiversity, and the provision of other ecosystem services for human well-being, including benefits to health. For this purpose, we applied the research framework proposed by Raymond et al. [6] that is further developed within the H2020 project CONNECTING Nature and is related to:

(1) Identification of either a problem or a challenge;
(2) Selection and assessment of an NBS and related actions;
(3) Classification and characterization of the NBS design implementation process (including different scales, types, scopes of NBS—see Table 1);
(4) Mapping of NBS potential;
(5) Analysis of governance models of NBS: initiators, actors, and stakeholders involved;
(6) Assessment of indicators for successes and failures when implementing NBSs [18,19].

First of all, we conducted a literature review on conceptualizations, approaches, and lessons about the co-creation, co-production, and co-development of NBSs, with a specific focus on application in cities. Then, we applied the analysis of environmental and urban development reports and white papers on NBS-related issues in Leipzig. This was followed by a focused review of the literature and reports on existing NBS governance models in cities, drawing on the governance work of other NBS-related projects such as Nature4Cities [20], Naturvation [10,21], and an analysis of dominant governance modes of NBS interventions presented in the Connecting Nature database [11,22].

In addition, semi-structured interviews were conducted using a questionnaire developed as part of the research work within CONNECTING Nature. We interviewed experts on emergent, innovative, and novel NBS experiments. The interviews were supplemented by site visits and participant observation including those during open public events, urban festivals, public lectures, guided excursions, and other events.

In order to conduct interviews, we applied templates (questionnaires) developed by the Connecting Nature WP2 lead partners from DRIFT (Dutch Institute for Transition Rotterdam). The aim of these interviews was to learn from experts on NBS experiments who were implementing NBS experiments in Leipzig (Figure 3). By analyzing their approaches to the co-production of the design and implementation of their NBS cases, we endeavored to identify lessons learned that will benefit other cities and stakeholders who are interested in designing, implementing, and stewarding NBSs.

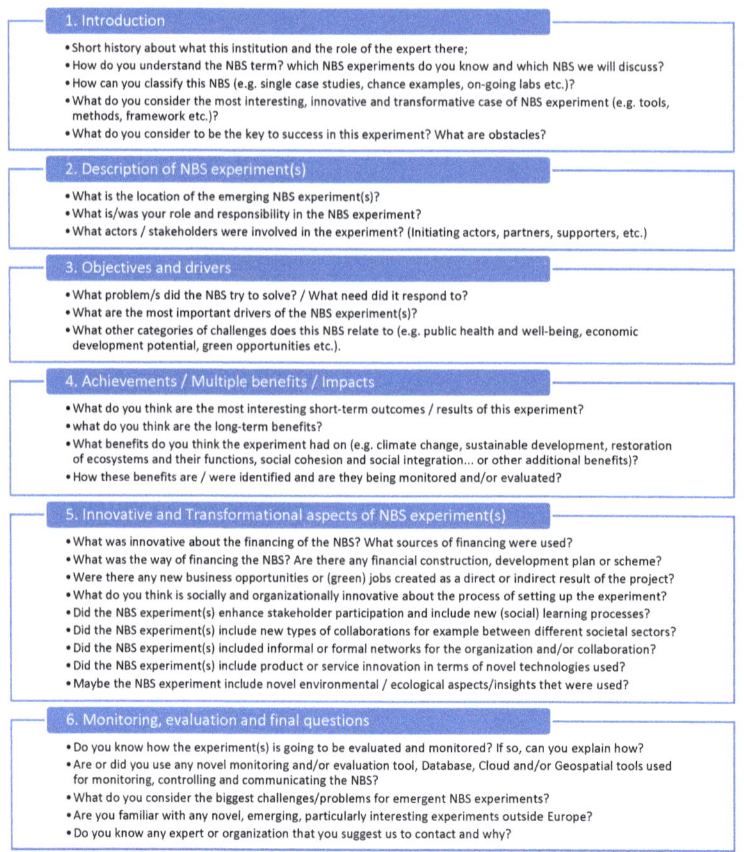

Figure 3. Interview template including the six-step iteration applied in the Leipzig study.

Table 1. Classification of NBS interventions and selected cases from Leipzig referred to each category (based on [23–25], with own additions). Photos: D. Dushkova.

Classification according to the Scale or Scope	Main Aim of the Interventions	Interventions Included in the Class	Examples from Leipzig Case	
			a	b
Building-scale interventions	Refurbishing of pre-existing buildings, design of new buildings	Actions on rooftops, facades (a), actions in community spaces of the buildings (b) (a) Greening facades within Kletterfix project; (b) Greening backyards in LWB municipal housing facilities		
Interventions in public spaces	Public space regeneration, urban land renewal, design of public living areas to increase social cohesion and integration	Actions in public living areas, urban parks, and public space (a), renaturing abandoned/post-industrial areas (b) (a) ANNALINDE intercultural garden; (b) Lene-Voigt Park—former Eilenburger Railway station		
Interventions in water bodies and drainage systems	Renaturing and recovery of river courses and wetlands, ponds, and lakes	Renaturing rivers and streams (a), restoration of ponds and lakes, sustainable urban drainage system (SUDs) (b), (a) Renaturing of Karl-Heine Channel highly polluted by textile industry; (b) SUDs in housing area		

Table 1. *Cont.*

Classification according to the Scale or Scope	Main Aim of the Interventions	Interventions Included in the Class	Examples from Leipzig Case	
			a	b
Interventions in linear transport infrastructures	Road projects, mobility plans, redevelopment, and greening streets	Naturing actions for both high capacity (i.e., roads–railways, etc. (a) and greening streets (b) (a) Greening the tram railways; (b) City's tree planting program: Baumstarke Stadt		
Interventions in natural areas and land management	Master plans to use/manage spaces, public space plans, Green Infrastructure strategies, agriculture, and forestry promotion plans	Natural protected areas, wetlands, peri-urban parks, rural land management (a) Leipziger Auenwald (Rosenthal park) (b) Leipziger Neueseenland recreation spot (former lignite mining area)		
Ecological education and awareness raising-related interventions	Awareness raising to environmental issues, stakeholders' and citizens' involvement, knowledge transfer	Ecological festivals, workshops, master classes (a) Umwelttage und Ökofete Leipzig (b) Project "Edible city" and Nutrition Council Leipzig		

In total, 24 semi-structured interviews were conducted with municipal officers, stakeholders from nongovernmental organizations, private owners from SMEs, and citizens in the period between February 2018 and August 2019 (Appendix A, List of interviews). The interviews were mostly conducted in German and, after being transcribed, were translated into English. The interviews lasted between 29 and 128 minutes. The recording device used was an OLYMPUS digital voice recorder (WS-853). The interviews were transcribed with notes taken for the remainder and were analyzed using content and thematic analysis. Photographs and notes were taken during participant observation of events, excursions, and workshops. These were also used to hold informal talks with participants and interview partners.

The information was also obtained during the events:

- Annual "Eco-Festival" (Umwelttage Leipzig, 5–24 June 2018 and 5–16 June 2019)
- Annalinde Saison opening (13 April 2018)
- Spring festival of Annalinde intercultural garden (8 May 2019)
- Annalinde Herbal festival (24–25 August 2019)
- Guided open public excursion from the Project "Kletterfix" (19 June 2018)
- Guided excursions in Leipziger Neuseenland (23 June 2018 and 31 August 2019)
- Workshop "Edible City" (3 February 2018).

We also used secondary information that was collected at these events, in public information centers, and by researching web-based reports, published plans, and documentation distributed at public consultation events. We also used plans, newsletters, briefings, online information, and press articles as well as the web sites of the NBS interventions mentioned and analyzed in this paper. Some papers that addressed related issues were discussed during the site visits, workshops, and expert discussions that were focused on the potential of NBSs to address specific urban sustainability challenges in Leipzig.

4. Results

4.1. History of Urban Greening and Implementation of NBS Interventions

Before we analyze how the NBS concept is perceived in the city of Leipzig by different stakeholders and actors (city government, planners, private companies, nongovernmental organizations (NGOs), civil society, etc.), we provide here an overview of the history of urban greening and examples of prior so-called NBS innovations. In the case of Leipzig, these include pre-war cooperative and late-socialist housing developments that emphasized the need for planned green corridors and communal spaces.

The main core of green infrastructure in the city was developed during the socialist period, mainly in the 1950s and 1960s. For example, the centrally located Clara-Zetkin Park was shaped according to a Moscow blueprint [13]. Housing estates erected in the socialist era were planned with green spaces. These green spaces—which comprised backyards and courtyards—were designed to satisfy the requirements of the socialist society and were well equipped with roads, parking places, pedestrian walkways, waste collection sites, and also vast green spaces with children playgrounds and sport grounds (Figure 4).

Many of these spaces, which were also equipped with artworks and fountains, typified the architectural qualities of the modernism of the second half of the 20th century. The green urban concept of housing estates from this period responded to the socialist idea of the "collective dream". The playgrounds and landscape design of the backyards and courtyards in the housing estates can be perceived as prior NBSs since they contributed to human health and well-being, shaped the attractiveness of place, provided recreational and cultural value, fostered social cohesion, and created places for communication. After the fall of socialism and the reunification of Germany, a large part of such socialist estates in Leipzig—especially in its central and northern parts—has survived, even if, in some cases, their values have been lost as a result of the densification of housing, civic amenities, and parking spaces. Today, their fragments are still legible in the open spaces of housing estates.

Figure 4. Greening the backyards and courtyards in socialist pre-fabricated housing estates of Leipzig Center as an essential element of socialist housing development and urban planning strategy. Photos: D. Dushkova.

As a consequence of intense industrial development, the negative environmental situation has been a crucial reason for out-migration from Leipzig since the 1960s, which shrank the city long before the political changes of the 1990s. Thus, from an environmental perspective, the post-socialist transformation offered a lot of relief, chances, and new opportunities for establishing novel nature-based solutions [13]. After 1990, a "green belt" around the city was created (Figure 5), which included the renaturation of former lignite coal mining sites as new water landscapes and also demanded active cooperation between urban and regional actors.

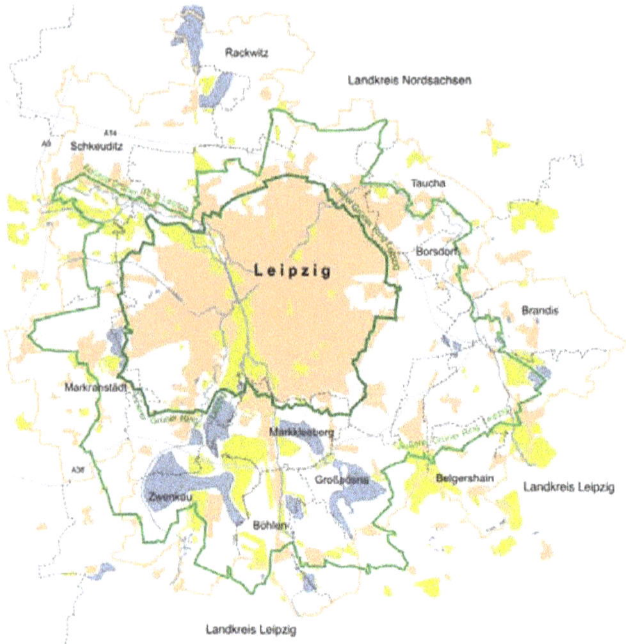

Figure 5. Green belt of Leipzig (dark green = 'Inner Green Belt'; light green = 'Outer Green Belt'). Source: Administrative Office of Green Ring Leipzig—https://gruenerring-leipzig.de/wp-content/uploads/2017/06/grünerring-kurzinhalt2.pdf.

The Karl-Heine Channel was completely restructured and made accessible for walkers, cyclists, and boaters [26]. Urban community gardens, playgrounds, and urban agriculture experiments are widely popular in the city of Leipzig. Leipzig is also the origin of the long-allotment tradition that started in 1864 with the "Schreber movement". Later, through a movement born in England during the 1960s, community gardens became popular in Germany and introduced alternative methods such as permaculture into urban gardening (Figure 6). With its 270 allotment gardens, Leipzig now has the highest density of urban gardens in Germany. Besides such important ecosystem services as food provision (self-supply) and recreation, these gardens also play an important role by providing local climate regulation and biodiversity conservation [27].

Figure 6. Allotment gardens of Leipzig-East. Photos: D. Dushkova.

Community gardens that contributed to the upswing of neighborhoods and attracted new inhabitants have to move constantly because of displacement. Traditional allotment gardens could also become endangered in the future by the demand for new building areas. At the moment, Leipzig finds itself in an interesting but complicated situation. The city benefits from a good image whereby green infrastructure plays an important role, but it is also under pressure because of new growth.

A novel concept of greening kindergarten areas within the NBS approach includes the nature-oriented playground for pre-school children that includes elements of nature alongside play equipment. These include sandy hills, live willow huts, and paths made from logs and stumps that are supplemented by green flowers, vegetable patches, herb beds, fruit bushes, and houses that allow children to grow their own plants or observe insects (Figure 7). By researching the incentivization of kindergartens in Leipzig, we discovered that a large number of developed and implemented projects de-sealed the concrete ground of playground areas by using native plants in the greening in order to enhance biodiversity with minimal maintenance. Designing the tree-based or meadow-like green spaces of kindergartens in this way can greatly contribute to adaptation to climate change, as air temperature is expected to decrease due to vegetation transpiration and canopy shade as well as due to de-sealing.

(**a**)   (**b**)   (**c**)

Figure 7. An innovative concept in greening areas near the kindergartens of Leipzig: (**a**) nature-oriented playground, (**b**) flower and vegetable beds, and (**c**) flower beds in water-sensitive design. Photos: D. Dushkova.

NBSs such as the greening the open spaces of kindergartens are a part of the overall development concept of kindergartens in Leipzig. As such, the users of these areas may alter and create the conditions of their everyday activity areas in a self-determined way. *"We organize the meetings with parents and children—users of our green spaces—in order to discuss the needs. Then, we can draw and even build miniature garden models in a box during the classes with our children. Such a participative approach excludes the hierarchies and is wonderful to hear and reflect every voice"* (Interview #20). In this sense, such newly created green areas of kindergartens can be defined as "play gardens" more than "playgrounds", giving children and their guardians the opportunity to experience nature directly and create an additional friendly greenspace in the city.

At the same time, solutions to the current social and environmental issues demand a set of tools and political will. Lene-Voigt Park, the Karl-Heine Channel, the Leipzig floodplains, and Leipzig New Lake Land are good examples of relevant problems and how greening cities can address them. These will be addressed in the following sections.

Under the conditions of climate change, it is crucial to protect and further develop the blue–green infrastructure network in Leipzig. In this context, NBSs can be responsible for the question of how activities should be improved in order to maintain important ecosystem services that are provided by the city's green and blue infrastructure. This will be discussed in the next sections of the paper.

4.2. NBS Approach: Environmental Aspects in Urban Planning Policy Related to NBS

In this section, we endeavor to answer two important questions: (1) Could many smart ideas that have been promoted in Leipzig (and described in the previous subsection as prior-NBS interventions) be included in and perceived as an NBS? (2) How is the NBS concept perceived in the city and by the city government, different stakeholders, and actors?

We find that the first question needs to be highlighted here in detail, since a great number of scientific debates and discussions with different stakeholders have revealed that the term is still unclear, and there is no precise differentiation between NBSs and prior interventions related to the environmental management, nature protection, greening measurements, etc. Nature conservation approaches and environment management initiatives have been carried out for decades and even for centuries as the case of Leipzig shows. What is new is that the multiple benefits of such NBSs to human well-being and society as a whole have been articulated well more recently. Even if the term itself is still being framed, examples of nature-based solutions can be found not only in Europe but also all over the world.

The concept of NBS itself was developed in order to propose a shift in nature conservation, environmental protection, and risk reduction from traditional 'hard' or 'gray' engineering solutions that exclusively involve structural features to 'softer', more eco-friendly solutions that simultaneously provide multiple benefits and are more cost-effective. In the last decades, a variety of new terms has started to be used to describe this type of solution, including the ecosystem-based approach, ecosystem services, green infrastructure etc. This variety of terms, coupled with a lack of specific details about the physical elements of the solution, can cause confusion. In our paper, we use the term nature-based solution (NBS) as the catch-all term for these approaches and according to the main stream of the CONNECTING Nature project, we are searching for innovative approaches by trying to identify what particular innovation can be found within particular NBSs to make it not just as a label but also a real tool to address the current societal challenges and achieve the sustainable development goals (www.undp.org).

As we learned from the work within the CONNECTING Nature project, one of the essential features that distinguishes NBSs from other environmental management-related interventions is using an innovation approach that brings together city governments, SMEs, academia, and civil society in order to co-produce usable and actionable knowledge, new business opportunities, and new governance models. By analyzing different NBS interventions presented in the database that we can name here, several innovations are coming through NBSs (presented in detail in the Appendix B for each NBS case):

- Improving the level of engagement of different actors participating in the development and implementation of NBSs (especially the business sector);
- Participation in decision making (public consultation, citizens budget);
- Inclusivity for multicultural society (e.g., a positive impact for as many social groups as possible);
- Accessibility to green and blue areas and providing environmental justice;
- Close collaboration between a wide range of stakeholders, which allows innovative problem solving and innovative organizational processes;
- Social and business innovations (innovations processes in organizational transformations (e.g., the people and structure of the organization have to be prepared for and capable of change);
- Fostering individual and collective innovative activity;
- Making use of innovative, interdisciplinary planning methods for green and blue space co-design and co-implementation, including the development of innovative social models for long-term positive management;
- Change in behaviors (awareness raising, promoting of eco-friendly lifestyle, etc.);
- Educational changes (workshops, events schools, pre-schools gardens, new gardens).

In order to answer the second question, we refer here to Cohen-Shacham et al. [24], who identified core principles for successfully implementing NBSs. Among these are synergy with other solutions, landscape-scale considerations, and policy integration. As the analysis of our NBS examples show, they can be implemented alone or in an integrated manner with other solutions. However, the successful cases should always be an integral part of the overall design of policies, measures, and actions taken to address societal challenges.

By discussing the features of current urban politics of Leipzig at the municipality level, we highlight the clear understanding that " ... *the city became aware of needs to focus on the topic of greenery and nature. We support the programs and events which educate our citizens and decision makers even more about the economic and social values of urban greenery and nature-based solutions—from historical, botanical, and private gardens to parks, roadside and water-space greenery, greening façades and rooftops, and other outdoor spaces*" (Interview #18). As stated by the City Office (Department of City green and waters, Department of Environment protection), " ... *partnerships and collaboration between different types of organizations and actors can produce previously unexpected solutions, especially when it comes to environmental issues*" (Interview #19). However, the problem is that members from different sectors of society often perceive NBS initiatives in different ways, pursuing their own interests. Thus, " ... *the current urban development strategy simply supports any kind of sustainable approaches somehow* ... " (Interview #19). This way, the city hopes to communicate and, subsequently, to harmonize the different ideas and visions regarding how to solve the social–environmental challenges on which this NBS paper is focuses.

The range of NBS initiatives in Leipzig "*expands from preventing flooding to responding to heat island effects in big cities*" (Interview #19). The governance of those initiatives involves public authorities, the private sector, civil society, NGOs, the green movement, and academia. One such prominent example is the Leipzig Charter on Sustainable European Cities (Die Leipzig-Charta) [28]: the National Initiative for Urban Development Policy. This initiative aims to provide high-quality urban development in terms of design and planning that enables growth with low carbon emissions, thus improving environmental quality and reducing carbon emissions.

As our analysis shows, the first and still broadly used governance model of NBSs in Leipzig includes state involvement in financial management. The second model of NBS governance is volunteerism. A third model refers to those that were initiated by the private sector and in which the main stakeholders are businesses (SME). Nevertheless, one of the main points in the governance of NBS concerns the sharing of costs and risks between the private sector and the state. However, the involvement of business can be expensive and risky, and sometimes, it leads to different forms of social exclusion. It is common to encounter the mixed governance model of NBS where private and public sectors manage the finances together. Economic and social relations, as well as gentrification pressures, could become a significant challenge for the governance of NBSs. The key to successful NBS initiatives is the joint reflection of and by all the stakeholders [21].

In the case of Leipzig in the 1990s, many urban green initiatives were initiated and governed by public authorities and state actors. However, in the past few decades, the vision of the role of the public sector in greening initiatives in cities has started to change. At the moment, local public administrations still emerge as the main source of the overall vision, planning, and management of green infrastructure in cities. However, " ... *state actors attempt to work with society more as more people are involved in the green movement when the environmental issues have become more crucial*" (Interview #18). Public administrations organize public education courses and activities and try to work with volunteers through planting events. Adding more parks, trees, green roofs, green walls, playgrounds, and allowing for urban agriculture, green spaces, and gardens, as well as replacing hard surfaces such as parking lots with green spaces, were reported to be the main types of successful NBSs in the city. "*The city administration is aware of the importance of its green and blue infrastructure for the city's economic development. We are welcoming new activities and pilot projects by providing financial incentives [and] supporting the cooperation with local stakeholders, local enterprises, and the various science and research institutes*" (Interview #18).

On the other hand, the idea of sharing opportunity costs and economic risks between the private sector and the state is not new, and it is certainly not unique to urban greening projects. Private sector and business should be involved in NBS projects, for example, in the context of shrinking state budgets and the long-term management of urban greening initiatives while benefits for the private sector fail with short-term interests [29]. Such arrangements are expected to deliver continuous economic growth, in the case of the EU, while avoiding irreversible and unpredictable changes to the global ecosystem [30].

Furthermore, state authorities and the private sector are not the only entities now actively involved in the governance of NBSs in Leipzig. Civil organizations and NGOs are also responsible for innovations and transitioning those green initiatives. They are seen as key actors for advocating for NBSs (providing applied evidence of their benefits) and re-establishing a green urban common [7]. As the case of Leipzig shows, " ... *local residents and community groups very often positively perceive new NBS interventions in their neighborhoods if they respond to the needs to distribute socio-ecological benefits in order to reach ecological justice*" (Interview #18).

Engaging all actors in the process of implementing NBSs is a potential solution in which all sides benefit and where innovation, economic gains, biodiversity protection, and climate change could go hand in hand [7,31,32]. The partnering of different actors in the governance of NBS is perceived as " ... *a way to reduce barriers/constraints to adopting NBS on a wider scale, which is especially important in terms of implementing NBS projects in cities*" (Interview #5).

The benefits of NBS only accrue when they are embedded in urban social–ecological systems—i.e., they need to reflect environmental conditions and needs, as well as socio-economic priorities, norms, and values, and human perceptions and institutional contexts that characterize specific local neighborhoods, the city as a whole, and regional connectivity [8,33]. While NBSs are often still initiated and financed by local governments, their long-term viability (e.g., through new business models) requires knowledge from multiple actors (e.g., citizens, NGOs, social innovation networks, businesses, and scientists) so that the solutions fit the needs and context of the city [32].

Today, a growing number of independent urban garden initiatives, intercultural community gardens, kindergartens, and school garden projects are revitalizing abandoned wasteland within the Leipzig city area. "*These projects are extending the traditions of gardens as a place of communication, creation, and learning [while] contributing at the same time to the conservation of biodiversity in urban areas and shaping the healthy environment for their inhabitants*" (Interview #4).

Maintenance and development plans (Pflege und Entwicklungspläne) play an important role, as they are an essential component of the overall green infrastructure site management developed by the city government. " … *[T]hey have primarily been used to safeguard the quality of valuable sites such as protected areas, but in [the] future, they should also include green spaces*" (Interview #19). These maintenance and development plans integrate innovative approaches such as *"protection through utilization"* projects or *"volunteer-based management interventions"* which are organized, for example, by the city of Leipzig as part of the Leipzig Garden Programme (Interview #4).

There are a number of current NBS projects that have been recognized as a novel collaborative mode of urban governance that use a mixed governance model based on co-creation and co-production. According to the European Union, these models are acknowledged as key mechanisms for dealing with sustainability challenges [34]. They allow for deep participation to leverage and weave together local, expert, and tacit knowledge and, ultimately, advance urban sustainability and resilience [3,7,8,32]. Co-creation and co-production promote collaborations and partnerships among diverse actors—including civil servants, citizens, planners, entrepreneurs, architects, scientists, and engineers—in the design, implementation, and eventual stewarding of NBS. As will be shown in the next sections of the paper, this mixed model has been successfully applied and works well in a variety of NBS examples.

4.3. Analysis of NBS Examples

In previous sections, we highlighted the urban challenge and how NBSs can meet them. We noted that the greatest benefits provided by NBSs include place making, air quality, urban heat mitigation, urban agriculture, acoustic control, storm water mitigation, biodiversity, aesthetic quality, human health and well-being, and social cohesion. In this section, we focus on the selected NBS examples that respond to the key sustainability challenges faced by the city of Leipzig (the NBS projects are listed in detail in Appendix B):

1. Greening façades (within the project Kletterfix)
2. ANNALINDE Intercultural Garden
3. ANNALINDE Urban Agriculture
4. ANNALINDE Academy
5. Project "Edible city"
6. Nutrition Council Leipzig
7. Lene-Voigt Park (renaturing a former industrial area to public green space)
8. Eco-food project from "Rosenberg Delikatessen"
9. Ecological festivals
10. City's tree-planting program
11. Leipziger Neuseenland (revitalization of post-mining area)
12. Karl-Heine Channel (revitalization of waterscape)
13. Leipziger Auwald (riverside/riparian forest)

These NBS examples represent the full range of NBSs by showing how social, economic, and environmental goals can be reached and which innovative approaches were implemented in order to provide a large variety of benefits such as climate change mitigation and adaptation, water management, quality of life, human health and well-being, attractiveness of place, social inclusion, and green job opportunities (Table 2).

Table 2. Classification of the selected NBS examples in Leipzig. NGOs: nongovernmental organizations.

NBS Case	Time Frame	Motivation	Initiator/Stakeholders	Benefits	Source of Financing	Project Budget
1. Green facades and walls for Leipzig–Kletterfix	2015–ongoing	to create and raise awareness about the multiple benefits of urban green and inspire the action of tenants and landlords on private property	NGO Ökolöwe/City Leipzig	improvement of urban microclimate and general living conditions, increase urban biodiversity, combat particulate matter pollution, generate economic savings for tenants	public local authority's budget, funds of NGOs	less than 50,000 EUR
2. ANNALINDE Interkultural Garden	2011–ongoing	to create places of exchange and learning on ecological growing of food, biodiversity, sustainable consumption, responsibility in resources, sustainable neighborhoodd, and urban development	civil society—neighbors and volunteers, NGO/civil society, citizens or community groups, ANNA-LINDE GmbH	public space regeneration, urban land renewal, and design of public living areas to increase social cohesion and integration, gardens on unbuilt areas open to all and for active engagement with social issues	EU funds, public national budget, public local authority's budget, funds provided by NGOs, crowdsourcing	n.a.
3. ANNALINDE Urban agriculture	2011–ongoing	to encourage, reactivate urban gardening, cultivate vegetables according to the standards of organic farming, to provide social interaction and mutual learning				
4. ANNALINDE academy	2011—ongoing	environmental awareness rising, knowledge transfer, stakeholders' and citizens' involvement through master classes, workshops in urban agriculture, gardening, recycling, composting		opportunities for joint and experience-based learning toward urban resilience and sustainable development, shaping environmental culture, contributing to sustainable urban living		
5. "Edible city"	2018—ongoing	to promote edible species within public spaces (parks, pedestrian streets, etc.) as a source of food with financial advantages for municipality and citizens, environmental awareness rising, knowledge transfer, stakeholders' and citizens' involvement	NGOs, civil society groups	good peri-urban relationships, sustainable regional agriculture, local food production, support of local food producers, gardening movement aiming to turn the city into an edible garden	NGO funds, civil society groups funds	n.a.
6. "Nutrition Counsil Leipzig"	2019—ongoing	to improve the food system in Leipzig and make it more ecological and more democratic; to ensure the long-term sustainable supply of the city's population with regional, seasonal, and healthy food	civil society groups, farmers, processors, retailers, restaurateurs, municipality representatives	better dialogue between different actors of food system in the city; the development of a sustainable regional agricultural system, biodiversity conservation; knowledge and education platform about the production, marketing, and the supply of the food in the region	civil society groups funds, private companies, municipality	n.a.
7. Lene Voigt Park	2001–2004	to renaturate a former industrial area to public green space, to connect eastern districts and the 'green lungs' of the city (park as a part of a green belt), to co-create with residents (through workshops), to offer more green space for the dense housing area	City of Leipzig	to initiate residential change of the surrounding areas and the development of the local urban infrastructure, e.g., cafes, shops. New residents moved there and housing vacancies started to decrease	City of Leipzig	n.a.

Table 2. *Cont.*

NBS Case	Time Frame	Motivation	Initiator/Stakeholders	Benefits	Source of Financing	Project Budget
8. Eco-food project "Rosenberg Delikates-sen"	2016—ongoing	to establish green business/company, to use regional food/products, to support regional fluxes of materials and energy balance	private initiative	respect to traditional processing/manufacturing of regional fruits, good networking that involves different actors, the promotion of regional agriculture and products	private initiative	n.a.
9. Ecological festivals (Umwelttage and Ökofete)	1990—ongoing	awareness raising on environmental issues, knowledge transfer, stakeholders' and citizens' involvement	NGO funds, City of Leipzig, private companies, etc.	environmental education, active engagement of citizens in the city greening and environmental activities	NGO funds, City of Leipzig, private companies, etc.	n.a.
10. City's tree planting program (Baumstarke Stadt)	1996—ongoing	to give private people and businesses the opportunity to donate money for the planting of one or more trees, active engagement of citizens in the city greening	City of Leipzig, citizens, and private companies	city greening, place attractiveness, air quality, promoting the special cultural meanings that many attach to trees, i.e., as a source/giver of life or a memory site or symbol, financial contribution of citizens to the city greening	public local authority's budget, corporate investment, crowdsourcing	n.a. (tree sponsorships of 250 EUR and more pro one tree)
11. Leipziger Neuseenland (Leipzig new lakes regions)	1994—ongoing	to create a new peri-urban landscape through revitalization of the post-mining landscape destroyed by vast open lignite mines (remaining craters are renaturalized and flooded with ground or rainwater to be revived as recreational lakes)	mining companies, scientific institutions, recreational business, regional and national bodies	renovation and development of a new concept for the former mines, including the creation of new employment opportunities, improvements in environmental conditions (air and water quality), recreational area	public national budget, public regional budget, corporate investment	above 4,000,000 EUR
12. Karl-Heine Channel	2005—onging	to renaturate and recover the river courses and wetlands, ponds, and lakes, sustainable redevelopment of the post-mining area into forestry and recreational areas and reintegrate it into the adjacent landscape	City of Leipzig, external experts, Forum for Leipzig West (business, NGOs, citizens)	installing waterside greenery, developing the green zones of former waterside embankments to interlink green areas in the whole city, a cycle path network along the canal, recreational areas, attractiveness of space, human well-being	City of Leipzig, national funds, EU funding	5,500,000 EUR
13. Leipziger Auwald (Riverside/Riparian forest)	1991–2011	environmental protection strategy and master plans to use/manage spaces, public space plans, Green Infrastructure strategies	Leipzig University, UFZ, iDiv, Naturkundemuseum, City of Leipzig, NGOs (Ökolöwe, NABU, etc.)	biodiversity conservation, air cooling, shading and transpiration, flood mitigation, carbon storage, habitat provision, legislation (nature reserve); workshops to promote the value of Auwald, awareness raising (cultural open-air events and concerts), recreation	public local authority's budget, funds provided by NGOs, etc.	n.a. (partly municipality 198,000 Euro

Note: n.a—the data was not available.

5. Discussion

5.1. Factors of Success of NBS Examples

The following aspects that contribute to the successful existence of selected NBS examples were discovered by looking at the history of their creation, their impact (environmental, social, and economic), governance models (initiator, stakeholders, beneficiaries), methods of implementation, design, and maintenance, and additional benefits, costs, and financing:

- **Multifunctional in the best sense**, having a wide spectrum of ecological and social activities and multiple benefits
- **"Formalize the NBS"—political support and willingness** (close collaboration with municipality: implementation of projects into the master planning of a city and municipality programs, regular dialogue with urban policy makers and planners to facilitate knowledge transfer, strong local commitment to protect implementation)
- **Community engagement** in the design and implementation process—a high degree of citizen participation and "accessibility" of the NBS
- **Close collaboration between a wide range of stakeholders** that allowed innovative problem solving
- **Participatory processes** that support stakeholder empowerment
- **Locally grown solutions** (e.g., a preference for using elements of traditional nature-based practices more than approaches that rely on bringing "new" interventions in from the outside)
- **Secured financing** for a longer period
- **Outcomes of the NBS are clearly** communicated, accessible, and attractive in regarding their benefits (environmental, economic, for health and well-being, joy, aesthetics, etc.) for local residents

As we see from all the selected NBS examples, at the local level, an NBS strategy (development) first requires the active consideration and combination of the above approaches to address sustainable development goals (SDGs) [35], especially when aiming to reduce climate risk on the ground. For example, among these SDGs are SDG3 (good health and well-being), SDG 11 (sustainable cities and societies), and SDG12 (responsible consumption and production). Second, to ensure their sustainability, NBS strategies must be implemented at the local, institutional, and inter-institutional levels. Third, the different measures and strategies only lead to sustainable change when combined. Finally, other cross-cutting issues (notably climate change mitigation) can create synergies and support progress.

By analyzing different NBS cases, we can state that the same NBSs may have a different impact in different local contexts. As our analysis shows, expensive greening and revitalization projects (e.g., Neuseenland) on former brownfield or post-industrial sites, for example, work only in contexts where the resulting costs can be paid by state and/or private actors. Projects related to urban gardening (e.g., ANNALINDE urban gardening and intercultural garden) may be successful, given the presence of a motivated and active local community, but will fail if there is apathy and disinterest or even a conflict of interests. The majority of selected NBS examples are largely initiated by citizens and with a strong presence of civil society groups in their creation, implementation, and maintenance. Considering the context in which NBSs are implemented is of great importance for political and power structures, ways of making decisions, and the inclusion of certain groups of inhabitants and/or actors.

As was pointed out in the interviews, factors that lead to success include state-driven initiatives that are supported by private companies and focus on enhancing the attractiveness of the area, providing new recreation facilities, and generating economic growth and employment through tourism (as in case of New Lake Land). Federal support for ecological revitalization was also one of the key drivers for the Karl-Heine Channel project, which aimed to harmonize living conditions during ecological transition as a part of welfare policy. However, as emphasized by several interview experts, federal funds would not be enough without the active participation of civil society. In order for this participation to occur, awareness of the importance of particular ecosystems and their functions is essential (ANNALINDE Academy, Ecofestivals). For example, in Leipzig, the creation of awareness of the floodplain and its related ecosystem services to both people and nature was a key driver of success. For the most part, a combination of different factors was responsible for the success of the NBS example. These included the continuous development of innovative strategies to produce healthy local food, respect for traditional farming of regional fruits and vegetables, effective networking that involves different actors (Rosenberg Delikatessen), co-creation with residents (through workshops), and the provision of more green space for dense housing areas, thus leading to an increase in the attractiveness of places and districts (Lene-Voigt Park).

5.2. Impact of NBS Examples on the Environment, Economics, Society, and the Sustainable Development of the City

In response to the research questions, several options can be outlined that utilize the functionality of NBSs to better face current societal challenges, especially the consequences of climate change in cities and urban regions. The following are the key observations:

- Only the right green projects in the right place (e.g., environmentally based, proven, and responding to the current needs of urban society) can have a positive and long-term effect.
- As our results show, NBSs have a far greater environmental, economic, and social value than generally assumed.
- Investing in restoring, protecting, and enhancing ecological functionality and ecosystem services through the implementation of different NBSs is not only ecologically and socially desirable [16,17,31], but also, as the contribution has shown, the multiple benefits obtained from NBSs have a great value for a large number of beneficiaries in the city of Leipzig.

The impacts of the selected NBS examples are presented in Table 3.

Table 3. Impact of NBS examples for environment, economics, society, and the sustainable development of Leipzig.

Social Value of NBS	Economic Value of NBS	Ecological Value of NBS
creating safe and welcoming spaces for residents, leisure and recreation with associated therapeutic values	improvements in residential quality and image of the area (through further upgrades to existing or creation of new green spaces), which contributes to its attractiveness for potential investors and new residents	climate change mitigation and adaptation (flood mitigation through local water retention and revitalization of water landscapes; heat and drought mitigation); microclimate improvements through raising the number and quality of green spaces
promoting integration and social cohesion	mobilization of private and public finances for the upgrading of structurally disadvantaged areas	biodiversity conservation and habitat provision
contribution to interaction between diverse ethnic and social groups, also with migration background	green job opportunities through the creation of green business models of NBSs	decrease in air pollution (including lower emission rates through promoting green mobility)
providing opportunities for active engagement	financial incentives when implementing NBSs	regeneration of derelict areas
raising a district's cultural vitality and image, thus contributing to the attractiveness of place	better infrastructure through the provision of safe pedestrian and cycle paths to connect different parts of the city	reduced land usage
enhancing public perception of the area	extension of the touristic infrastructure and better connection to transregional tourism networks	sustainable living (through responsible consumption and production, increased resource efficiency, etc.)

During the analysis, we found the follow types of NBS interventions in Leipzig:

- Type 1: NBSs that involve making better use of existing natural or protected ecosystems (e.g., measures to increase ecosystem services supply)—the Karl-Heine Channel, Leipziger Auwald
- Type 2: NBSs based on developing sustainable management protocols and procedures for managed or restored ecosystems (e.g., re-establishing traditional agro or forestry systems, etc.)—Baumstarke Stadt, Leipziger Auwald, Nutrition Counsil, and Edible City

- Type 3: NBSs that involve creating new ecosystems from existing abandoned, brownfields, or neglected area (e.g., establishing green buildings, green walls, green roofs)—Lene-Voigt Park, Neuseenland, and ANNALINDE intercultural garden
- Type 4: NBSs that involve creating new ecosystems (e.g., establishing green buildings, green walls, green roofs)—Kletterfix project
- Type 5: NBSs that aim for responsible consumption and production based on generating awareness, ecological education, and sustainable actions—ANNALINDE Academy, Rosenberg Delikatessen, Ecofestivals.

5.3. Limitatations Associated with Trade-Offs and Conflicts around the NBS

The implementation of effective and durable NBSs face potential barriers, as identified by [7,8,21]. They include "strong stakeholders" or private business (such as housing associations, investors, or developers) with whom a city or municipality has to enter into contractual obligations. Another associated barrier is the risk of "overselling nature" [9,36] or encouraging a perception of ecosystems as entirely substitutable by other assets used by humans. In addition, financial limitations can occur, leading to the inadequate development and maintenance of NBSs (e.g., financial stress, changes in funding priorities within a new political cycle, other barriers in financial mechanisms). In addition, insufficient incentives and regulations to encourage the private sector to include NBSs in developments may hinder the implementation process of NBSs. For instance, the private sector often lacks incentives that would make the business cases for the NBS better defined and attractive.

As we learned from the CONNECTING Nature project discussions and interviews with different stakeholders, many positive effects are not taken into account when implementing NBSs, such as social impacts, which simply cannot be monetarized. The scientific community still experiences a lack of tools or methods to describe the positive effects of NBS, which require a unified methodology and framework for the assessment and evaluation of NBSs. There are still gaps and barriers in the impact assessments of NBSs—in most of the NBS projects, the long-term monitoring of the impact of NBSs is not foreseen; however, it would be an essential part of promoting the NBS concept and approach.

As stated by variety of stakeholders, the long-term benefits are not clearly defined for them and for citizens, which pose an additional task for the city to visualize the effects of projects regarding the long-term benefits [7,9,11,13,17,19,37]. As our previous research shows, very few trade-offs were identified so far, such as eco-gentrification, ecosystem disservices, etc. [7,13,37].

Coming back to our case of Leipzig, we reveal the following actions, which were developed and introduced by a variety of different actors in order to overcome these limitations:

- Development of shared visions (as we noted by different NBS cases presented in the Appendix B such as Nutrition Council, ANNALINDE Academy, etc.).
- Implementation of environmental and sustainability management systems (Leipzig has developed and introduced different strategies such as the Leipzig Charter on Sustainable European Cities, the Leipzig Garden program, maintenance and development plans, the city's tree planting program, ecological festivals etc., which highlight the embedding of NBSs in existing plans and strategies).
- Different campaigns to raise awareness for different stakeholders, including citizens (such as Edible Cities, Nutrition Council, Ecological festivals, etc.).
- Supporting formal and informal networks between different actors involved in the creation and implementation of NBS (for further details, please see the Appendix B).
- Supporting public–private partnership (for further details, please see the Appendix B).
- Use financial incentives and compensation schemes (e.g., when implementing green walls, as shown by the Kletterfix project).

In addition, powerful socio-economic interests can dominate greening initiatives and be placed above other social equity needs or priorities [37]. For instance, this could involve the generation of economic value from deprived locations through the redesign of parks and the capture of a rent gap, which can in turn cause gentrification—possible trade-offs of greening projects—in [16,17,29]. This can be illustrated by the current situation in Lene-Voigt Park, which has been redesigned following a growth-oriented strategy that involved targeting the rising attractiveness of nearby housing estates and neighborhoods. This caused rents to rise and poses the risk of displacement for vulnerable social groups, e.g., low-income households and migrants.

Another barrier can arise because of short-term subcontracting to non-profit and private actors for the management of green spaces, which is a practice that is currently actively used in Leipzig. On the one hand, as a positive outcome, the neglected or abandoned areas are managed through public budgets and as a public responsibility, which helps the city to save scarce municipal funds. On the other hand, it can result in transforming the green area into a market-based and a quasi-private space [21,29].

As we have learned, local residents and community groups might either welcome or contest NBS interventions in their neighborhoods, depending on how the interventions (re)distribute socio-ecological benefits. In this regard, the issue of justice in the distribution of benefits and power—between genders as well as ethnic and cultural groups—can arise. Therefore, existing research on environmental justice in cities provides a distinct pathway toward—and source of data and analysis on—inequities produced by urban greening initiatives and strategies [37,38]. In addition, [39] conclude that access to green space is still highly stratified according to income, race, ethnicity, age, gender, (dis)abilities, and various other axes of difference. As concluded by [37], if green or natural areas do not meet the culturally and socially defined needs of potential visitors, the authors conclude that they may be used less frequently or be less effective for recreation. Recently, much research has addressed the process of making older urban areas inhabited by the lower-income population more livable and attractive with greening projects. The resulting processes have been termed ecological-, green-, environmental-, or eco-gentrification [40–43].

In our previous research [11,13,22], we concluded that even when NBSs are expected to contribute to the overall improvement of the living and health standards of urban citizens, the trade-offs that have resulted from several NBS interventions can be connected with a number of unexpected and undesired issues. For instance, urban greening might not automatically provide social and environmental sustainability in cities when we consider maintenance costs resulting from dry summers and the need to invest additional money, as we learned from the case of Lene-Voigt Park or the concept of the Tree-City (Baumstarke Stadt). Another dimension of justice in the planning and management of greening projects in urban areas is the sense of community and refuge that certain types of natural areas in cities tend to provide. The improved environmental conditions are often related to emotional needs and the search for community, belonging, protection, and safe spaces [40]. This emotional need reflects the importance of providing intangible values for some individuals such as place attachment or place of belonging, for example, through social urban gardens (e.g., ANNALINDE projects).

6. Conclusions

In summary, there is a broad spectrum of NBS types present in Leipzig that includes the creation of wildness patches and recreational areas in former mining landscapes and central parts of the city, the renaturation of rivers (especially those that were highly polluted during the time of industrialization), and the creation of urban green and blue infrastructure that is interconnected by cycling pathways (green roads, green walls, etc.) in order to connect the green and blue spaces of the city center with those in the suburban areas. Along with these NBS interventions, which are mostly initiated and supported by the government, many different initiatives have occurred or appeared in which the main initiators and drivers are citizens and NGOs (intercultural and neighborhood gardens, projects aiming to create awareness of environmental issues, awareness raising, and citizen involvement, etc.). They all addressed the issues of urban regeneration, coping with former industrial and neglected areas (e.g., brownfields), demanding green space for several urban districts that are currently referred to as the biggest challenges for several cities in Europe and beyond.

The idea of NBSs is now increasingly being implemented in finding a solution for biodiversity loss, climate change, the sustainable use of natural resources, and air pollution together with public health, social justice, and green economic opportunities in the cities. In the era of urbanization—and its associated sustainability challenges and business opportunities for those involved in the construction sector and urban planning—the concept of NBSs turns out to be highly topical.

Most of the NBS examples analyzed in this paper are valued recreational areas. They provide many benefits for urban dwellers such as fresh air, moisture, oxygen, and biogenic essentials as well as many cultural and place-based values. They are very efficient spaces for climate change mitigation, water and matter regulation, pollutants fixation, and flood water retention [16,17]. Thus, they represent perfect nature-based solutions for almost all current sustainable developments goals, particularly in the case of dense urban areas of Leipzig. These goals include risk mitigation and adaptation to the effects of climate extremes such as flood and drought, the disruption of food provision, social justice, and more. Moreover, they can serve as a buffer against high air temperatures and provide moisture during heatwaves. However, urban wetlands and riparian forests are often endangered by different land-use conflicts that have resulted from urbanization pressure such as land take for construction purposes and pollution. This paper has argued that urban NBSs now present a great opportunity with multiple benefits for a great number of beneficiaries in cities that are facing the effects of climate change. Further, we argue that NBSs present design options for expanding existing and even creating innovative new tools for sustainable development policy.

However, we did not mention the NBS approach as a panacea for urban sustainable development, since as mentioned above, it has a variety of limitations (e.g. financial mentioned above), trade-offs (such as eco-gentrification [37], ecosystem disservices [44]), and barriers [7–9,21,36]. We did not deny the importance of other concepts such as ecosystem services (ES) and urban green infrastructure (UGI), and share the ideas of Badiu et al. ([45], p. 3) that UGI plays an essential role in the NBS concept by tackling environmental problems and societal challenges in urban settings. Moreover, as highlighted by Artmann et al. [46], concepts from smart growth and green infrastructure are found to be compatible and, in our opinion, they fit with the main approaches of NBS such as functionality, smart governance and planning, compact cities, and sustainable networking. In this regard, the concept of NBSs can provide a better science–policy integration of UGI and ES approaches for addressing societal challenges, as was shown by the selected NBS intervention from Leipzig. Here, we agreed with the authors [45,46] that both UGI and NBS (which involves UGI as an essential part of its concept) demonstrate their powerful tools to make cities more livable, resilient, sustainable, and better adapted to current societal challenges.

There are a number of other concepts such as the concept of a "nurtured landscape" presented by Yang and Lay [47] and proposed for mediating between the natural ecosystem and the urban/industrial environment, which corresponds with the concepts of NBS, UGI, and ES. In this regard, we can perceive the several selected NBS cases in Leipzig (e.g., Leipzig New Lakes Region, Karl-Heine Channel) to be framed into the concept of the nurtured landscape, which provides a basis for the development of new ecological technology using landscape to ameliorate the polluting effects of the urban/industrial neighbourhood for the possible remediation and revitalization of former industrial areas [47].

By examining the role of selected NBSs in the urban landscape, we can conclude that human intervention and governance can make nature an increasingly essential part in the development of cities toward the use of more resilient systems. This can be one while using the benefits that ecosystems provide and co-designing them with manufactured and technological solutions [48,49] that are also needed in order to create cities that are fit for a sustainable future.

Author Contributions: Conceptualization, D.D. and D.H.; methodology, D.D. and D.H.; validation, D.H. and D.D.; formal analysis D.H. and D.D.; investigation, D.D.; resources D.D.; data curation and interpretation of results, D.D.; writing–original draft preparation, D.D. and D.H.; writing–review and editing, D.D. and D.H.; visualization, D.D.; supervision, D.H.; project administration, D.H.; funding acquisition, D.H. and D.D. All authors have read and agree to the published version of the manuscript.

Funding: This research was funded by the Horizon 2020 Framework Programme of the European Union, research and innovation project "CONNECTING Nature—COproductioN with NaturE for City Transitioning, Innovation and Governance", Grant Agreement No 730222.

Acknowledgments: We are thankful to the CONNECTING Nature project team for their fruitful discussions regarding the NBS issues and the opportunity to learn about NBSs from a more theoretical side as researchers. We are also thankful for the opportunity to discover all the milestones, from the creation of an NBS idea to its implementation, impact assessment, and monitoring, together with different actors of NBS. Our special thanks go to the stakeholders, namely the decision makers and the active civil society of the city of Leipzig, especially from the organizations listed in the paper and their representatives who took a part in our interviews and supported us with their valuable expertise and advices. The authors also extend great gratitude to the anonymous reviewers and editors for their helpful reviews and critical comments.

Conflicts of Interest: The authors declare no conflict of interest.

Appendix A. List of Interviews

Interview 1   workshop "Essbare Stadt" (Edible City), expert from NGO (February 2018)

Interview 2   officer of Ernährungsrat (Nutrition Council) Leipzig (February 2018)

Interview 3   head officer of Rosenberg Delikatessen private company (February 2018)

Interview 4   head officer of ANNALINDE GmbH (March 2018)

Interview 5   EU NBS project expert (April 2018)

Interview 6   Kletterfix project expert (June 2018)

Interview 7   restaurant owner in response to Kletterfix project (June 2018)

Interview 8   Kletterfix project participant (house resident) who implemented a green façade in his house in the Connewitz district of Leipzig (June 2018)

Interview 9   expert from the NABU (Naturschutzbund—Nature and Biodiversity Conservation Union) NGO by Ecofestival "Ökofete" Leipzig (June 2018)

Interview 10   expert from Solar-energy Producer by Ecofestival "Ökofete" Leipzig (June 2018)

Interview 11   expert from Grüne Liga Leipzig by Ecofestival "Ökofete" Leipzig (June 2018)

Interview 12   visitor of Ecofestival "Ökofete" Leipzig (June 2018)

Interview 13   expert of Naturkundemuseum Leipzig during excursion to Leipzig Auwald (June 2018)

Interview 14   staff of Center for Environmental Information of City Leipzig—Umweltinformationszentrum Leipzig (April 2019)

Interview 15   inhabitant of prefabricated LWB housing estate in the center of Leipzig (April 2019)

Interview 16   user of allotment garden cooperative in Leipzig East (April 2019)

Interview 17   expert from NuKLA e.V./Naturschutz und Kunst—Leipziger Auwald (May 2019)

Interview 18   municipal officer of the Environment Protection department of the City of Leipzig (May 2019)

Interview 19   municipal officer of Green Space and Waters department of the City of Leipzig (May 2019)

Interview 20   staff of Kindergarden Leipzig East (June 2019)

Interview 21   visitor 1 of Lene-Voigt Park (July 2019)

Interview 22   user of ANNALINDE Gemeinschaftsgarten (August 2019)

Interview 23   staff of Neseenland Bergbautechnikpark (August 2019)

Interview 24   visitor of Neseenland area (August 2019)

Appendix B. NBS Projects in Detail

Appendix B.1. Greening Façades (within the Project Kletterfix)

Challenge: Raising awareness of the benefits of urban green spaces, financing incentives from the municipality for action of tenants and landlords on private property.

Scale of impact (Macro—region, Meso—city and district, Micro-level—street, building): micro.

Relationship with Sustainable Development Goals—SDG [35]: SDG 3 (good health and well-being), SDG 11 (sustainable cities and communities), SDG 13 (climate action).

Temporal scale of impact (short versus long-term): long-term.

Method of implementation: Financial incentives, green infrastructure development, social networking.

ES provided: Climate mitigation through improvement of urban microclimate, air purification (combat particulate matter pollution), increasing quality of life and general living conditions, aesthetic value, increase urban biodiversity (green walls as ecological corridors for flora and fauna), economic savings for tenants.

Design and main drivers: In 2015, the city of Leipzig, with the support of the Ökolöwe ecological organization, initiated a project called "Kletterfix". The goal of this initiative is to encourage citizens to become more active in terms of "urban planting" by greening their walls and façades. The implementation of green walls and façades in a growing city is a good opportunity to extend green surface by using space that is available at almost every street corner, thereby contributing to a more livable city. The Ökolöwe ecological organization provides free courses and services about greening walls and façades, and five types of climbing plants are provided free of charge as well. The project "Kletterfix—Green Walls for Leipzig" is a part of the public greening that is implemented by Ökolöwe together with the property owner. Due to the "Kletterfix" initiative, the concept of a wall-filling and easy-care greening was developed. Green walls contribute to a healthy environment in the city, and their implementation improves living conditions and microclimates. As experts of Kletterfix stated: *"Not only urban parks act as a 'green lung' and cooling spot. Evaporation of the plants on the 850 square meter green façade corresponds to a cooling capacity of about 45 air conditioners. The winter heat loss of the building can also be reduced by 50%. Green façades filter out air pollutants and fine dust, especially when implemented along roads with high traffic. Moreover, along with the air pollution control, green façades contribute to noise reduction and play the important role of being green corridors (especially for insects and small animals), which connect different green spaces within the urban green infrastructure"* (Interview #6). The campaign has already shown positive functions of façade greening, which turned out to be more and more popular amongst the population of Leipzig. For example, the owner of the Italian restaurant from the district Connewitz (Leipzig) mentioned that green façades are not only good for a greener city but are also a luxury that is unique and beautiful to look at; he says: *"[The] implementation of green façades has an impressive result. It is much cooler and therefore more pleasant in the restaurant. Guests feel comfortable in the green ambience"* (Interview #7). They decided to plant wine because it expresses the Italian lifestyle and grows rapidly. *"The advantage is that the desired effect of the shading has set in in a very short time"*, claims the restaurant manager (Interview #7). *"We also harvest the grapes in the summer and make schnapps from it. The dense green is also popular among birds"* (Interview #7). Thus, greening walls and façades can not only be good for the climate in the city but also profitable for property owners: *"Such green walls not only look attractive, but also have many other advantages, including protecting the surface of buildings from direct sunlight and, as a result, reducing the cost of air conditioning, reducing noise, and increasing the oxygen content in the air. Moreover, vertical landscaping allows increases in the service life of buildings"* (Interview #8). Both sides, the city and the people, generally have a positive perception of the greening of walls and buildings, since they believe they only profit from the green façades.

Appendix B.2. ANNALINDE Intercultural Garden and Urban Agriculture

Challenge: Public space regeneration, the redevelopment of abandoned land, the renewal and redesign of public living areas to increase social cohesion and integration, raising awareness through active citizen engagement, food provision.

Scale of impact (Macro—region, Meso—city and district, Micro-level—street, building): meso.

Relationship with Sustainable Development Goals—SDG [35]: SDG 3 (good health and well-being), SDG 10 (reduced inequalities), SDG 11 (sustainable cities and communities), SDG 12 (responsible consumption and production), SDG 13 (climate action), SDG 14 (life on land).

Temporal scale of impact (short versus long-term): long-term.

Method of implementation: Sharing knowledge, social networking, change in physical infrastructure, green infrastructure.

ES provided: Climate-change mitigation and adaptation through a range of provisioning, regulating, and cultural ecosystem services: providing food, water regulation through unsealed soils, improved air circulation and cooling through plant transpiration and shading, mitigation and adaptation to the urban heat island effect, habitat for wildlife and genetic diversity, cultural ecosystem services (leisure and recreation, promoting health and well-being, as well as a sense of place, cultural identity, and social cohesion).

Design and main drivers: Similar to the other German cities (e.g., Cologne, Berlin, Göttingen, Munich, etc.) Leipzig has developed an innovative approach to urban food growing. Amongst the best examples are its intercultural gardens, which refer today to the integral part of the German community gardens (they number more than 100 located across Germany). Their main goal is intercultural exchange, which attracts people from diverse backgrounds (e.g., social, national, cultural, etc.) and helps them better integrate into the ever more multicultural society by discovering and sharing different aspects of food culture. As a proportion of inhabitants, Leipzig is the city with the greatest number of allotments in Germany, which is due to the active allotment movement that had already been established in the 1860s. In the 20th century, a great number of bottom–up initiatives have greatly contributed to the appearance of nationally and socially oriented urban agricultural projects, mostly focusing on regional food growing. Significantly, the first intercultural garden was established in Göttingen in 1995 (Internationale Gärten), and the second one was founded in Leipzig in 2001 (Bunte Gärten) in the former footprint of an unused municipal horticultural nursery [50]. *"Beginning in 2011, the ANNALINDE community garden has initiated a number of projects related not only to the food growing site but to the broad spectrum of issues related to the future development of Leipzig (such as the participatory use of urban space, environmental education programs)"* (Interview #4).

Established in 2011, ANNALINDE originated in the cooperative initiative and creativity of young people having fun on green business activities in the western districts of Leipzig. *"Our goal is to create places of exchange and learning on issues of local and ecological food growing, biodiversity, sustainable consumption, responsibility in resource use, and a sustainable neighborhood and urban development. Thus, ANNALINDE provides an ideal combination of all aspects such as urban ecology, land management, urban gardening, and so on."* (Interview #4). Since 2018, ANNALINDE has operated in the eastern districts of Leipzig as well; in particular, Gärtnerei Ost (an urban gardening and nursery project in Eastern Leipzig) was created near the S-Bahnhof Station Anger-Crottendorf. Briefly explained, ANNALINDE includes a community garden and was the first area. The second one was the Gärtnerei West (urban gardening and nursery project in Western Leipzig), which was established in 1870 and is now a cultural monument of the Municipality of Leipzig.

The urban gardening of ANNALINDE can be assessed as a multifunctional urban nature-based solution. Besides providing food, it contributes to water regulation through unsealed soils, improved air circulation and cooling through plant transpiration and shading, and offers microclimate oases to many users, visitors, and neighbors. It also can help mitigate the urban heat island effect. Another important ecosystem service provided by urban gardening is a habitat for wildlife and genetic diversity. Cultural ecosystem services include opportunities for leisure and recreation, promoting health and well-being, as well as a sense of place, cultural identity, and social cohesion, all of which are important factors for societies that are adapting to change.

Appendix B.3. ANNALINDE Academy

Challenge: To develop joint and experience-based learning toward urban resilience and sustainable development, shaping environmental culture, contributing to sustainable urban living.

Scale of impact (Macro—region, Meso—city and district, Micro-level—street, building): Meso.

Relationship with Sustainable Development Goals—SDG [35]: SDG 3 (good health and well-being), SDG 4 (quality education), SDG 11 (sustainable cities and communities), SDG 12 (responsible consumption and production), SDG 13 (climate action), SDG 14 (life on land).

Temporal scale of impact (short versus long-term): long-term.

Method of implementation: Sharing knowledge and knowledge transfer, social networking, stakeholders' and citizens' involvement, workshops.

ES provided: Cultural (information on environmental issues).

Design and main drivers: The ANNALINDE Academy was established in 2014. It was created from the community garden, firstly as a discussion platform and educational program for different aspects of sustainable urban design under the leitmotif "cultivating tomorrow". Their topics range from environmental psychology, the cyclical use of resources, and renewable energies to alternative economies. The academy is supported by the Leipzig garden program and the gallery of Contemporary Art. *"Since 2015, the ANNALINDE Academy has been the interface between our three project areas—community garden, gardening, orchard—our environmental education and environmental policy work, as well as our commitment to urban and neighborhood development"* (Interview #4).

Other fields of educational activity include locally adaptable organic agriculture, urban gardening, biodiversity, sustainable urban development, food sovereignty, environmental justice, the campaign for an Edible City in Leipzig, lectures and workshops for multipliers and interested parties, and the regular environmental education programs in schools and daycare centers.

"The ANNALINDE Academy perceives itself as a place of experience-based and action-oriented learning through a co-constructive learning approach, including co-creation and co-development and acting live on-site. […] we can experience our topics through touch, taste, and participation and convey creative competence" (Interview #4). This is an approach in which theoretical knowledge is linked with practical activities and concrete implementation possibilities, making the academy an open platform for knowledge exchange.

Appendix B.4. Project "Edible City" (Essbare Stadt)

Challenge: To support sustainable regional agriculture and local food production, gardening movement aiming to turn the city into an edible garden.

Scale of impact (Macro—region, Meso—city and district, Micro-level—street, building): Macro.

Relationship with Sustainable Development Goals—SDG [35]: SDG 2 (zero hunger), SDG 3 (good health and well-being), SDG 10 (reduced inequalities), SDG 11 (sustainable cities and communities), SDG 12 (responsible consumption and production), SDG 13 (climate action), SDG 14 (life on land).

Temporal scale of impact (short versus long-term): long-term.

Method of implementation: Sharing knowledge, social networking, workshops.

ES provided: Regulating and cultural ecosystem services.

Design and main drivers: "Edible City" (*"Essbare Stadt"*) appeared in 2014 as an initiative of young and cooperative people [51]. The concept came out of a series of workshops on social agriculture. The organization responsible for its development is ANALINDE (NGO). *"The initiative 'Edible City' is about food cultivation in public spaces. On the one hand, it is about those spaces where people can garden and harvest. On the other hand, many fruits and herbs are already growing in public spaces; thus, this project includes the mapping of and informing about harvesting possibilities. It also aims to make Leipzig an edible city and create and implement certification for [producing] and marketing the food"* (Interview #1). Community gardens and market gardens are being used for cooperative learning about local ecological food production, biological diversity, sustainable consumption, and the economy, which can empower the idea of a sustainable society. The city council has allocated 150,000 Euros for the promotion of community garden projects for the years 2019 and 2020.

The "Edible City" concept includes two main strategies: revitalizing urban communally owned brownfield sites and replacing decorative planting in public parks and other non-food-growing green facilities (such as pocket parks, backyards, etc.) with edibles. *"It not only contributes to increasing the biodiversity and improving the quality of green spaces, but also helps to engage the public (broad community), which can greatly contribute to the lowing of maintenance costs for communally owned open spaces"* (Interview #1).

One such workshop, named "Essen für alle" ("Good food for all"), was organized in February 2018. The questions discussed during the workshop ranged from "how we can feed ourselves in the city more sustainably and independently than today" to "what trends we can observe now in terms of newer developments in the food scene?" The presenters highlighted the value of the community and neighborhood gardens, as well as the classical allotment gardens, community-supported agriculture, the città slow (the slow food movement is an international network of small towns that originated in Italy less than a decade ago that is aimed at promoting slowness against the fast life and includes improving the quality of life in towns by having a cleaner environment, eating wholesome food, participating in a rich social life, and taking time out to think about what we should be doing and how we should be doing it) etc. by underlying the role of local politics and other actors in urban governance.

Appendix B.5. "Nutrition Council Leipzig" (Ernährungsrat Leipzig)

Challenge: To improve the food system in Leipzig and make it more ecological and more democratic; to ensure the long-term sustainable supply of the city's population with regional, seasonal, and healthy food.

Scale of impact (Macro—region, Meso—city and district, Micro-level—street, building): Macro.

Relationship with Sustainable Development Goals—SDG [35]: SDG 2 (zero hunger), SDG 3 (good health and well-being), SDG 10 (reduced inequalities), SDG 11 (sustainable cities and communities), SDG 12 (responsible consumption and production), SDG 13 (climate action), SDG 14 (life on land).

Temporal scale of impact (short versus long-term): Long-term.

Method of implementation: Social networking, sharing knowledge, workshops.

ES provided: Provisioning, supporting, cultural ecosystem services.

Design and main drivers: The "Nutrition Council Leipzig" was set up in 2019 in order to improve the food system in Leipzig and make it more ecological and more democratic. The first nutrition councils in Germany were founded in 2016 in Berlin and Cologne. The "Nutrition Council Leipzig" is an association of civil society groups, farmers, processors, retailers, and restaurateurs from Leipzig and the region, as well as representatives of the municipality. *"It was set up for better dialogue between different stakeholders for a better meal" (Interview #3).* One of the biggest global challenges in the 21st century is the production and availability of healthy food. Against the background of climate change and the constant decline of biodiversity, it is also necessary to develop a sustainable agricultural system at the regional level. Therefore, *"Leipzig wants to emphasis a local political agenda and ensure the long-term sustainable supply of regional, seasonal, and healthy food to the city's population"* (Interview #2). In doing so, the Nutrition Council Leipzig is involved in three main activities: (1) it connects different actors within the food system in the city; (2) it collects knowledge about the production, marketing, and supply of the food in the region; and (3) it performs an advisory function and appears as a platform for education and informational exchange.

"We have to highlight one important role of such nutrition councils, namely that they are a tool for food policy 'from the bottom'" (Interview #2). The main idea for the foundation of the Nutrition Council, its goals and visions, as well as important milestones for its successful implementation, are the following: *"More and more people today are committed to sustainable change in agriculture, food production, and consumption style: you participate in community gardens or solidarity, agriculture, cooking for others, sharing your food, organizing discussion events, or going for a different agricultural policy on the road. These diverse civil society initiatives and actors come together in nutrition councils and jointly develop policies for the renewal of our nutrition systems"* (Interview #2). One of the indicators of success is the city's support: *"In this regard, the City of Leipzig makes its great contribution in giving financing support for such kind of initiatives and networks"* (Interview #1).

Another set of activities promoted by the Nutrition Council include those that refer to the sharing and learning initiatives for a different food culture: *"Our eating habits are very diverse and, at the same time, characterized by social norms, habits, and incentive structures. This initiative is aimed to education at sustainable nutrition in the sense of fair food culture"* (Interview #2).

"I am interested in the work of this organization (Nutrition Committee Leipzig) and will attend the next meetings for sure. I think different aspects will be discussed that relate to my business. Especially how projects like mine could be supported. And I would like to understand the processes from inside in order to understand in depth what the driving forces for transformations in society are, for shift of life-models and method of product consumption" (Interview #3).

Appendix B.6. Lene-Voigt Park (Renaturing of Former Industrial Area to Public Green Space)

Challenge: To renature the post-industrial area to public green space, to connect eastern districts and the 'green lungs' of the city (park as a part of green belt), to initiate residential change of the surrounding areas and the development of the local urban infrastructure.

Scale of impact (Macro—region, Meso—city and district, Micro-level—street, building): meso.

Relationship with Sustainable Development Goals—SDG [35]: SDG 3 (good health and well-being), SDG 11 (sustainable cities and communities), SDG 13 (climate action), SDG 14 (life on land).

Temporal scale of impact (short versus long-term): Long-term.

Method of implementation: Change in physical infrastructure, green infrastructure development, social networking.

Indicators of success: Co-creation with residents (through workshops), offering more green space for the dense housing area, increasing in the attractiveness of place and district.

ES provided: Climate mitigation and adaptation, habitat provision, recreation

Design and main drivers: Lene-Voigt Park is a good example of the successful renaturing of the former heavily disturbed area of the Eilenburg railway station, which was formerly located on this site. Now, it is a very popular place of recreation for people of all ages that was created in 2001 and officially opened in 2004. From the time of its completion in 1874, the approximately 11 ha area Eilenburger Bahnhof (Railway Station) together with other four railway stations made Leipzig one of the important railway junctions of the former German Empire. After the construction of the Central Railway Station in 1915, passenger traffic was largely relocated there. In 1942, the operation at the Eilenburg railway station was completely discontinued, and the site was mostly neglected. An 800 m long and 80–130 m wide area remained behind in a prominent location close to the city center [12–14].

In an area characterized over several decades by vacant buildings and extensive brownfields, the park has become part of a green belt connecting the eastern districts and the "green lungs" of the city. Residents' opinions and desires were included in the process of its creation through workshops. The main aim of the creation of this park was to offer more green space for the dense housing area and to create playgrounds for kids. The park was well received and became popular shortly after it was completed. It helped to initiate the residential change of the surrounding areas and the development of the local urban infrastructure, e.g., cafés and shops. New residents moved there, and housing vacancies started to decrease. Among the new residents, there were more young families with children and higher incomes. As a consequence, rents started to rise from 4.5 Euro per square meter in 2000 to almost 7 Euro per square meter as of today. Lene-Voigt Park is the heart of an increasingly expensive housing area for young and educated residents. Former residents of the area, including elderly and less affluent households, had to leave due to increasing housing costs.

"The aim of the redevelopment of this area was not only to preserve and restore the historic quality and complexity of the district, but also to create a district (neighborhood) park that invites you to play and recover and that continues in a green radius towards the outskirts of the city" (Interview #19).

An interesting detail is that park was created by the active involvement of local inhabitants in the project planning. As a result, a number of sport facilities (table tennis tables, playing fields for ball sports, as well as opportunities for boules, inline skating, and mountaineering), children's playgrounds (including a fairytale forest and sand island for tasting and creating), and places for picnics make this park popular among young people and families with children.

A direct connection from Lene-Voigt-Park to the "Grünen Ring" (Green Belt) around Leipzig forms the walking and cycling path and is an important element of Leipzig's green infrastructure. In 2002, Lene-Voigt Park received the European prize for landscape architecture, as the concept deliberately provided a lot of space for civic engagement and participation. There are a number of citizen's forums and workshops that are organized here yearly and which contribute to the raising of awareness of all the related sustainable goals. These also provide a platform where the critical opinions of different urban actors can be stated out loud, and the creativity of local residents can be shown.

One of the important elements of design of the Lene-Voigt Park includes spaces for nature experiences (Naturerfahrungsräume). This concept offers spaces for nature experiences and was developed in the 1990s. These are near-natural sites that provide opportunities for children to play in an unregimented and largely unsupervised way.

Appendix B.7. Eco-Food Project "Rosenberg Delikatessen"

Challenge: To establish a green business/company, to use regional food/products, to support regional variations of materials and energy balance.

Scale of impact (Macro—region, Meso—city and district, Micro-level—street, building): Macro.

Relationship with Sustainable Development Goals—SDG [35]: SDG 3 (good health and well-being), SDG 11 (sustainable cities and communities), SDG 12 (responsible consumption and production), SDG 13 (climate action), SDG 14 (life on land).

Temporal scale of impact (short versus long-term): Long-term.

Method of implementation: Social networking, sharing knowledge, financial incentives.

Indicators of success: Continuous development of innovative strategies to make the products better, respect to traditional processing/manufacturing of regional fruits and vegetables, good networking that involves different actors.

ES provided: Provisioning, supporting, cultural.

Design and main drivers: Rosenberg Delicatessen was founded in 2016 as a joint private company driven by the idea of producing bio(eco)products from the fruits and vegetables growing and cultivated in the region of Leipzig, in close cooperation with bio-shops of eco-food (orchards). The private company is involved in a network of actors who are dealing with the distribution, landscaping, and landscape conservation as well as environmental protection aspects and the processing of fruits and further production.

"As for me, the term NBS is first of all related to human well-being and health [...] and also to the aspects such as establishing a green business/company, using regional food/products, supporting regional variations of materials, and energy balance [...] In this sense, my case fits well" (Interview #3).

The project *"[...] first of all [...] aims to strengthen regional material cycles as well as the production and processing of fruits and vegetables in the Leipzig region. One of the important goals is raising the awareness of such topics as the Edible City and healthy food in the population"* (Interview #3).

"[...] A network of people dealing with the subject of the Edible City is very important for my company, with ideas on regional development, local production, landscaping, fruit, local food producers, landscape care, and nature conservation. I am glad to be involved in the network of producers of local foods and to better understand what moves the people, what they offer, and what they consume [...] How are citizens engaged with the aspects of healthy food, green space, [and] healthy nutrition, and what do they do for that? For me, these are the important areas of experience" (Interview #3).

"In fact, Rosenberg Delikatessen can be perceived as an NBS experiment that refers to the continuous process started in 2016 and that will further develop. Last week (February 2018), I met people and we discussed the idea of new orchards for the city and who will handle, process, and maintain them. Thus, it is the networking that involves developing and maintaining connections with different individuals, the process of establishing a mutual beneficial relationship between them" (Interview #3).

By highlighting the factors of success, the following can be identified as the most prominent: *"[...] I have monthly bookkeeping and accounting. This is a good thing. Thanks to this, I notice how dynamic everything is […] Also, a website was established and it helps me […] to reflect on all activities within the project that I am working on and, especially, to develop new strategies. […] Currently, I am developing a web shop. So, I had learned and [I] apply different strategies on how to start a private company, how to make the products and the regions well known on the market, and [how] to promote them"* (Interview #3).

Another key to the success of the experiment, as identified by the company chair, is *"[…] not only the love of my work ('die liebevolle Handarbeit') and the respect for traditional processing/manufacturing of regional fruits, but also good networking that involves different actors"* (Interview #3). Thus, the goal of Rosenberg Delicatessen's establishment was not primarily connected to total the profit-maximization objective, but first of all that *"[…] regional products made from the fruits growing and cultivated in the region of Leipzig will be presented and well-known on the market. […] Then, they will be processed in the regional products and finally will be sold and consumed also locally. So, it uses less energy [and] transportation resources and is healthy for people"* (Interview #3).

If the city and surroundings get more orchards, this means that it will provide more habitat for animals and plants and biodiversity support. It also means more space for leisure and recreation by citizens, the development of green infrastructure, and support for local food organizations and private companies. In addition, it provides a healthy lifestyle for citizens in the sense of the Edible City, where people are aware of bio-products and the importance of local food consumption as well as participate in the development of intercultural/community gardens. For instance, the Foundation of Leipzig's Citizens ("Stiftung Bürger für Leipzig") promotes this initiative, and it will be decided in the near future where such areas will be established.

Appendix B.8. Ecological Festivals (Umwelttage and Ökofete)

Challenge: Raising awareness of environmental issues, stakeholders' and citizens' involvement.
Scale of impact (Macro—region, Meso—city and district, Micro-level—street, building): Macro.
Relationship with Sustainable Development Goals—SDG [35]: SDG 3 (good health and well-being), SDG 10 (reduced inequalities), SDG 11 (sustainable cities and communities), SDG 12 (responsible consumption and production), SDG 13 (climate action), SDG 14 (life on land).
Temporal scale of impact (short versus long-term): Long-term.
Method of implementation: Sharing knowledge and knowledge transfer, social networking.
ES provided: Cultural.
Design and main drivers: The annual Leipzig Environment Days ("Umwelttage") have offered a colorful program since 1990 with over 100 environmental events in and around Leipzig. They normally commence on 5 June (World Environment Day) and include a variety of events around the topics of environment, sustainability, and nature conservation. They allow citizens of all ages to explore, experience, and discover related issues with films, excursions, discussions, guided tours, lectures, concerts, and much more. As a highlight of the green event, at the end of the Environment Days, the Ecofestival is organized in Clara Zetkin Park, which has a rich program dedicated to environmental protection, sustainable, and environmentally friendly consumption, and eco-friendly products. The festival offers an opportunity for everybody to communicate with representatives of the government, various environmental organizations, political units, and producers of eco products regarding their activities and the current environmental issues that the city of Leipzig faces. *"The Leipzig Environment Days show how sustainable and environmentally friendly our life in the city can be. A variety of eco-products are presented there, including apples from the local market place, noodles unpacked into the glass, T-shirts made of fair-trade organic cotton, and other sustainable products"* (Interview #9).

The program includes different activities and excursions. One of these is the excursion to the green paradise in the middle of the city, the Leipziger Auwald, where participants can find out which animals and plants are at home here (native species) and which were introduced but have now found their habitat. A large portion of the activities and events is devoted to the issue of whether gardeners in a big city such as Leipzig can be close to nature.

"You can learn how easily and ecologically the fruits and vegetables can be grown in your own garden or backyard. This we got thanks to the Leipzig Environment Days" (Interview #12).

"Here, you can not only buy organic products such as food and beverages, but also learn something about the current environmental issues, smart solutions, and about those in Leipzig (associations, offices, companies, individuals) who are contributing to making the environmental situation better" (Interview #9).

Appendix B.9. City's tree Planting Program (Baumstarke Stadt)

Challenge: Achieve more green in the city, improve air quality, and reduce noise, citizen and stakeholder involvement.

Scale of impact (Macro—region, Meso—city and district, Micro-level—street, building): Meso.

Relationship with Sustainable Development Goals—SDG [35]: SDG 3 (good health and well-being), SDG 11 (sustainable cities and communities), SDG 13 (climate action), SDG 14 (life on land).

Temporal scale of impact (short versus long-term): Long-term.

Method of implementation: Green infrastructure development, change in legislation or regulation, social networking.

ES provided: Climate change mitigation and adaptation, air purification (increasing air quality and reducing noise), cultural.

Design and main drivers: The city of Leipzig initiated the project Tree-Strong City ("Baumstarke Stadt") in 1996–1997 [52]. This NBS initiative has been implemented by the city authorities in order to attract people's attention to modern ecological issues and increase the popularity of ecological movements among the population by reforesting streets in the city. The aim of this project is to increase the tree stock in the city by planting new trees in public areas such as parks, streets, urban cemeteries, etc. The initiative consists of tree sponsorship bids starting from 250 Euros paid by citizens or companies who are going to be a sponsor for the particular tree. The sponsorships include an oak stele with a sign on it that indicates tree species and a personal message. This personal dedication can commemorate any special occasion or event (birthday, anniversary, graduation, etc.). Thus, the Tree-Strong City initiative in Leipzig is a creative and beneficial way to reforest the streets, which ultimately improves the quality of life since trees provide fresh air and contribute significantly to a livable, green city. This project contributes to a greener city landscape and strengthens the attachment of citizens to their urban environment.

Appendix B.10. Leipziger Neuseenland (Leipzig New Lakes Region)

Challenge: Revitalization of the post-mining area, regeneration of a vast mining landscape.

Scale of impact (Macro—region, Meso—city and district, Micro-level—street, building): Macro.

Relationship with Sustainable Development Goals—SDG [35]: SDG 3 (good health and well-being), SDG 9 (industry, innovation, and infrastructure), SDG 11 (sustainable cities and communities), SDG 13 (climate action), SDG 14 (life on land).

Temporal scale of impact (short versus long-term): Long-term.

Method of implementation: Change in physical infrastructure, change in legislation or regulation (federal program), green infrastructure development, networking.

Indicators of success: State-driven initiative, supported by private companies, focusing on enhancing the attractiveness of the area, providing new recreation facilities, and generating economic growth and employment through tourism.

ES provided: Climate adaptation and mitigation, biodiversity and habitation provision, climate and water regulation, recreation, cultural ES.

Design and main drivers: The region has a long tradition of mining activity, which has constituted the most important sector of the regional economy. The first open pit appeared in the area in 1900. During the existence of the GDR, the coal industry was actively developed, as there was no access to the West German deposits. The consequence was a significant deterioration of the environment, so the air in Leipzig and its surroundings was considered to be the most polluted in all of Europe. Sprawling quarries and empty rock dumps occupied a huge territory, and tens of thousands of people were relocated to other regions. Active land reclamation started in the 1990s, since after the reunification of Germany, the demand for coal fell, and work at many quarries ceased. The question was what to do with such a large area of land from the former developments (the total area of the former quarries was 18,000 ha, the depth of some reached 100 m). As part of the remediation, new reservoirs have been formed and continue to be created. A plan exists to connect these together (at the moment, 15 lakes have been created with different purposes, including recreation or environmental protection). The total water surface area of the plans is 70 square km.

In recent decades, many traditional European centers of the mining industry have been rendered unprofitable, which has led to the closure of the pits. In addition, such regions as middle Germany, in which Leipzig is centered, started to face environmental issues in the form of persistent pollution of the water, soil, and air. Those environmental issues led to the closure of the coal industry in the region and required the revitalization of the ground, which led to the foundation of a new recreational area called Neuseenland. Nowadays, this region has become one of the most popular destinations for relaxation, sunbathing, fishing, and yachting. There is an industrial museum founded here to show the history of the coal mining in the region and its transformation after the reunification of Germany.

"In the 1990s, it was decided that a museum would be established that would tell the history of the coal industry in the region of Leipzig. Different types of big industrial machines, which had been used for mining, are exhibited in our open-air museum. Mostly, people, visiting this museum are interested in the coal industry and the history of the Leipzig region, but there are not that many visitors. Our exhibits are hard to miss on the horizon. They are huge. Those machines are visible from far away, from Lake Markkleeberg where people come to swim and relax during the summer. Kanupark is also there, which is very popular among locals and tourists. Some people visit just because they want to know what is here. And, even though most of the people don't reach the museum, they still get involved in the mining history of the region because they swim or yacht in the lakes created on the site of the former coal mines. School groups visit for educational purposes during the academic year. It seems to me that our museum is very interesting and unusual" (Interview #24).

At the moment, recreational nature management is developing in the former industrial territory. The functional use of the territory has transformed from an industrial one to a recreational one. One such famous touristic object is Canoe Park, which was created on Lake Markkleeberger and Lake Cospudener. The tourist infrastructure of the park is well developed, and there are visual maps of the functional zoning of the park at the entrance so that all guests can easily find what they need. On the banks of beaches, there are cafés, restaurants, developing infrastructure for water sports, and an amusement park located near the technical Park Museum that is dedicated to the mining industry. At the entrance, there is a map of the geological structure of the territory of the former quarries. The park for water sports activities was built as part of the promotion of Leipzig as a candidate for the 2012 Summer Olympics. In this area, there is a hotel consisting of cottages and apartments, which is equipped with a developed tourist infrastructure for a comfortable year-round stay. The development of infrastructure continues. Many people, including children, visit for water sports and just relax on the shores of the lakes. In addition to recreational activities, the former industrial area is used for power generation, and a large area is occupied by solar panels located on the slopes of the waste-rock dumps.

In general, it should be noted that this method of reclamation of the former industrial territory was chosen and executed very successfully. The transportation infrastructure within the park is constantly developing and is maintained at a high level (both personal and public transport). This makes the park even more popular among tourists and vacationers of all ages. *"Everyone can find something that he/she likes, from professional athletes who have the opportunity to train (water sports) to a quiet family holiday with children in a comfortable hotel on the lake"* (Interview #25).

Appendix B.11. Karl-Heine Channel

Challenge: To deal with the consequences of industrialization and deindustrialization, to renature and recover the river courses and wetlands, ponds and lakes, sustainable redevelopment of the post-industrial landscape.

Scale of impact (Macro—region, Meso—city and district, Micro-level—street, building): Meso.

Relationship with Sustainable Development Goals—SDG [35]: SDG 3 (good health and well-being), SDG 6 (clean water and sanitation), SDG 11 (sustainable cities and communities), SDG 13 (climate action).

Temporal scale of impact (short versus long-term): Long-term.

Method of implementation: Change in physical infrastructure, change in legislation or regulation, green infrastructure development, monitoring.

Indicators of success: Support from federal policy and programs, ecological revitalization as one of the main goals of policy after German reunification in order to harmonize living conditions in the east and the west (e.g., ecological transition as a part of welfare policy), federal money stemming from subsidies to the labor market, active participation of civil society.

ES provided: Improvement of the environmental situation, redevelopment of an area that makes the residential area attractive for new restaurants and businesses.

Design and main drivers: The Karl-Heine Channel is a 3.3-km long artificial water route west of Leipzig (it terminates near Lindenau port). The channel was created in 1856–1898 on the Initiative of the Leipzig lawyer and industrial pioneer Carl Heine as the first part of a projected shipping channel from the White Elster up to the Saale. In the 1990s, the canal was revitalized. In 1996, a cycle path was set up on the northern canal bank. In 2007, the municipality decided to extend the canal to the port in order to enable the long-planned connection to Leipzig [26]. By 2015, 18 million Euro had been invested in the project, including funding from the municipality, land sales, and the EU Urban Development Fund. Another source indicates that the total costs of the construction project will come to 9.997 million Euro, including planting and landscaping works. Thus, having originally followed the natural course of the river, the canal was redesigned with a strong emphasis on urban development. The revitalization of the channel also contributed to the attractiveness of the whole western district of Leipzig, transforming it from the one of the most polluted and deteriorated into the so-called "Leipzig's Venice". *"This manmade canal doesn't have to be explored by boat. A cycle lane leads from the Nonnenbrücke (bridge) to the Luisenbrücke (bridge) through several interesting sightseeing spots such as the 'Riverboat', the MDR's former talk-show studio. In between, you will spot grand villas, ultramodern industrial lofts, and lovely places to stop and enjoy"* (Interview #23).

Appendix B.12. Leipziger Auwald (Riverside/Riparian Forest)

Challenge: Development and conservation of the local floodplains, to restore natural habitats and provide ecological benefits as well as opportunities for environmental learning and citizen participation.

Scale of impact (Macro—region, Meso—city and district, Micro-level—street, building): Macro.

Relationship with Sustainable Development Goals—SDG [35]: SDG 3 (good health and well-being), SDG 11 (sustainable cities and communities), SDG 13 (climate action), SDG 14 (life on land).

Temporal scale of impact (short versus long-term): Long-term.

Method of implementation: Change in legislation or regulation, green infrastructure development, monitoring, social networking, sharing knowledge.

Indicators of success: Raising awareness of the importance of the floodplain and its related ecosystem services to both people and nature.

ES provided: Mitigation of flood risk, cleaning water by nutrient retention, delivering oxygen, fixating carbon dioxide in the production of floodplain forests, enrichment of quality of life for citizens by providing people with an enjoyable place of rest, relaxation, and recreation.

Design and main drivers: In Leipzig, the development of the city and the forest are closely interrelated, not least because one of the largest floodplain forests areas of Central Europe is situated in the urban territory of Leipzig. Indeed, the urban floodplain forest is the backbone of the city's multifaceted green network that includes lots of parks, gardens, and allotments and that contributes to the high quality of life in our city. This green network is complemented by a system of rivers and water courses that cross the city and shape the cityscape [53]. Awareness of the city's blue infrastructure is now emerging due to the reconstruction and rehabilitation of canalized water courses, the improvement of water quality, and the river system. In recent years, it has developed into a significant attraction for Leipzig's citizen as well as for the tourism business in the region [54].

In order to revitalize the floodplains of Leipzig and Schkeuditz in 2012, and especially the river and riparian forest landscapes, the project Live Luppe River ("Lebendige Luppe") was launched. A variety of negative changes and consequences for the wetlands appeared due to intensive human activities in the past such as flood and stream controls, embankment sealing, and the drainage of arable land and meadows. One of these activities was the regulation of the "New Luppe" river, which embanked in the 1930s, causing ecological damage due to droughts and missing groundwater contact in the floodplains [16,17]. It resulted in the situation that a riverine landscape that had once regularly flooded now suffered from a marked reduction in groundwater and inundation. This developed alongside the progressive drying of habitats and poses a threat to the floodplain biodiversity and related ecosystem services.

The afforestation of urban forests on derelict land introduces a new type of urban green space and offers an ecologically, economically, and socially viable alternative to costly designed green spaces [53]. In the long term, the urban forests shall be managed similar to rural forests and, at the same time, serve as neighborhood parks. Through a thorough process, suitable sites for urban forests in the inner-city area were selected. By now, two of them have been afforested. The implementation of the project is accompanied by comprehensive research design, which will clarify the effects of urban forests on the urban climate, biodiversity, recreation, population, and urban redevelopment.

References

1. Albert, C.; Schröter, B.; Haase, D.; Brillinger, M.; Henze, J.; Herrmann, S.; Gottwald, S.; Guerrero, P.; Nicolas, C.; Matzdorf, B. Addressing societal challenges through nature-based solutions: How can landscape planning and governance research contribute? *Landsc. Urban Plan.* **2019**, *182*, 12–21. [CrossRef]
2. Scott, M.; Lennon, M.; Haase, D.; Kazmierczak, A.; Clabby, G.; Beatley, T. Nature-based solutions for the contemporary city. *Plan. Theory Pract.* **2016**, *17*, 267–300. [CrossRef]
3. European Commission. *Towards an EU Research and Innovation Agenda for Nature-Based Solutions and Re-Naturing Cities*; CEC: Brussels, Belgium, 2015.
4. European Commission. Policy Topics: Nature-Based Solutions. 2016. Available online: https://ec.europa.eu/research/environment/index.cfm?pg=nbs (accessed on 15 August 2019).
5. Urban Agenda for the EU. Sustainable Use of Land and Nature-Based Solutions Partnership. Draft Action Plan. 2018. Available online: https://ec.europa.eu/futurium/en/sustainable-land-use/introduction-draft-action-plan-sustainable-use-land-and-nature-based-solutions (accessed on 23 July 2019).
6. Raymond, C.M.; Berry, P.; Frantzeskaki, N.; Kabisch, N.; Breil, M.; Nita, M.R.; Geneletti, D.; Calfapietra, C. A framework for assessing and implementing the co-benefits of NBS in urban areas. *Environ. Sci. Policy* **2017**, *77*, 15–24. [CrossRef]
7. Kabisch, N.; Frantzeskaki, N.; Pauleit, S.; Naumann, S.; Davis, M.; Artmann, M.; Haase, D.; Knapp, S.; Korn, H.; Stadler, J.; et al. Nature-based solutions to climate change mitigation and adaptation in urban areas: Perspectives on indicators, knowledge gaps, barriers, and opportunities for action. *Ecol. Soc.* **2016**, *21*, 39. [CrossRef]
8. Kabisch, N.; Korn, H.; Stadler, J.; Bonn, A. *Nature-Based Solutions to Climate Change Adaptation in Urban Areas: Linkages between Science, Policy and Practice*; Springer International Publishing: New York, NY, USA, 2017. [CrossRef]
9. Nesshoever, C.; Assmuth, T.; Irvine, K.J.; Rusch, G.M.; Waylen, K.A.; Delbaere, B.; Haase, D.; Jones-Walters, L.; Keune, H.; Kovacs, E.; et al. The science, policy and practice of Nature-Based Solutions: An interdisciplinary perspective. *Sci. Total Environ.* **2017**, *579*, 1215–1227. [CrossRef] [PubMed]

10. Almassy, D.; Pinter, L.; Rocha, S.; Naumann, S.; Davis, M.; Abhold, K.; Bulkeley, H. Urban Nature Atlas: A Database of Nature-Based Solutions Across 100 European Cities. Report of H2020 Project Naturvation. 2018. Available online: https://naturvation.eu/sites/default/files/result/files/urban_nature_atlas_a_database_of_nature-based_solutions_across_100_european_cities.pdf (accessed on 15 June 2019).

11. Haase, D.; Dushkova, D. Learning from existing nature-based solutions experiences in Europe. In *Connecting Nature: Bringing Cities to Life, Bringing Life into Cities*; Deliverable 1.1.2018 of H2020 Project Connecting Nature; Unpublished document; 2018.

12. Rink, D. Stadt der Extreme. *Leipz. Bl. Sonderh.* **2015**, *1000*, 4–7.

13. Haase, D.; Dushkova, D.; Haase, A.; Kronenberg, J. Green infrastructure in postsocialist cities: Evidence and experiences from Russia, Poland and Eastern Germany. In *Post-Socialist Urban Infrastructures*; Tuvikene, T., Sgibnev, W., Neugebauer, C.S., Eds.; Routledge: London, UK, 2019; pp. 105–122.

14. Rink, D.; Behne, S. Grüne Zwischennutzungen in der wachsenden Stadt: Die Gestaltungsvereinbarung in Leipzig. *Stat. Quart.* **2017**, *1*, 39–44.

15. Wolff, M.; Haase, A.; Haase, D.; Kabisch, N. The impact of urban regrowth on the built environment. *Urban Stud.* **2016**, *54*, 2683–2700. [CrossRef]

16. Haase, D. Urban ecosystem, their services and town planning. Critical reflections of selected shortcomings. *Urbanistica* **2017**, *159*, 90–94.

17. Haase, D. Urban wetlands and riparian forests as a nature-based solution for climate change mitigation and adaptation in cities and their surroundings. In *Nature-Based Solutions to Climate Change in Urban Areas: Linkages of Science, Society and Policy*; Kabisch, N., Bonn, A., Korn, H., Stadler, J., Eds.; Springer International Publishing: New York, NY, USA, 2017; pp. 111–122. [CrossRef]

18. Connop, S.; Vandergert, P.; Eisenberg, B.; Collier, M.J.; Nash, C.; Clough, J.; Newport, D. Renaturing cities using a regionally focused biodiversity-led multifunctional benefits approach to urban green infrastructure. *Environ. Sci. Policy* **2016**, *62*, 99–111. [CrossRef]

19. Connop, C.; Dushkova, D.; Haase, D.; Nash, C. *Contributions to the European Handbook on Impact Assessment: Indicators Factsheets*; Connecting Nature Project (Internal Report); Unpublished document; 2019.

20. Nature4Cities. NBS Multi-Scalar and Multi-Thematic Typology and Associated Database; Project Deliverable D1.1. 2018. Available online: https://docs.wixstatic.com/ugd/55d29d_8813db2df690497e80740537b6a8a844.pdf (accessed on 15 June 2019).

21. Sekulova, F.; Anguelovski, I. *The Governance and Politics of Nature-Based Solutions*; Naturvation, Deliverable 1.3: Part VII; Ashgate Publishing: Farnham, UK, 2017.

22. Haase, D.; Dushkova, D. Naturbasierte Lösungen für Umweltprobleme in Städten—Ein Anwendungsfall für Die Stringente Umsetzung von Urbanen Ökosystemleistungen? Innovationsnetzwerk Ökosystemleistungen Deutschland (ESP-DE). 2019. Available online: https://www.esp-de.de/naturbasierte-loesungen-fuer-umweltprobleme-in-staedten-ein-anwendungsfall-fuer-die-stringente-umsetzung-von-urbanen-oekosystemleistungen/ (accessed on 2 August 2019).

23. Cohen-Shacham, E.; Walters, G.; Janzen, C.; Maginnis, S. *Nature-Based SOLUTIONS to Address Global Societal Challenges*; IUCN: Gland, Switzerland, 2016.

24. Cohen-Shacham, E.; Andrade, A.; Dalton, J.; Dudley, N.; Jones, M.; Kumar, C.; Maginnis, S.; Maynard, S.; Nelson, C.R.; Renaud, F.G.; et al. Core principles for successfully implementing and upscaling Nature-based Solutions. *Environ. Sci. Policy* **2019**, *98*, 20–29. [CrossRef]

25. Gutiérrez, L.; García, G.; García, I. *Nature-Based Solutions for Local Climate Adaptation in the Basque Country: Methodological Guide for Their Identification and Mapping*; Donostia/San Sebastián Case Study; Environmental Management Agency, Ministry of the Environment, Territorial Planning and Housing—Basque Government; Ihobe: Bilbao, Spain, 2017.

26. Stadt Leipzig. Machbarkeitsstudie zur Anbindung des Lindenauer Hafens an Dem Saale-Elster Kanal. Einleitung Stadt-Umland-Konferenz des Grünen Ringes Leipzig. 7 November 2018. Available online: https://gruenerring-leipzig.de/wp-content/uploads/2018/11/zabojnik-liha-1.pdf (accessed on 5 September 2019).

27. Cabral, I.; Keim, J.; Engelmann, R.; Kraemer, R.; Siebert, J.; Bonn, A. Ecosystem services of allotment and community gardens: A Leipzig, Germany case study. *Urban For. Urban Green.* **2017**, *23*, 44–53. [CrossRef]

28. Bundesministerium für Umwelt, Naturschutz, Bau und Reaktorsicherheit (BMUB). *Die Leipzig Charta zur Nachhaltigen Europäischen Stadt [The Leipzig Charter on Sustainable European Cities]*; BMUB: Berlin/Heidelberg, Germany, 2007. Available online: https://www.bmi.bund.de/EN/topics/building-housing/city-housing/national-urban-development/leipzig-charter/leipzig-charter-artikel.html (accessed on 10 August 2019). (In German)

29. Perkins, H.A. Green spaces of self-interest within shared urban governance. *Geogr. Compass* **2010**, *4*, 255–268. [CrossRef]

30. Maes, J.; Jacobs, S. Nature-Based Solutions for Europe's Sustainable Development. *Conserv. Lett.* **2015**, *10*, 121–124. [CrossRef]

31. Elmqvist, T.; Andersson, E.; Frantzeskaki, N.; McPhearson, T.; Olsson, P.; Gaffney, O.; Takeuchi, K.; Folke, C. Sustainability and resilience for transformation in the urban century. *Nat. Sustain.* **2019**, *2*, 267–273. [CrossRef]

32. Frantzeskaki, N.; Hölscher, K.; Bach, M.; Avelino, F. *Co-Creating Sustainable Urban Futures: A Primer on Applying Transition Management in Cities*; Springer International Publishing: New York, NY, USA, 2018. [CrossRef]

33. Pauleit, S.; Zölch, T.; Hansen, R.; Randrup, T.B.; van den Bosch, C.K. Nature-based Solutions and Climate Change—Four Shades of Green. In *Nature-Based Solutions to Climate Change in Urban Areas: Linkages of Science, Society and Policy*; Kabisch, N., Bonn, A., Korn, H., Stadler, J., Eds.; Springer International Publishing: New York, NY, USA, 2017; pp. 29–49. [CrossRef]

34. Biggs, R.; Kizito, F.; Adjonou, K.; Ahmed, M.T.; Blanchard, R.; Coetzer, K.; Handa, C.O.; Dickens, C.; Hamann, M.; O'Farrell, P.; et al. Current and future interactions between nature and society. In *The IPBES Regional Assessment Report on Biodiversity and Ecosystem Services for Africa*; Secretariat of the Intergovernmental Science-Policy Platform on Biodiversity and Ecosystem Services: Bonn, Germany, 2018; pp. 297–352.

35. United Nations. *Transforming Our World: The 2030 Agenda for Sustainable Development*; Division for Sustainable Development Goals, Department of Economic and Social Affairs, United Nations: New York, NY, USA, 2015. Available online: https://sustainabledevelopment.un.org/content/documents/21252030%20Agenda%20for%20Sustainable%20Development%20web.pdf (accessed on 23 July 2019).

36. Wright, H. Understanding green infrastructure: The development of a contested concept in England. *Local Environ.* **2011**, *16*, 1003–1019. [CrossRef]

37. Kabisch, N.; Haase, D. Green justice or just green? Provision of urban green spaces in Berlin, Germany. *Landsc. Urban Plan.* **2014**, *122*, 129–139. [CrossRef]

38. Laszkiewicz, E.; Kronenberg, J.; Marciaczak, S. Attached to or bound to a place? The impact of green space availability on residential duration: The environmental justice perspective. *Ecosyst. Serv.* **2018**, *30*, 309–317. [CrossRef]

39. Byrne, J.; Wolch, J. Nature, Race, and Parks: Past Research and Future Directions for Geographic Research. *Prog. Hum. Geogr.* **2009**, *33*, 743–765. [CrossRef]

40. Anguelovski, I.; Connolly, J.; Masip, L.; Pearsall, H. Assessing green gentrification in historically disenfranchised neighborhoods: A longitudinal and spatial analysis of Barcelona. *Urban Geogr.* **2018**, *39*, 458–491. [CrossRef]

41. Checker, M. Wiped out by the "greenwave": Environmental gentrification and the paradoxical politics of urban sustainability. *City Soc.* **2011**, *23*, 210–229. [CrossRef]

42. Dooling, S. Ecological gentrification: A research agenda exploring justice in the city. *Int. J. Urban Reg. Res.* **2009**, *33*, 621–639. [CrossRef]

43. Gould, K.A.; Lewis, T.L. *Green Gentrification Urban Sustainability and the Struggle for Environmental Justice*; Routledge: London, UK, 2017.

44. Döhren, P.; Haase, D. Ecosystem disservices research: A review of the state of the art with a focus on cities. *Ecol. Indic.* **2015**, *52*, 490–497. [CrossRef]

45. Badiu, D.L.; Nita, A.; Ioja, C.I.; Nita, M.R. Disentangling the connections: A network analysis of approaches to urban green infrastructure. *Urban For. Urban Green.* **2019**, *41*, 211–220. [CrossRef]

46. Artmann, M.; Kohler, M.; Meinel, G.; Gan, J.; Ioja, I.C. How smart growth and green infrastructure can mutually support each other—A conceptual framework for compact and green cities. *Ecol. Indic.* **2019**, *96*, 10–22. [CrossRef]

47. Yang, P.P.J.; Lay, O.B. Applying ecosystem concepts to the planning of industrial areas: A case study of Singapore's Jurong Island. *J. Clean. Prod.* **2004**, *12*, 1011–1023. [CrossRef]

48. Eggermont, H.; Balian, E.; Azevedo, J.M.N.; Beumer, V.; Brodin, T.; Claudet, J.; Fady, B.; Grube, M.; Keune, H.; Lamarque, P.; et al. NBS: New Influence for Environmental Management and Research in Europe. *GAIA* **2015**, *24*, 243–248. [CrossRef]

49. McPhearson, T.; Pickett, S.T.A.; Grimm, N.; Niemelä, J.; Alberti, M.; Elmqvist, T.; Weber, C.; Haase, D.; Breuste, J.; Qureshi, S. Advancing Urban Ecology toward a Science of Cities. *BioScience* **2016**, *66*, 1–15. [CrossRef]

50. Viljoen, A.; Bohn, K. *Second Nature Urban Agriculture: Designing Productive Cities*, 1st ed.; Routledge: London, UK, 2014. [CrossRef]

51. Essbare Stadt. Available online: http://www.leipziggruen.de/de/2016_EssbareStadt_Uebersicht.asp (accessed on 1 October 2019).

52. Stadt Leipzig. Straßenbaumkonzept: Leipziger WüNschen Sich GrüNere Strassen. 2016. Available online: https://www.leipzig.de/news/news/strassenbaumkonzept-leipziger-wuenschen-sich-gruenere-strassen/ (accessed on 1 October 2019).

53. Klotz, S. Auenwald—Ein einzigartiges Ökosystem. In *Leipzig: Eine Landeskundliche Bestandsaufnahme im Raum Leipzig Landschaften in Deutschland: Werte der deutschen Heimat 78*; Denzer, V., Dix, A., Porada, H.T., Eds.; Böhlau: Vienna, Austria, 2015; pp. 136–137.

54. Stadt Leipzig; Bundesamt für Naturschutz (BfN); Naturschutzbund Deutschland (NABU); Landesverband Sachsen E.V. Erhalt Leipziger Auen: Stadt Leipzig und Landesverband Sachsen des NABU e.V. Erhalten Fördermittel aus Bundesprogramm. 2012. Available online: https://www.leipzig.de/news/news/erhalt-leipziger-auen-stadt-leipzig-und-landesverband-sachsen-des-nabu-e-v-erhalten-frdermittel-aus/ (accessed on 21 December 2019).

 *land*

Article

Designing a Blue-Green Infrastructure (BGI) Network: Toward Water-Sensitive Urban Growth Planning in Dhaka, Bangladesh

Sanjana Ahmed [1], Mahbubur Meenar [2,*] and Ashraful Alam [3]

[1] Department of Urbanism, Faculty of Architecture and the Built Environment, Delft University of Technology, Julianalaan 134, 2628 BL Delft, The Netherlands
[2] Department of Geography, Planning, and Sustainability, School of Earth and Environment, Rowan University, Glassboro, NJ 08028, USA
[3] School of Geography, University of Otago, Dunedin 9054, New Zealand
* Correspondence: meenar@rowan.edu

Received: 20 August 2019; Accepted: 13 September 2019; Published: 16 September 2019

Abstract: In a warming world, urban environmental stresses are exacerbated by population-increase-induced development of grey infrastructure that usually leaves minimal scope for blue (and green) elements and processes, potentially resulting in mismanagement of stormwater and flooding issues. This paper explores how urban growth planning in the megacity of Dhaka, Bangladesh can be guided by a blue-green infrastructure (BGI) network that combines blue, green, and grey elements together to provide a multifunctional urban form. We take a three-step approach: First, we analyze the existing spatial morphology to understand potential locations of development and challenges, as well as the types of solutions necessary for water management in different typologies of urban densities. Next, we analyze existing and potential blue and green network locations. Finally, we propose the structural framework for a BGI network at both macro and micro scales. The proposed network takes different forms at different scales and locations and offers different types of flood control and stormwater management options. These can provide directions on Dhaka's future urban consolidation and expansion with a balance of man-made and natural elements and enable environmental, social, spatial, financial, and governance benefits. The paper concludes with some practical implications and challenges for implementing BGI in Dhaka.

Keywords: blue-green infrastructure; water-sensitive planning; urban growth planning; stormwater management; flooding; urban morphology; space syntax; spacematrix; Dhaka

1. Introduction

The city is a dynamic landscape characterized by natural (blue and green) and man-made (grey) elements. These elements, accumulated over time, shape the urban form and influence the behavior of residents. Over densification and unplanned urbanization leave little room for interaction among blue, green, and grey elements, and as a result, the natural elements (e.g., water, green space) and natural characteristics (e.g., topography) are deprioritized in many cities. One manifestation of this is that water—a vital structuring element—can become a challenge for the urban environment during extreme weather events such as heavy rainfall. The compact urban fabric often does not possess porous surfaces for water permeability, causing historically unprecedented flooding events.

Despite guaranteeing the protection of water bodies and actively managing water resources, many cities across the world are encountering complicated issues related to population growth, water shortage, and climate change [1]. This is more common in megacities that often confront water-related challenges [2,3]. Megacities are frequently considered as enlarged forms of cities, but in reality, the city-scale imposes immense pressure on ecology and infrastructure with added complications such as fragmented societies, higher inequality, and rising informality [4,5]. Megacities, characterized by numerous decision-making authorities, often struggle to achieve collaboration in solving urban sprawl, land use conversion, and water management problems [6].

The rapid process of urbanization in megacities is causing environmental, economic, and social problems. Development has been accompanied by negative consequences for many river systems, including changes in their hydrology and ecology. In recent decades, the increasing frequency of disaster events—including hydro-meteorological disasters—has threatened human lives and infrastructure. One of the most common water-related disasters frequently affecting urban social life, particularly in Asian regions, is flooding [7,8]. Specifically, cities are experiencing pluvial flooding due to increased urbanization and climate change [9–11]. Sources of water bodies capable of capturing a significant volume of floodwater are slowly disappearing as the volume of impervious surfaces rapidly escalates. As a result, urban areas are experiencing increasing high pick flow and stormwater runoff incidents that have noticeable adverse effects on social and economic lives [10,12]. Simultaneously, the remaining receiving water bodies have been polluted by mixed stormwater and wastewater, degrading their water quality.

In order to address escalating events of environmental dilapidation, resource susceptibilities, booming urban population, and other uncertain impacts from climate change, cities worldwide are reconsidering conventional urban water management systems [13,14]. Researchers have suggested numerous approaches for renovating urban systems, such as integrated water resources management [15], sustainable urban water management [15–17], and water sensitive cities [13,18]. Several concepts and technologies have emerged in recent years to manage urban stormwater with alternative solutions, including low impact development and green stormwater infrastructure in the USA [19–23], sustainable urban drainage systems in the UK [24,25], water sensitive urban design in Australia [26,27], and the 'sponge city' in China [28–30].

Within this context, the purpose of this paper is to explore if and how an urban blue-green infrastructure (BGI) network can be proposed—both at macro and micro scales—as part of a future urban development strategy for the megacity of Dhaka, the capital of Bangladesh. We attempt to answer the following research question: How can the urban structure be adapted to mediate between the need to accommodate population growth and at the same time give room for water and green spaces? We start with a literature review on BGI and discuss the current context of Dhaka. Next, we describe the three-step methodology applied in our research and accompanying results. Finally, we discuss how a BGI network can be implemented in Dhaka and offer concluding remarks.

2. Literature Review

2.1. Understanding Blue-Green Infrastructure (BGI) and Current Trends

BGI, albeit used less frequently, is an umbrella term used in planning (often in landscape planning) and is closely related to the concept of "green infrastructure" (GI) [31]. It combines the concept of green (including blue) networks [32] and ecological networks [33]. Scholars mostly draw on a standard definition given by the European Commission –

> *"a strategically planned network of natural and semi-natural areas with other environmental features designed and managed to deliver a wide range of ecosystem services. It incorporates green spaces (or blue if aquatic ecosystems are concerned) and other physical features in terrestrial (including coastal) and marine areas. On land, green infrastructure is present in rural and urban settings."* [34]

Compared to its earlier predecessors, more notably GI, BGI is relatively a new term which was first used for describing the planning efforts in Sao Paolo, Brazil to create a network of "green and blue" infrastructures in response to flood risks [35]. A systematic review of literature on blue, natural, ecological, and green infrastructure from 1989 to 2015 [36] suggests that there is a shift of focus observed in the transitioning from green, or natural, infrastructure to BGI. The later seeks an integrated approach, utilizing different types of eco-systems and associated eco-system services. It is a shift away from a comparatively simple "land use view" [37], towards recognizing more flexible eco-system service-based solutions [38,39] that do not only include green and blue elements and processes but also take into account man-made interventions, such as permeable pavements, bioswales, retention basins, and constructed wetlands as an integrated whole. The present paper aligns with the later body of works and the definition provided by Ghofrani that BGI is

> *"an interconnected network of natural and designed landscape components, including water bodies and green and open spaces, which provide multiple functions such as: (i) flood control, (ii) water storage for irrigation and industry use, (iii) wetland areas for wildlife habitat or water purification, among many others."* [40] (p. 499)

This later approach to BGI moves away from earlier engineering discourses that proliferated through the conceptualization of green infrastructure mostly for the management of water in urban areas. A range of nomenclature and acronyms were used to define different, or sometimes very much the same, elements, designs, and purposes [41]. Studies have used different nomenclature and acronyms for water management in literature since 1980s [14]. The approach to BGI in the later period, more closely evolves from the concept of water sensitive urban design (WSUD) [13] which shows a clear emphasis on the 'blue' elements (e.g., rainfall and flood) to describe the infrastructure [41]. Liao et al. [42] further clarifies that BGI is a particular type of green infrastructure involving a network of landscape systems, which often combine both natural and artificial materials, and is purposefully designed and managed to provide (storm) water-related ecosystem services (see also Fletcher et al. [14], for a review of the concept).

By adopting the term BGI, the present paper shifts away from an earlier fragmented and to some extent elective approach to blue elements as indicated by European Commission's definition ("or blue if aquatic ecosystems are concerned" [34]). Through the use of the term BGI, this paper attempts to recognize the active agencies of and adequate sensitivity to the water elements in planning urban infrastructure. In this regard, the paper also acknowledges the notion of "intentional landscapes" [43] (p. 133) in encapsulating BGI, not just comprising of natural landscapes but also made out of man-made elements. This approach to BGI helps a greater recognition of an array of ecosystem services (e.g., water purification, heat retention, as well as cultural and economic benefits) with a higher sensitivity towards human interventions, i.e., planned or designed urban spaces [44]. BGI, in this paper, is approached as purposeful and intentional, not just the remnant or leftover landscapes but designed and deployed primarily for social, economic, and environmental benefits [45,46], yet without compromising but sustaining natural processes [47].

Many studies have focused on BGI, but most of them seek to highlight the benefits of BGI [48]. In countries such as the Netherlands and Sweden, BGI is well accepted; however, the theoretical considerations of how the concept of BGI can inform the practice of planning remain undeveloped. BGI represents a conceptual shift from conventional approaches to water management, emphasizing the natural landscape [49] to provide "resilient and adaptive measure to deal with flooding by mimicking pre-development hydrology" [50] (for example, detention and retention techniques). This approach reduces stress on 'grey' infrastructure in urban areas [51]. Thus, planning for BGI is significantly different from conventional planning that historically relied too much on 'hard' built infrastructures, such as streets, sewage and drainage systems, and utility lines [31]. From the experiences of the Malmo City Council in Sweden [52], it is recommended that a careful balance is needed between hard and soft elements for a successful BGI through the incorporation of blue and green infrastructure among existing land uses where sealed or paved surfaces can be effectively minimized. Research also reinforces that planning for BGI should aim for a "symbiotic" relationship between the city and its region [53]; the emphasis should be on enticing more polycentric development and multiple densities across the entire region. Overall, there is a call for a more integrated mix of blue, green, and grey elements in which "the boundary between the natural and the technical networks is blurred" [53] (p. 9).

In the rest of the paper, a range of literature informs the investigation, analysis, and the overall approach of designing the BGI networks in the context of Dhaka, Bangladesh. Aligning with Ahern's [54] recommendation, the paper adopts a 'multi-scalar' planning approach to implement BGI integrating both micro (i.e., land parcel and neighborhood) and macro levels (i.e., urban and peri-urban region). Scholars have suggested the 'scale' in terms of regional/urban and private /public component in a given urban area [55]. Some key studies [56,57] propose for three spatial relationships that need to be taken into account when implementing BGI: site-specific elements, linkages, networks and connectivity [58], and other broader (regional) scale landscape elements. This paper also follows Hansen et al. [59], who recommend promoting 'multifunctional' BGI for high-density urban areas. BGI is well suited for dense urban areas as it reduces the need to upgrade or expand conventional stormwater/drainage system, both spatially and financially [42]. Following Hansen et al. [59], the first step of designing a BGI network involves a systematic "spatial assessment" of the urban morphology to identify all blue and green spaces at the 'site-level' as well as 'city-wide' that could be strategically organized in the future to meet multiple functions (p. 99). The second step focuses on the 'connectivity' of blue and green corridors as a critical component. The third step combines the two to envisage a multi-scalar solution that incorporates Dhaka's urban morphology and the dynamics of water.

2.2. Context: Water and Dhaka's Urbanization

Dhaka—one of the world's most densely populated megacities with 17 million people—is geographically situated in the deltaic plain of three major rivers: Padma, Brahmaputra, and Meghna, and surrounded by tributaries of these major rivers (see Figure 1). To describe Dhaka's historical context, in this section, we mostly rely on Rahman's edited collection [60] that collate facets of Dhaka's urban social and spatial transformations. Dhaka traces back its origin in the 17th century as a trading hub of the Mughals. As water played a principal role in transportation at that time, the city was not only structured by the bodies of water that flowed through it, but also culturally constructed as 'the river city'. At this time, the city's boundaries were clearly defined by the tributaries: Buriganga to the south, Turag to the north and west, and Balu to the east, as well as the swampy lands all around.

The story of Dhaka is similar to many highly urbanized cities of contemporary time such as Delhi and Shanghai that neglected their water systems to find room for growth, if seen from a morphological point of view. The city underwent different phases of development throughout history. In the 16th century Mughal rule, the development followed the river Buriganga in the east-west axis. During the British rule of the 17th century, the city started to grow toward the north because of the relocation of the industrial zones and the connection with railway lines. The 18th century was all about the coexistence of the city and the river where the city development respected the river course and used it as the network of the main transportation system. After the independence of Bangladesh in 1971, the city expanded in all directions, maintaining the tradition of a centralized capital city with a concentration of all governmental and private developments. The historical flooding event of 1988 put 75% of the city under water (see Figure 2). As a result, the western riverbank was protected by dam construction, which in the later phases caused dense urbanization process along the dam, keeping minimum room for the river. With the development of other grey infrastructures, waterways eventually became a forgotten backyard of the city.

The phenomenon of rural-urban migration has worsened the situation as the city doubled in size from 1990 to 2005 [61]. The U.N. predicts, by 2025, the population of Dhaka will cross 20 million making the city larger than Jakarta, Mexico City, or Shanghai today [62]. New residents move to Dhaka every day, seeking residence in the low-lying lands—by filling up water retention basins, river beds, or even portions of a river itself [63].

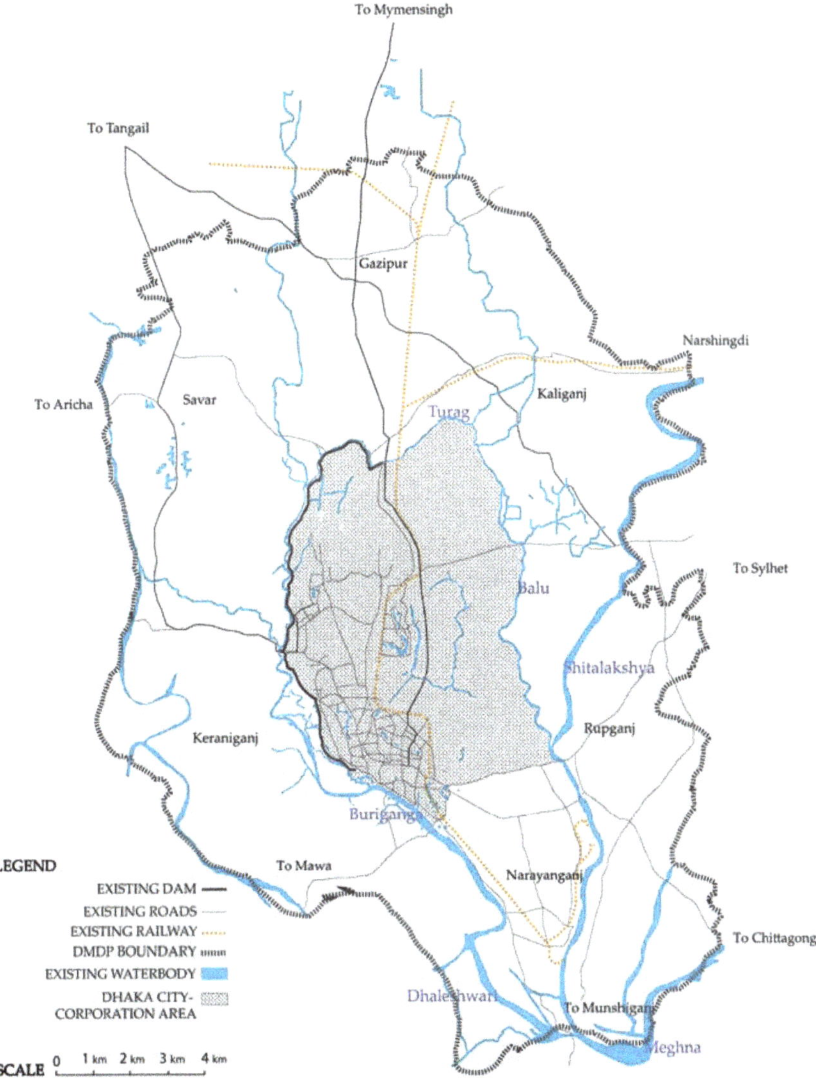

Figure 1. Location of Dhaka at the territorial scale, at the confluence of three major rivers.

The process of neglecting water in urban planning and development brought about both physical and metaphorical consequences. People started perceiving water as an element to neglect, polluting and encroaching its banks indiscriminately without any concern of environmental impacts on the overall city landscape. The waterways that used to clearly define the limit of the urban–rural condition of the city are blurred today. The city continues growing in all directions, keeping bare minimum porosity in the urban tissue. The city's green landscape is also a distant memory with more than 1000 ponds, canals, and parks replaced by houses, workplaces, and markets; now, only 5 percent in old parts and 12.5 percent in new parts of the city is composed of green areas [64]. The most common trend in Dhaka is to build multi-storied buildings on smaller parcels with minimal open space preserved. Private developers are filling up low-lying areas, including wetlands and marshlands, specifically on the eastern part of the city (see Figure 2). Such development patterns are resulting in severe scarcity of pervious land for rainwater to permeate and recharge the ground. Since the surface runoff cannot accumulate on the eastern lower side of the city, it causes high amounts of water clogging all around the city during major rain events. The inner part of the city that used to have water channels connecting the rivers on both sides has been occupied over time. Subsequently, there is not sufficient surface water corridors to channel the water from the inner part of the city to the downstream areas. High rainfall coinciding with a high-water level in the river quickly floods the city because stormwater cannot be naturally drained through the remaining water system that is in part due to the insufficient drainage system. As a result, frequent flooding events have increased in the last decade. The city now struggles relentlessly for the survival of its few remaining water courses and very often fails to withstand the pressure of the urbanization process that either completely ignores or mostly replaces the natural system with asphalt surfaces.

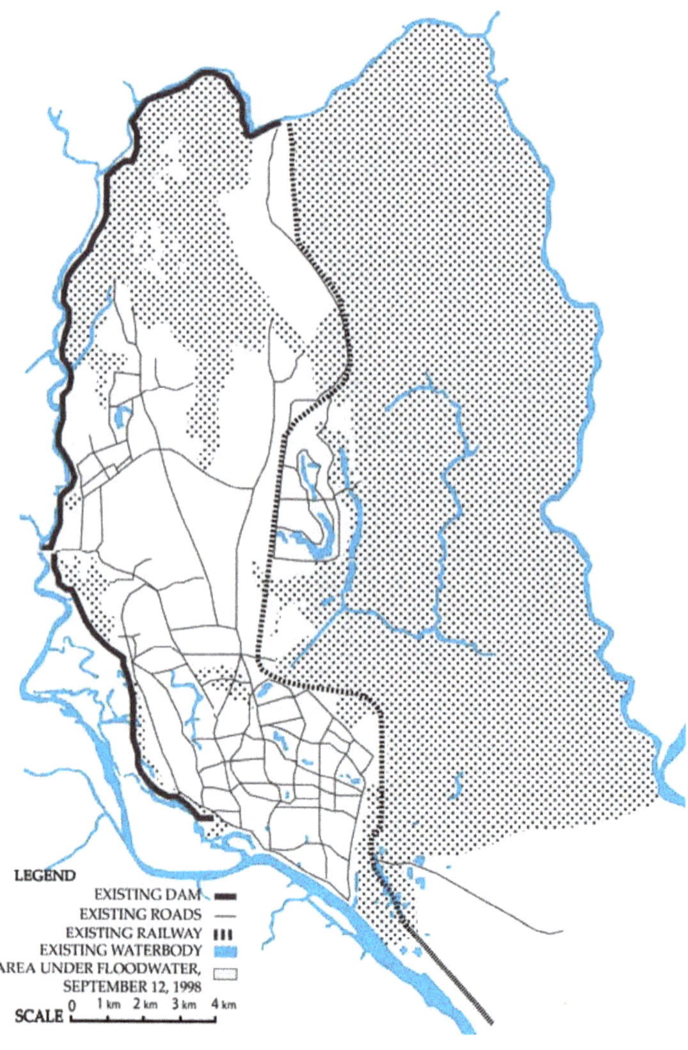

Figure 2. Flood-affected areas during the historic 1998 flood in Dhaka.

This paper addresses the opportunity to rethink how to accommodate future inhabitants of the city without necessarily compromising the blue (and green) infrastructural elements (e.g., water and green/open spaces) while possibly reviving those that have been lost. The recognition of the importance of blue and green natural elements as a structural tool is evident in the recent Draft Dhaka Structure Plan 2016–2035 [65], where significant spatial strategies have been premeditated to re-organize the future growth development pattern. The structural plan proposes to identify special zones for protecting and preserving natural areas for both flood management and recreational purposes. Notably, the plan identifies strategies for reclaiming the illegally occupied flood-flow zones in the city to be implemented in the coming years. In the context of the future urban development strategy, the following sections explore the potential of BGI networks—both at macro and micro scales.

3. Materials and Methods

We addressed our research question—how the urban structure can be adapted to mediate between the need to accommodate population growth while preserving natural resources—by taking a three-step approach described below (see Figure 3). The process of designing a BGI network, including the methodology and results of this three-step approach, is presented in this section. We collected GIS data from Dhaka City Corporation (administrative boundaries, parcel boundaries, building footprints, building heights, number of stories per building), Dhaka Water Supply and Sewerage Authority (rivers and waterways, wetlands, topography, and flood plains), Geodash (parks and open spaces), and Bengal institute (streets and other infrastructure).

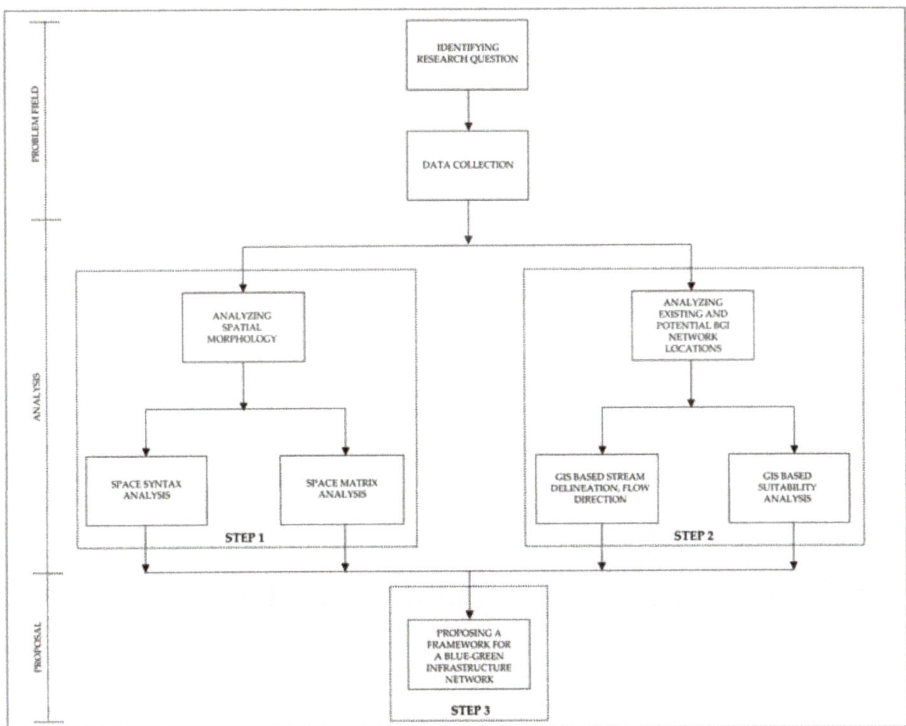

Figure 3. A graphical framework to represent our methodological approach.

Step 1—Analyzing Spatial Morphology: The primary author conducted a city-wide spatial morphological analysis to understand potential locations of development and challenges. This included a space syntax analysis and a spacematrix analysis (described below). The analysis was done based on an understanding that Dhaka's morphological structure differs across the city and, therefore, the same development strategy should not be applied to the whole city. The southern part of the city from the 1600s, for example, still has its narrow organic web-like street pattern with smaller parcels. The upper northern and eastern new towns, on the other hand, also have small parcel sizes but with wider streets following a grid-iron pattern. The morphological patterns of the rest of the city were either grown organically, owned by the army, or built illegally and later blended with the other parts of the city.

Space syntax is built on quantitative analysis and geospatial computer technology and provides a set of theories and methods for the analysis of spatial configurations of various kinds and at all scales [66]. The global axial integration analysis with a radius of "N" expresses the vehicular movement or street connectivity in the context of the whole city, whereas a radius (e.g., 2, 3, 4, 5) represents connectivity/integration at the local level. Radius 2, for example, stands for two steps away from each element, where a step stands for a change in directional turns from each element in a system. The higher the value of the radius, the closer it is to global axial analysis. In axial analysis, topological depth is the change in direction between one axial line and another that has no geometric value. Integration represents potential distance in a system and these destinations are highlighted from red to blue representing most integrated and segregated areas respectively [67].

We used the space syntax software to understand the integration of street network in Dhaka (or depth map analysis) and also to use the outputs of street integration as inputs in the Spacematrix analysis (described below) in order to understand the relationship between the levels of street integration with types of urban densities. First, a global axial analysis was conducted by editing the street network GIS shapefile collected from Dhaka City Corporation. Each vector line of the street network was edited in AutoCAD to intersect and thus prepared for a depth map analysis. After importing the converted dxf file into depth map, the map was converted to an axial map.

The method of Spacematrix is used to efficiently calculate urban density and understand the correlation of density and urban form—as well as the composition of built and unbuilt areas—by taking the street block as the smallest aggregation unit [68]. Spacematrix has the following parameters: floor space index or FSI (gross floor area/plan area), ground space index or GSI (built area/plan area), and network density (N), which can be calculated from a space syntax analysis (described above). These three measures are represented in a three-dimensional diagram. The open space ratio (OSR), the average numbers of floors (L), and the size of urban blocks (W) can be understood from that diagram. The readings of FSI, GSI, and OSR inform us how dense urban blocks are and what the amount of pressure is in the open space on those blocks.

We conducted this analysis using building and parcel shape files from Dhaka City Corporation. Parcel size and building footprints were calculated using a GIS. The values were used to calculate FSI, GSI, and OSR. All values were plotted on the graph using red dots. Each red dot was overlapped with blue, green, or yellow dots to represent low-, moderate-, or high-integrated streets respectively—resulting from their corresponding streets from the Space syntax/depth map analysis.

Step 2—Analyzing Existing and Potential Blue and Green Network Locations: The purpose of this step was to identify possible options for new blue-green corridors combining the existing water network, hidden streams, potential stormwater infrastructure locations, and existing green spaces—these can work together as a system and can drain stormwater runoff during heavy rainfall events. Using data on topographical configuration and existing water bodies of the city, we ran a GIS-based stream delineation model to identify the hidden streams that are buried under current development (e.g., buildings, streets, other impervious features). We also delineated the flow directions of rainwater on the ground based on their elevation level. We mapped existing green spaces (e.g., parks, open spaces) within the city and finally conducted a GIS-based suitability analysis to identify suitable locations for potential green stormwater management projects. Criteria used for the suitability modeling included elevation (areas with lower elevation are highly suitable), waterways (areas closer to waterways are highly suitable), rivers (areas closer to rivers are highly suitable), hidden streams (areas closer to hidden streams are highly suitable), and parks and open spaces (areas within parks and open spaces are highly suitable).

Step 3—Proposing a Framework for a Blue-Green Network: We overlaid all the outputs from steps 1 and 2 so that the framework for a BGI network followed a bilateral process to address two major concerns: morphological characteristics and water dynamics. These two concerns, however, needed a multi-scalar solution—territorial (macro) and local (micro). Therefore, we reviewed government proposals for the macro-level structure plan of greater Dhaka that, among other topics, focused on population growth management and flood control. As the final step, we proposed a framework of a BGI network.

4. Results and Recommendations

Step 1—Spatial Morphology: Figure 4 shows the result of space syntax analysis; the color range from blue to red indicates the level of street integration from lower to higher values. This result helped us identify streets that are not globally highly connected, and which can be converted to water carrying structures to create an integrated water system.

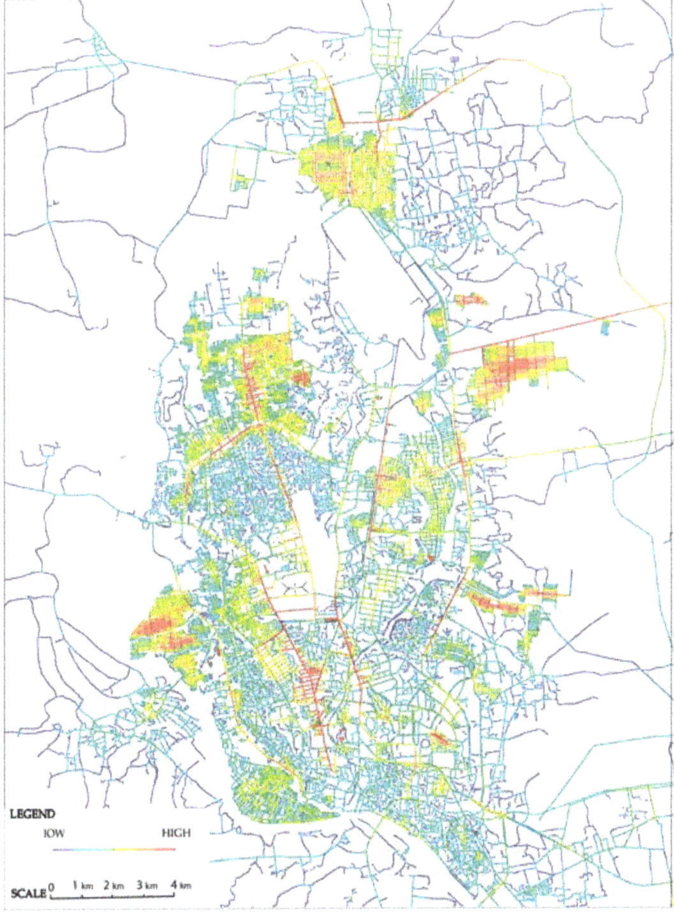

Figure 4. Results of Space Syntax analysis. Blue to red represents lower to higher street integration.

Figure 5 shows the results of the Spacematrix analysis for an area in Keraniganj as an example. FSI on the Y axis gives an impression of the intensity of the built environment of a particular area and GSI on the X axis shows the compactness or openness of that area. OSR represents the spaciousness or the pressure on the unbuilt area, and the L represents the average number of floors. The street blocks having GSI values from 0.30 to 0.90 with FSI values from 1.00 to 6, are located in the areas where streets are comparatively less integrated. On the other hand, the street blocks having GSI values from 0.35 to 1.80 with FSI values from 2.00 to 6.00 are located in the areas where the streets are more integrated. Overall, the majority of the blocks of Keraniganj show higher ground space intensity (GSI) values which indicates the high compactness of this area. On the other hand, the majority of the street blocks showing low floor space intensity (FSI) values indicate the possibility of future densification of this area.

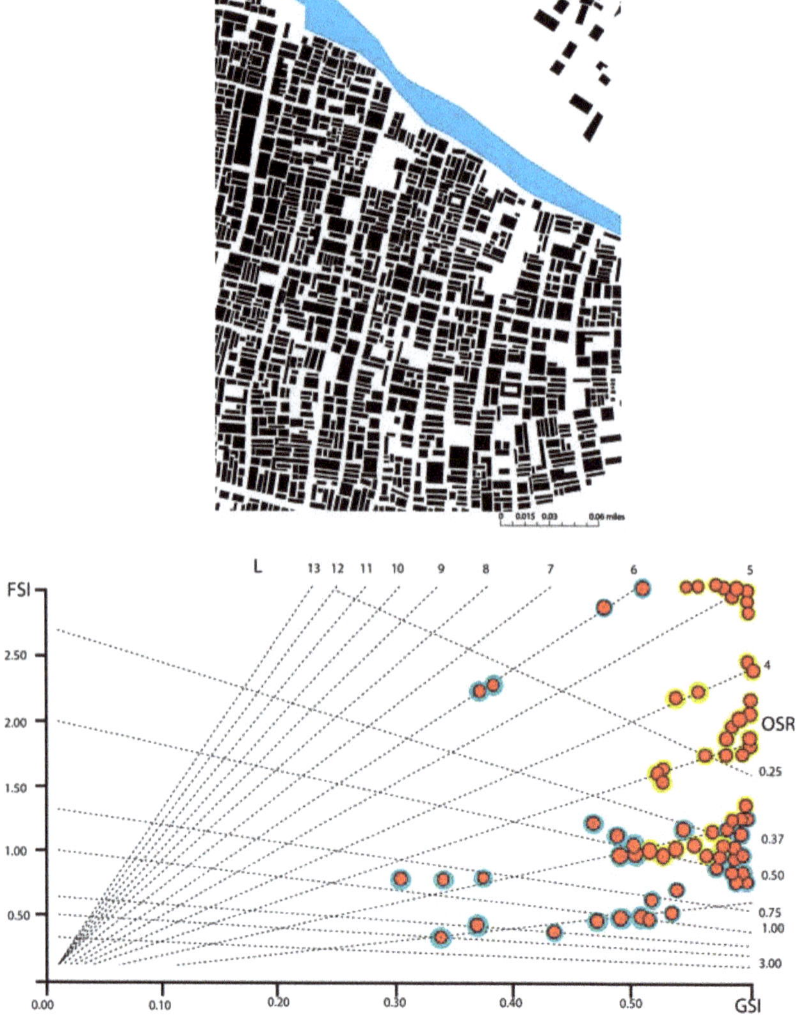

Figure 5. Results of Spacematrix analysis for an area in Keraniganj as an example.

Table 1 lists the results of the morphological analysis of nine selected areas in Dhaka, representing different types of development patterns (as shown in Figure 6, see Figure 5 for Keraniganj) and different historical, political, and economic influences. The major phases of socio-economic and political change, including the Hindu period, the Mughal period, the Pre-colonial period, the Colonial period, the Pakistan period, and finally the Bangladesh period shape the growth of different morphological typologies in different parts of the city, constituting the driving force of the selection of these nine locations.

Table 1. Morphological analysis results of nine selected areas in Dhaka.

Sample Location	GSI Values	Floors Per Building	Street Grid Integration	FSI Values	OSR Values	Comment
Bangshal	Moderately high 0.38–0.90	2–6	Low	on both the lower and higher end 1.00–5.00	2.0–6.0	Location under transition. Potential of managing storm water runoff on the ground floor.
Gulshan	Moderately high 0.30–0.70	6–15	High	On the higher end 2.70–6.00	6.0–16.0	The OSR which is also on the higher end of the graph suggests the potential of using the building volume for managing water since space is scarce on the ground level.
Lalmatia	Moderately high 0.35–0.80	2–10	Moderate	Vary from lower to higher 0.50–6.00	2.0–10.0	Location under transition. Option for micro scale water management (e.g., roof gardens or micro scale rain water harvesting systems)
Mirpur	Lower to moderately high 0.27–1.00	3–10	High	On the higher end 1.00–5.80	3.0–9.0	Potential of applying micro scale water management strategies using the transitioning building blocks (from midrise to high-rise apartments)
Pirerbagh	Lower to moderately high 0.40–0.70	1–6	Low	Mostly low 0.25–4.20	1.0–6.0	Possibility of future densification
Rampura	Lower to moderately high 0.40–0.80	2–9	Low	Mostly low 1.00–4.60	2.0–8.0	Possibility of future densification
Rayerbazar	Low0.30–0.79	3–10	Low	Relatively higher 0.90–5.50	2.5–10.0	Opportunity for managing stormwater runoff on the ground level
Uttara	Moderately high 0.38–0.70	2–9	High	Mostly on the mid to higher end 0.70 – 5.00	2.0–8.0	Possibility of future densification
Keraniganj	Very high 0.35–1.80	1–6	Moderate	Relatively low 1.00–6.00	1.5–10.0	Opportunity for future densification to meet the pressure of population increase

Overall, this spatial morphology analysis allowed us to categorize the blocks in Dhaka according to different qualities to be densified or restricted for development in the future. It also helped us understand what kind of solutions would be necessary for water management in different typologies of urban densities.

Step 2—Existing and Potential Blue and Green Network Locations: Figure 7a shows the delineated hidden streams along with existing water network, Figure 7b shows existing green areas, and Figure 7c shows the results of the suitability analysis (darker red represents highly suitable areas). Overall, this analysis assisted us in step 3 (proposal phase) to identify the areas following hidden streamlines where the peripheral water network could be connected through the existing urban fabric in order to better manage stormwater runoff. It also helped us identify the locations for stormwater infrastructure in upstream, midstream, and downstream to slow down the process of the natural drainage system.

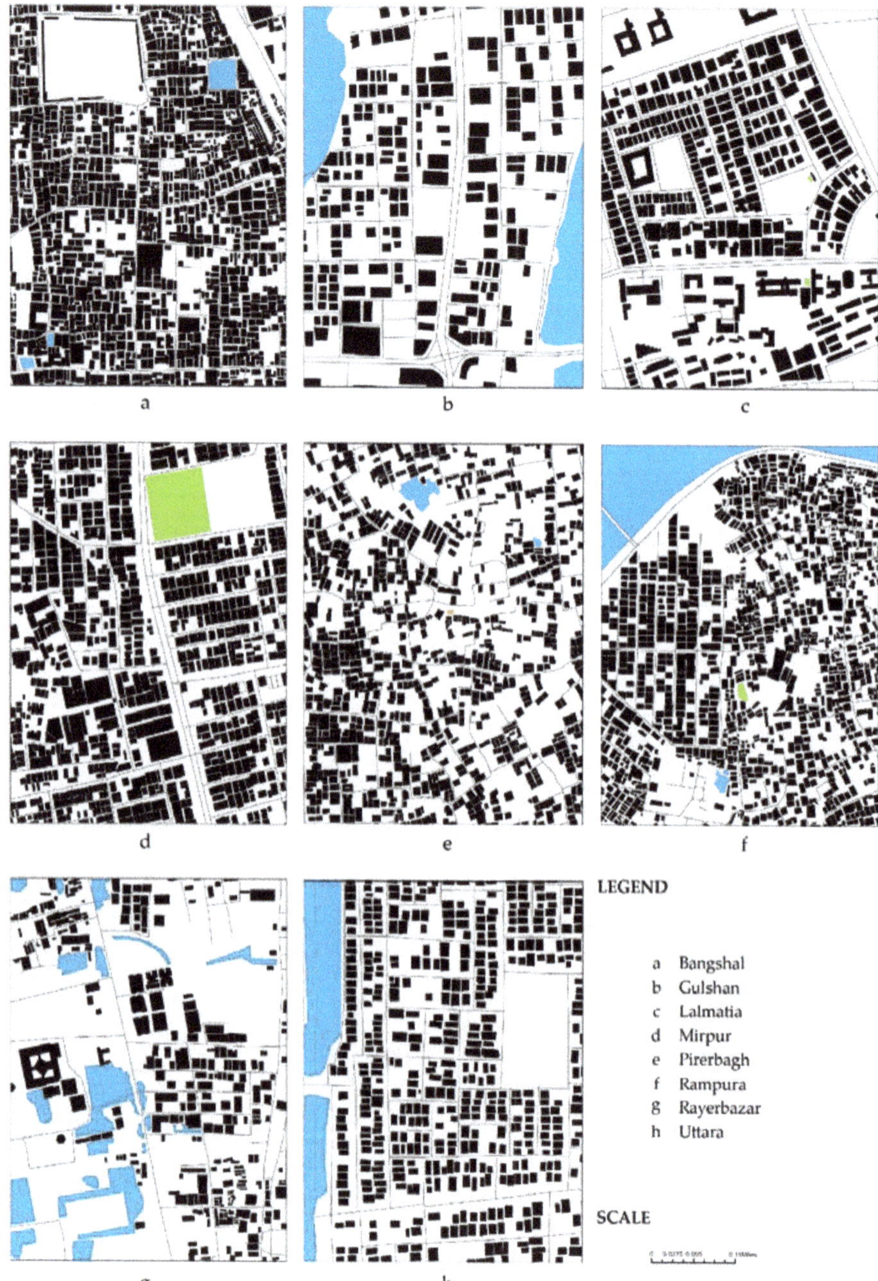

Figure 6. Ground space intensity and spatial pattern of eight selected areas in Dhaka.

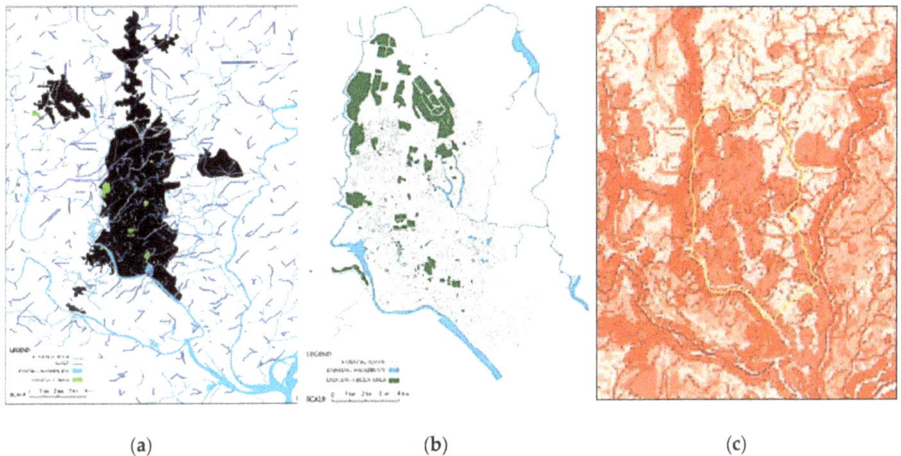

Figure 7. (a) Hidden streams and existing water network. (b) Existing green areas. (c) Suitable locations for green stormwater infrastructure (darker red represents highly suitable areas).

Step 3—A Framework for a BGI Network: A review of Dhaka Structure Plan 2016–2035 suggested long term strategies for macro-scale decentralization in order to reduce pressure on the city core. Government proposals included strategic development zones (e.g., satellite towns) with varied density outside the core area, circular/ring roads to serve the flow of traffic outside the core area, and a park system to improve the quality of the urban environment. We conceptualized that a BGI network at the macro scale could guide the development of future satellite towns, by channeling the natural water system of the whole territory [65]. Based on our proposal, a BGI network can provide both a direction on urban expansion zones and limitation on urban control zones by focusing on nature as the main guiding and limiting factors respectively.

Figure 8 shows the water system as a combination of the existing ones and newly proposed corridors—based on re-surfaced/retrofitted hidden streams and existing groundwater recharge areas—for the BGI network. The new water corridors while crossing existing highly dense urban areas would follow different types of patterns—sometimes following the less integrated streets, channeling underground, or incorporating existing smaller but open water bodies to accommodate excess water in extreme cases. While crossing upstream, midstream, or downstream areas, the network would slow down stormwater runoff, store water in identified appropriate locations, work as usable water reservoirs, or drain the excess runoff to the downstream. Multi-scalar rainwater storage areas would supply water in the urban areas and reduce pressure on the groundwater resource. Detention areas in the downstream would be integrated with the highly urban tissue. Loose urban tissue in the upstream area would turn into bigger retention areas and be preserved against unplanned urbanization. In some cases, the streets with low integration values would be retrofitted as water carrying structures/blue-ways with different strategies appropriate in different locations, including both subsurface and surface flow systems. The green network following the water channels would have multi-functionality. Sometimes, in the case of stormwater overflow during heavy rainfall, it might work as a fluvial park, agriculture land, energy park, or meadow. This macro-scale blue-green parkway, while crossing urban areas, would become fragmented at different scales following existing characteristics of different urban densities. Sometimes, in areas closer to the urban core, it would serve as a series of smaller-size neighborhood parks and accommodate collective recreational activities in order to bring people closer to rapidly diminishing green spaces. The BGI network, therefore, would not only manage the water system of the city but also limit the development of environmentally sensitive areas where more productive and recreational spaces can be activated with different social or cultural programs along its trajectory.

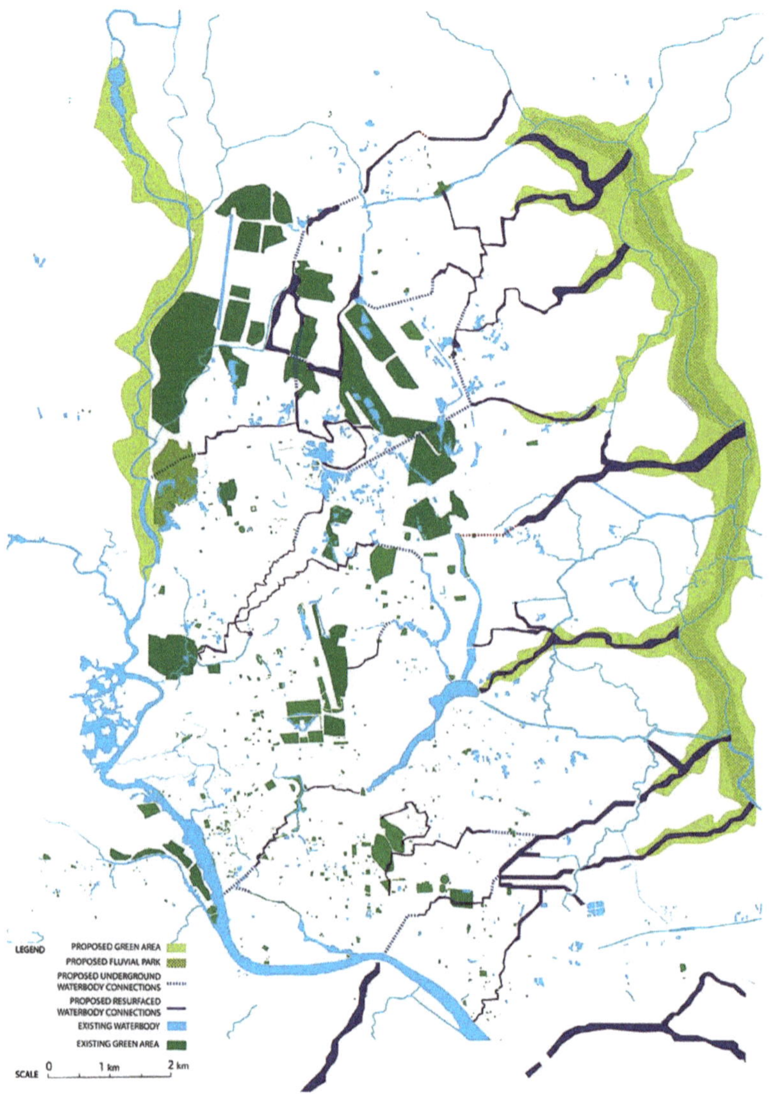

Figure 8. The proposed blue-green infrastructure (BGI) network at macro scale.

By setting the basic parameters for future development at the macro-scale, it is possible to delve further into the local or micro-scale strategies, which vary by locations. Different morphological configurations and levels of the street integrity at local levels opened up the requirement of multi-scalar approaches for stormwater management as well as different types of densification or controlling strategies for areas with a different FSI and GSI index. The BGI network would be an appropriate solution for water management in highly dense urban areas with high FSI and low GSI index. Densification would be needed in the areas with low FSI and development control in the areas with high FSI index. The micro-scale interventions of the BGI network would include various types of green stormwater infrastructure projects at different scales—buildings, parcels, blocks, or districts—to recharge groundwater, circulate rainwater runoffs, and connect the missing links of waterways to the drainage system. The key concept would be to promote composite water management structures that encourage social lives and activities and systematically drain excess water (see examples of dual-functionality of green stormwater infrastructure projects in Meenar [22]). Examples of such interventions include green roofs at the building level, bio-swales and rain gardens at the parcel or block level, and stormwater detention sites and blue-ways at the district level.

5. Discussion

The proposed BGI network may have significant practical (and planning) implications in achieving a sustainable urban form for Dhaka. First, the BGI network addresses two major challenges of Dhaka: (i) As the center of economic, educational, productive, and administrative sectors of the whole country, Dhaka faces extreme pressure of population influx and rapid urbanization and (ii) severe flooding and stormwater management issues have created challenges for current and future urbanization. The proposed water corridors include existing water features and ground water recharge areas, as well as re-surfaced (and retrofitted) hidden streams. The corridors take different forms at different scales and locations—upstream, midstream, or downstream—and offer different types of flood control and stormwater management options. Green 'corridors' [58] following the water channels may ensure flexible and diverse eco-system services [38,39], such as fluvial parks, agriculture land, energy parks, meadows, and small-size neighborhood breathing spaces. These serve as 'multifunctional' [46,57,59] spaces providing the much-needed room for green spaces in a dense urban context and recreational opportunities for residents. At the same time, they become an integrated part of the water retention and channeling system during extreme rainfall events. Overall, the proposed BGI network provides directions on the regeneration of the existing urban fabric, future urban growth direction, and puts restrictions on development in environmentally sensitive areas.

Among the many possible ways of implementing a BGI network in an urban setting, we have shown a way of structuring BGI with specific foci on tackling the realities of high urban density; namely, the continuing negligence to blue (and green) elements of the city and future risks from climate change and overwhelming urban growth. We have provided some principles and guidelines for the megacity of Dhaka that would allow different stakeholders to follow a conceptual and structural background for its future development. The strategies at the macro-scale would help in protecting and limiting development in water-sensitive areas, whereas the strategies at the micro-scale would encourage new water-sensitive developments in the fringes of the city. Together, these can produce more compatible morphological pattern respecting the existing landscape topology while simultaneously accommodating water and people in harmony.

It is often evident in the context of Dhaka that a higher degree of pessimism surrounds the planning and implementation of large infrastructure projects because of weak governance regimes leading to a lack of citizen trust in public institutions. This is compounded by a lack of citizen participation in the conception, planning, and implementation phases; lack of political will for collaboration among stakeholders and multiple overlapping bodies (e.g., Capital Development Authority or Rajdhani Unnayan Kartripakkha, Dhaka Electric Supply Company Limited, Dhaka Water Supply and Sewerage Authority, Public Works Department, Urban Development Directorate); and the complexities surrounding ownership of land. Notably, the implementation of comprehensive citywide infrastructural solutions become difficult because of budgetary constraints. The culture of multiple organizations doing the same work "affects the development plans having the absence of role casting principle" [69] (p. 331).

Our approach to implementing a BGI network can address such difficulties in a number of ways. First, the proposal emphasizes the adoption of a multi-scalar approach that combines both micro-level (private lot) and macro-level (city-wide) interventions within an integrated system. Such a scalar approach creates opportunities for engagement by different stakeholders at different scales. For example, community engagement may be leveraged to restore blue and green spaces at the parcel or block level through green stormwater management practices such as rainwater harvesting, bio-swales, and porous ground covers; public-private partnership may be leveraged at the district level multi-functional stormwater detention sites. Second, the proposed corridors can accommodate fluvial park, community garden projects, agricultural lots, and energy parks. These spaces could enable successful public-private partnership and citizen communing projects. Previous studies [70] have already proven the value of multi-scalar BGI network that provides opportunities for developing different urban forms ranging from privatized spaces (e.g., roof garden, backyard garden) to more community-owned and cooperative spaces (e.g., power transmission corridors, easements, and alleyways).

In a warming world, urban environmental stresses (e.g., flood, heatwave, etc.) are exacerbated by population-increase-induced development of grey infrastructure. As the BGI network, in principle, strives to incorporate both natural (blue and green) and man-made (grey) elements [37], it can be a more financially and spatially efficient solution for Dhaka. As advised by Liao et al. [42], BGI, with a greater carrying capacity than conventional infrastructure, can serve a greater population size and, would reduce the need for implementing new grey infrastructure. Furthermore, increasing reliance on the natural elements as an infrastructural strategy could reduce 'stress' [51] on the existing and underperforming grey infrastructure of Dhaka which are evidently failing due to unprecedented environmental forces caused by climate change impacts.

BGI holds the promise of achieving more sustainable urban form for Dhaka. With a philosophical underpinning of 'multifunctionality', our proposal can be considered a more appropriate alternative to the traditionally designed single-functioned infrastructural solutions. BGI provides a framework for integrating those conventionally isolated and independent blue and green initiatives [42] (pp. 221–222). With appropriate design consideration of ecological, economic, and social factors, BGI networks can be a valuable strategy for guiding other traditional infrastructure sectors, such as housing and transportation.

Hansen et al. [59] describe that the implementation of BGI can create spaces for cooperation between sectors and public departments. Overall, our proposed BGI network suggests for an integration of forms, structures, and processes across multiple scales, from the block level to larger urban setting of Dhaka; therefore, it may eliminate the problem of short-termism and piecemeal urban development that are common in Dhaka's institutional culture and political economies. BGI will encourage more strategic and collaborative urban planning and environmental management discourse.

In addition to the tangible impact on Dhaka, our approach can help rethink the implementation of BGI in other megacities. The proposed BGI approach is bottom-up—beginning with an understanding of the attributes (e.g., density, the mix of existing natural and grey elements) at the parcel or block level before integrating these micro-level strategies with district-level and city-wide topographical considerations. This novel approach presents a means of addressing the heterogeneous urban dynamics of megacities, promising a means of implementing BGI networks throughout informal urban settlements across the megacities of the Global South.

6. Conclusions

This paper has proposed a structural framework for designing and implementing BGI for future development in the megacity of Dhaka. The proposed BGI network addresses the city's morphological characteristics and water dynamics and provides multi-scalar and multi-functional solutions. The proposal considers a range of micro and macro levels dynamics, including population density, existing grey elements through FSI, GSI, OSR, and natural elements (e.g., hidden and existing water elements, green corridors). We also discussed the social, spatial, environmental, financial, and governance benefits of implementing BGI network in the existing political economic context of Dhaka.

Nevertheless, the implementation of a BGI network must be done cautiously. Since the inception of BGI, concepts are being drawn from countries such as the Netherlands and Sweden. The political, social, environmental and resource dynamics in the developed world are different from the context of Asian megacities like Dhaka. Very little is understood on BGI in the third world mega-urban context.

Learning on "the development of, and agreement on, conceptual understandings as well as practical processes" [71] (p. 77) should be taken from the Global South (e.g., Brazil) [35]. The socio-cultural dynamics in implementing BGI in the third-world megacity context warrants future research.

Unlike typical infrastructural projects, the implementation of BGI takes on a different approach by blending hard and soft elements, which take time to mature. Therefore, a BGI network may also suffer from a lack of visibility due to the slow pace of development and an absence of 'development-as-usual' grey elements within it; as a result, a greater community buy-in may be challenging at times. To overcome these issues, a greater understanding of the synergies, trade-offs, and the capacity of the existing blue and green spaces is required (see Hansen et al. [59] for critical recommendations). There are needs of adequate non-regulatory interventions (e.g., education) to develop awareness among communities and partners about the components of BGI and the multiple benefits in personal (e.g., health and wellbeing) and public life (e.g., community interactions, better amenities, climate resilience) of a city. Planning for a BGI network can be greatly benefited from more participatory processes. There are opportunities of learning from creative and participatory flood management experimentations that successfully avoid top-down, technology-heavy, and large-scale infrastructural solutions; for example, see "flood apprenticeship" projects [72].

Despite the proven and foreseeable advantages, widespread implementation of a BGI network can sometimes be hampered by uncertainties regarding hydro-logical performances, negative perceptions by residents about BGI [73], service delivery, and lack of confidence by decision-makers [74]. Such uncertainties reinforce the idea that BGI should be understood as a combination of civic culture, social equity, and environmental awareness as part of the greater urban climate change adaptation response. Greater multi-level collaboration is required to recognize the BGI network as part of the climate change response agenda. Future research should focus on the climate adaptation aspects of cities in relation to BGI networks. Proposals should be backed by adequate strategic elements with carefully chalked out contingency plans, rather than being specific to professional chauvinism and siloed institutional views.

Finally, since the last decade, there has been an ideological shift in the approach of implementing BGI from engineering and system-oriented discourses to a more integrated spatial planning approach that advocates for a symbiotic relationship among different natural and man-made elements [53]. Therefore, planning for BGI requires a more holistic vision to recognize cities as "socio-ecological" entities [43] where social and natural elements co-exist. BGI researchers have already started to recognize the collective socio-natural and processual aspects [75]. There is a need for recognition of the 'controlled' as well as 'flexible' options [76] in the structure of a BGI network based on which a livelier urban morphological matrix may mature over time. In the context of megacity Dhaka, where the demographic densities, climate vulnerabilities, deterioration of urban nature all are pervasively at play, there is a greater need for future research initiatives to understand the ways in which a more flexible and process-focused BGI network can be sustained.

Author Contributions: Conceptualization, S.A. and M.M.; methodology, S.A.; software, S.A.; formal analysis, S.A.; resources, S.A., A.A. and M.M.; writing—original draft preparation, M.M., A.A. and S.A.; writing—review and editing, M.M., A.A. and S.A.; visualization, S.A.

Funding: This research received no external funding.

Acknowledgments: The analytical part of this paper was originally conceptualized as part of the thesis completed by primary author S.A. under the European Post-master in Urbanism (EMU) program at the Faculty of Architecture and the Built Environment at Delft University of Technology in 2017. S.A. acknowledges the guidance and feedback of her mentors and examiners Han Meyer, Birgit Hausleitner, Paola Pelligrini, and Leo van den Burg. The authors also acknowledge Kyle Hearing for technical assistance.

Conflicts of Interest: The authors declare no conflict of interest.

References

1. Wong, T.H.; Brown, R.R. The water sensitive city: Principles for practice. *Water Sci. Technol.* **2009**, *60*, 673–682. [CrossRef] [PubMed]
2. OECD. *Water Governance in Cities*; OECD Publishing: Paris, France, 2016.
3. Varis, O.; Biswas, A.K.; Tortajada, C.; Lundqvist, J. Megacities and water management. *Water Resour. Dev.* **2006**, *22*, 377–394. [CrossRef]
4. Ahmed, S.; Meenar, M. Just Sustainability in the Global South: A Case Study of the Megacity of Dhaka. *J. Dev. Soc.* **2018**, *34*, 401–424. [CrossRef]
5. Kraas, F.; Mertins, G. Megacities and global change. In *Megacities*; Springer: Berlin, Germany, 2014; pp. 1–6.
6. Kim, J.H.; Keane, T.D.; Bernard, E.A. Fragmented local governance and water resource management outcomes. *J. Environ. Manag.* **2015**, *150*, 378–386. [CrossRef] [PubMed]
7. Centre for Research on the Epidemiology of Disasters-United Nations Office for Disaster Risk Reduction (CRED-UNISDR). *The Human Cost of Weather-Related Disasters 1995–2015*; Centre for Research on the Epidemiology of Disasters (CRED): Brussels, Belgium; United Nations Office for Disaster Risk Reduction (UNISDR): Geneva, Switzerland, 2015.
8. Guha-Sapir, D.; Vos, F.; Below, R.; Ponserre, S. *Annual Disaster Statistical Review 2011: The Numbers and Trends*; Centre for Research on the Epidemiology of Disasters (CRED): Brussels, Belgium, 2012.
9. Cousins, J.J. Infrastructure and institutions: Stakeholder perspectives of stormwater governance in Chicago. *Cities* **2017**, *66*, 44–52. [CrossRef]
10. Meenar, M.; Fromuth, R.; Soro, M. Planning for watershed-wide flood-mitigation and stormwater management using an environmental justice framework. *Environ. Pract.* **2018**, *20*, 55–67. [CrossRef]
11. Petit-Boix, A.; Sevigné-Itoiz, E.; Rojas-Gutierrez, L.A.; Barbassa, A.P.; Josa, A.; Rieradevall, J.; Gabarrell, X. Floods and consequential life cycle assessment: Integrating flood damage into the environmental assessment of stormwater Best Management Practices. *J. Clean. Prod.* **2017**, *162*, 601–608. [CrossRef]
12. Dhakal, K.P.; Chevalier, L.R. Urban stormwater governance: The need for a paradigm shift. *Environ. Manag.* **2016**, *57*, 1112–1124. [CrossRef] [PubMed]
13. Brown, R.R.; Keath, N.; Wong, T.H. Urban water management in cities: Historical, current and future regimes. *Water Sci. Technol.* **2009**, *59*, 847–855. [CrossRef]
14. Fletcher, T.D.; Shuster, W.; Hunt, W.F.; Ashley, R.; Butler, D.; Arthur, S.; Trowsdale, S.; Barraud, S.; Semadeni-Davies, A.; Bertrand-Krajewski, J.-L. SUDS, LID, BMPs, WSUD and more–The evolution and application of terminology surrounding urban drainage. *Urban Water J.* **2015**, *12*, 525–542. [CrossRef]
15. Mitchell, B. Integrated water resource management, institutional arrangements, and land-use planning. *Environ. Plan. A* **2005**, *37*, 1335–1352. [CrossRef]
16. Gleick, P.H. Global freshwater resources: Soft-path solutions for the 21st century. *Science* **2003**, *302*, 1524–1528. [CrossRef] [PubMed]
17. Pahl-Wostl, C. Requirements for adaptive water management. In *Adaptive and Integrated Water Management*; Springer: Berlin, Germany, 2008; pp. 1–22.
18. Floyd, J.; Iaquinto, B.L.; Ison, R.; Collins, K. Managing complexity in Australian urban water governance: Transitioning Sydney to a water sensitive city. *Futures* **2014**, *61*, 1–12. [CrossRef]
19. Ahiablame, L.M.; Engel, B.A.; Chaubey, I. Effectiveness of low impact development practices in two urbanized watersheds: Retrofitting with rain barrel/cistern and porous pavement. *J. Environ. Manag.* **2013**, *119*, 151–161. [CrossRef] [PubMed]
20. Christman, Z.; Meenar, M.; Mandarano, L.; Hearing, K. Prioritizing suitable locations for green stormwater infrastructure based on social factors in Philadelphia. *Land* **2018**, *7*, 145. [CrossRef]
21. Gogate, N.G.; Kalbar, P.P.; Raval, P.M. Assessment of stormwater management options in urban contexts using Multiple Attribute Decision-Making. *J. Clean. Prod.* **2017**, *142*, 2046–2059. [CrossRef]
22. Meenar, M.R. Integrating placemaking concepts into Green Stormwater Infrastructure design in the City of Philadelphia. *Environ. Pract.* **2019**, *21*, 4–19. [CrossRef]
23. Pyke, C.; Warren, M.P.; Johnson, T.; LaGro, J., Jr.; Scharfenberg, J.; Groth, P.; Freed, R.; Schroeer, W.; Main, E. Assessment of low impact development for managing stormwater with changing precipitation due to climate change. *Landsc. Urban Plan.* **2011**, *103*, 166–173. [CrossRef]

24. Ellis, J.B.; Lundy, L. Implementing sustainable drainage systems for urban surface water management within the regulatory framework in England and Wales. *J. Environ. Manag.* **2016**, *183*, 630–636. [CrossRef]

25. Fryd, O.; Dam, T.; Jensen, M.B. A planning framework for sustainable urban drainage systems. *Water Policy* **2012**, *14*, 865–886. [CrossRef]

26. Bach, P.M.; Mccarthy, D.T.; Deletic, A. Can we model the implementation of water sensitive urban design in evolving cities? *Water Sci. Technol.* **2015**, *71*, 149–156. [CrossRef] [PubMed]

27. Wella-Hewage, C.S.; Alankarage Hewa, G.; Pezzaniti, D. Can water sensitive urban design systems help to preserve natural channel-forming flow regimes in an urbanised catchment? *Water Sci. Technol.* **2016**, *73*, 78–87. [CrossRef] [PubMed]

28. Chan, F.K.S.; Griffiths, J.A.; Higgitt, D.; Xu, S.; Zhu, F.; Tang, Y.-T.; Xu, Y.; Thorne, C.R. "Sponge City" in China—A breakthrough of planning and flood risk management in the urban context. *Land Use Policy* **2018**, *76*, 772–778. [CrossRef]

29. Jiang, Y.; Zevenbergen, C.; Fu, D. Understanding the challenges for the governance of China's "sponge cities" initiative to sustainably manage urban stormwater and flooding. *Nat. Hazards* **2017**, *89*, 521–529. [CrossRef]

30. Wang, Y.; Sun, M.; Song, B. Public perceptions of and willingness to pay for sponge city initiatives in China. *Resour. Conserv. Recycl.* **2017**, *122*, 11–20. [CrossRef]

31. Ghofrani, Z.; Sposito, V.; Faggian, R. A Comprehensive Review of Blue-Green Infrastructure Concepts. *Int. J. Environ. Sustain.* **2017**, *6*, 15–36. [CrossRef]

32. Ahern, J. Greenways as a planning strategy. *Landsc. Urban Plan.* **1995**, *33*, 131–155. [CrossRef]

33. Jongman, R.H.; Pungetti, G. *Ecological Networks and Greenways: Concept, Design, Implementation*; Cambridge University Press: Cambridge, UK, 2004.

34. European Commission. *Green Infrastructure (GI)—Enhancing Europe's Natural Capital*; European Commission: Brussels, Belgium, 2013.

35. Frischenbruder, M.T.M.; Pellegrino, P. Using greenways to reclaim nature in Brazilian cities. *Landsc. Urban Plan.* **2006**, *76*, 67–78. [CrossRef]

36. Silva, J.M.C.d.; Wheeler, E. Ecosystems as infrastructure. *Perspect. Ecol. Conserv.* **2017**, *15*, 32–35. [CrossRef]

37. Scott, M.; Lennon, M.; Haase, D.; Kazmierczak, A.; Clabby, G.; Beatley, T. Nature-based solutions for the contemporary city/Re-naturing the city/Reflections on urban landscapes, ecosystems services and nature-based solutions in cities/Multifunctional green infrastructure and climate change adaptation: Brownfield greening as an adaptation strategy for vulnerable communities?/Delivering green infrastructure through planning: Insights from practice in Fingal, Ireland/Planning for biophilic cities: From theory to practice. *Plan. Theory Pract.* **2016**, *17*, 267–300. [CrossRef]

38. Haase, D. Reflections about blue ecosystem services in cities. *Sustain. Water Qual. Ecol.* **2015**, *5*, 77–83. [CrossRef]

39. Haase, D.; Haase, A.; Rink, D. Conceptualizing the nexus between urban shrinkage and ecosystem services. *Landsc. Urban Plan.* **2014**, *132*, 159–169. [CrossRef]

40. Ghofrani, Z.; Sposito, V.; Faggian, R. Designing resilient regions by applying blue-green infrastructure concepts. *Wit Trans. Ecol. Environ.* **2016**, *204*, 493–505.

41. Everett, G.; Lamond, J.; Lawson, E. Green infrastructure and urban water management. In *Handbook on Green Infrastructure: Planning, Design and Implementation*; Edward Elgar: Gloucester, UK, 2015; pp. 50–66.

42. Liao, K.-H.; Deng, S.; Tan, P.Y. Blue-green infrastructure: New frontier for sustainable urban stormwater management. In *Greening Cities*; Springer: Berlin, Germany, 2017; pp. 203–226.

43. Byrne, J.A.; Lo, A.Y.; Jianjun, Y. Residents' understanding of the role of green infrastructure for climate change adaptation in Hangzhou, China. *Landsc. Urban Plan.* **2015**, *138*, 132–143. [CrossRef]

44. Lovell, S.T.; Taylor, J.R. Supplying urban ecosystem services through multifunctional green infrastructure in the United States. *Landsc. Ecol.* **2013**, *28*, 1447–1463. [CrossRef]

45. Beer, A.R. *Greenspaces, Green Structure, and Green Infrastructure Planning. American Society of Agronomy*; Crop Science Society of America, Soil Science Society of America: Madison, WI, USA, 2010; pp. 431–448.

46. Brears, R.C. Blue-Green Infrastructure in Managing Urban Water Resources. In *Blue and Green Cities*; Springer: Berlin, Germany, 2018; pp. 43–61.

47. Wade, R.; McLean, N. *Multiple Benefits of Green Infrastructure*; Booth, C.A., Charlesworth, S.M., Eds.; Wiley Blackwell: Oxford, UK, 2014; pp. 319–335.

48. Lennon, M.; Scott, M.; Collier, M.; Foley, K. The emergence of green infrastructure as promoting the centralisation of a landscape perspective in spatial planning—the case of Ireland. *Landsc. Res.* **2017**, *42*, 146–163. [CrossRef]

49. Hoyer, J.; Dickhaut, W.; Kronawitter, L.; Weber, B. *Water Sensitive Urban Design: Principles and Inspiration for Sustainable Stormwater Management in the City of the Future*; Jovis: Berlin, Germany, 2011.

50. O'Donnell, E.C.; Lamond, J.E.; Thorne, C.R. Recognising barriers to implementation of Blue-Green Infrastructure: A Newcastle case study. *Urban Water J.* **2017**, *14*, 964–971. [CrossRef]

51. Lawson, E.; Thorne, C.; Ahilan, S.; Allen, D.; Arthur, S.; Everett, G.; Fenner, R.; Glenis, V.; Guan, D.; Hoang, L. Delivering and evaluating the multiple flood risk benefits in blue-green cities: An interdisciplinary approach. *Wit Trans. Ecol. Environ.* **2014**, *184*, 113–124.

52. Kruuse, A. The green space factor and the green points system. *Grabs Expert Pap.* **2011**, *6*, 12.

53. Bogunovich, D. Urban sustainability: Resilient regions, sustainable sprawl and green infrastructure. *Sustain. City Vii Vol* **2012**, *1*, 3–10.

54. Ahern, J. Green infrastructure for cities: The spatial dimension. In *Cities of the Future: Towards Integrated Sustainable Water and Landscape Management*; Novotny, V., Brown, P., Eds.; IWA Publishing: London, UK, 2007; pp. 267–283.

55. Pötz, H.; Bleuze, P. *Urban Blue-Green Grids for Sustainable and Dynamic Cities*; Coop for Life: Delft, The Netherlands, 2012.

56. Davies, C.; MacFarlane, R.; McGloin, C.; Roe, M. Green infrastructure planning guide. 2006.

57. Hansen, R.; Pauleit, S. From multifunctionality to multiple ecosystem services? A conceptual framework for multifunctionality in green infrastructure planning for Urban Areas. *Ambio* **2014**, *43*, 516–529. [CrossRef] [PubMed]

58. Ahern, J. From fail-safe to safe-to-fail: Sustainability and resilience in the new urban world. *Landsc. Urban Plan.* **2011**, *100*, 341–343. [CrossRef]

59. Hansen, R.; Olafsson, A.S.; van der Jagt, A.P.N.; Rall, E.; Pauleit, S. Planning multifunctional green infrastructure for compact cities: What is the state of practice? *Ecol. Indic.* **2019**, *96*, 99–110. [CrossRef]

60. Rahman, M. (Ed.) *Dhaka: An Urban Reader*; The University Press Limited: Dhaka, Bangladesh, 2016.

61. Sheth, K. Biggest Cities In Bangladesh. Available online: https://www.worldatlas.com/articles/biggest-cities-in-bangladesh.html (accessed on 20 August 2019).

62. Global Post. Dhaka, Bangladesh: Fastest Growing City in the World. Available online: https://www.cbsnews.com/news/dhaka-bangladesh-fastest-growing-city-in-the-world/ (accessed on 28 August 2019).

63. Burkart, K.; Gruebner, O.; Khan, M.; Staffeld, R. Megacity Dhaka-informal settlements, urban environment and public health. *Geogr. Rundsch.* **2008**, *4*, 4–10.

64. Islam, T. Environment-Bangladesh: Saving Dhaka's Remaining Green Charm. Available online: http://www.ipsnews.net/2000/08/environment-bangladesh-saving-dhakas-remaining-green-charm/ (accessed on 20 August 2019).

65. RAJUK. *Draft Dhaka Structural Plan 2016–2035*; Dhaka Metropolitan Development Planning, Rajdhani Unnayan Kartripakkha Dhaka: Dhaka, Bangladesh, 2015.

66. Hillier, B. *Space is the Machine: A Configurational Theory of Architecture*; Space Syntax: London, UK, 2007.

67. Al-Sayed, K.; Hillier, B.; Penn, A.; Turner, A. *Space Syntax Methodology*; Bartlett School of Architecture, UCL: London, UK, 2018.

68. Berghauser Pont, M.; Haupt, P. *Spacemate: The Spatial Logic of Urban Density*; Delft University Press Science: Delft, The Netherlands, 2004.

69. Rahman, M.A.U. Coordination of urban planning organizations as a process of achieving effective and socially just planning: A case of Dhaka city, Bangladesh. *Int. J. Sustain. Built Environ.* **2015**, *4*, 330–340. [CrossRef]

70. Mell, I.C.; Henneberry, J.; Hehl-Lange, S.; Keskin, B. Promoting urban greening: Valuing the development of green infrastructure investments in the urban core of Manchester, UK. *Urban For. Urban Green.* **2013**, *12*, 296–306. [CrossRef]

71. Mell, I.C. Green infrastructure: Concepts and planning. *Forum Ejournal* **2008**, *8*, 69–80.

72. Whatmore, S.J.; Landström, C. Flood apprentices: An exercise in making things public. *Econ. Soc.* **2011**, *40*, 582–610. [CrossRef]

73. Williams, J.B.; Jose, R.; Moobela, C.; Hutchinson, D.J.; Wise, R.; Gaterell, M. Residents' perceptions of sustainable drainage systems as highly functional T blue green infrastructure. *Landsc. Urban Plan.* **2019**, *190*, 103610. [CrossRef]

74. Thorne, C.R.; Lawson, E.C.; Ozawa, C.; Hamlin, S.L.; Smith, L.A. Overcoming uncertainty and barriers to adoption of Blue-Green Infrastructure for urban flood risk management. *J. Flood Risk Manag.* **2018**, *11*, S960–S972. [CrossRef]

75. Murphy, M.; Hedfors, P. Systems theory and Landscape Architecture. In Proceedings of the Urban Nature, Nature, CELA, Los Angeles, CA, USA, 30 March–2 April 2011.

76. Berg, P.G.; Ignatieva, M.; Granvik, M.; Hedfors, P. Green-Blue Infrastructure in Urban-Rural Landscapes-Introducing Resilient Citylands. *Nord. J. Archit. Res.* **2013**, *2*, 11–42.

 *land*

Article

Predicting Land Use Changes in Philadelphia Following Green Infrastructure Policies

Charlotte Shade * and Peleg Kremer

Villanova University, Department of Geography and the Environment, Villanova, PA 19085, USA;
peleg.kremer@villanova.edu
* Correspondence: charshade17@gmail.com

Received: 20 December 2018; Accepted: 30 January 2019; Published: 1 February 2019

Abstract: Urbanization is a rapid global trend, leading to consequences such as urban heat islands and local flooding. Imminent climate change is predicted to intensify these consequences, forcing cities to rethink common infrastructure practices. One popular method of adaptation is green infrastructure implementation, which has been found to reduce local temperatures and alleviate excess runoff when installed effectively. As cities continue to change and adapt, land use/landcover modeling becomes an important tool for city officials in planning future land usage. This study uses a combination of cellular automata, machine learning, and Markov chain analysis to predict high resolution land use/landcover changes in Philadelphia, PA, USA for the year 2036. The 2036 landcover model assumes full implementation of Philadelphia's green infrastructure program and past temporal trends of urbanization. The methodology used to create the 2036 model was validated by creating an intermediate prediction of a 2015 landcover that was then compared to an existing 2015 landcover. The accuracy of the validation was determined using Kappa statistics and disagreement scores. The 2036 model successfully met Philadelphia's green infrastructure goals. A variety of landscape metrics demonstrated an overall decrease in fragmentation throughout the landscape due to increases in urban landcover.

Keywords: landcover change; green infrastructure; spatial modeling; TerrSet; policy; GEOMOD; Land Change Modeler; nature-based solutions

---

1. Introduction

On a global scale, urbanization is predicted to continually harm important ecosystem services far into the future [1], causing continuous challenges for governments, policymakers, and urban planners in resource reallocation [2]. Many communities are combating the consequences of urbanization through policies focused on nature-based solutions (NBS). NBS policies encourage actions that help societies address a variety of environmental, social, and economic challenges in sustainable ways [3], using the science–policy–practice interface [4]. Green infrastructure (GI) is a common example of NBS. The concept of GI describes the interdependence of land conservation and land development, and refers to a contiguous, interconnected green network consisting of a range of natural environments [5]. Green infrastructure can enhance ecosystem function in fewer, larger areas compared to numerous, small patches [5], but connectivity cannot be achieved by solely enlarging the total area of GI.

Modeling potential urban scenarios and solutions has emerged as a useful tool to explore uncertain futures in complex urban systems and to further understand the impacts from land use/landcover changes (LULCCs). Although scenarios usually lack quantified probabilities [6,7], they instead function as alternative narratives that present important possibilities about the future [6–8]. Using satellite images from the past and present via remote sensing techniques [9,10], researchers can calculate patterns of urbanization, apply drivers of change, and extrapolate LULCC trends into the future.

To assess the best course of action, different land use policies can be applied to models by adding constraints and/or incentives.

To evaluate the consequences of urbanization and the validity of possible NBS, social and environmental scientists are increasingly using highly detailed LULCC models [11,12]. Landcover models have been used to address general questions of landcover change and urbanization around the world [2,9,10,13–27]; however, only one other study models LULCCs under GI policies [28]. To predict precise landcover transitions and to answer specific questions of policy, future LULCCs need to be modeled at finer scales. Urban modeling studies are conducted at a variety of resolutions depending on the satellite imagery available. Most landcover models are created at a 30 m resolution using Landsat imagery [2,9,13,15–18,27]; yet, small landcover features, like GI, require modeling at a much higher resolution, as GI projects can be smaller than 30 m. Similarly, urban models have been created at different levels of detail with varying numbers of landcover classes. Some studies present a broad overview of urbanization with only two landcover classes [10,19,21–27], usually "urban" and "nature" or "nonurban." Other studies present more realistic models with seven to ten landcover classes representing many of the features in the urban system [13,15,16,18], such as buildings, roads, trees, and grass. A specific landcover class, such as GI, must be modeled with a larger number of landcover classes, as to accurately represent the landscape and the specific variables that effect GI's location.

A variety of different LULCC modeling tools exists today, all allowing for the prediction of socio-environmental changes in a study area over time and projecting these changes into the future in a way that it relates to measured land change [29]. Modeling LULCCs involves historical estimates of landcover combined with biophysical and socioeconomic information to create estimates of future change [30]. Selecting the method to model LULCC is an important first step, based on a study's purpose and available data [27]. This study uses a hybrid modeling approach, utilizing two different tools to model future GI growth and continued urbanization.

One common method in land change prediction is cellular automata. Cellular automata (CA) models are based on the interaction of several components: the grid space can be a one, two, or multidimensional space; and the "cell" or the "automaton" is a discrete variable that represents the structural units of the grid [28]. The "cell state" describes the characteristics of the cell which are subject to change [28]. The change occurs according to specified transition rules, which are mathematical expressions that govern changes of the cell state [28]. Cellular automata methods have been widely used in various modeling tools, such as SLEUTH [31], the CA-Markov module in TerrSet [32], and GEOMOD [33]. GEOMOD was chosen in this study to model future growth in GI, as GEOMOD allows the user to define the quantity of change over time.

Machine learning is a newer approach that is gaining popularity in LULCC studies. Machine learning describes the automated procedures with which the knowledge can be acquired [34]. This study utilizes the machine learning process of the multilayer perceptron (MLP) neural network in the Land Change Modeler (LCM) tool in TerrSet 18.3 [32]. The MLP uses a back-propagation learning algorithm, one of the most widely used neural network models [35], to calculate transition potentials over time. The transition potential models are then combined in a Markov chain process to determine the overall quantities of change over time [35]. LULCC models created in the LCM have been found to be more accurate than LULCC models created in other tools, such as SLEUTH and FUTURES [27]. The LCM was used to model continued urbanization in this study.

Using detailed landcover models at a 1 m resolution, this study aims to model GI policies in Philadelphia, Pennsylvania, USA, and the resulting LULCCs from new GI and continued urbanization for the year 2036. A hybridization of GEOMOD and the LCM was used to model future changes in GI and urbanization, respectively. The model is driven by patterns of historical change, conservation and development constraints, the physical landscape, and distance variables. The model is validated using Kappa statistics and disagreement scores, and the landcovers are compared using a variety of spatial metrics.

2. Study Area

The city of Philadelphia, Pennsylvania, USA is located towards the eastern coast of the United States at 39.95° N, 75.17° W. The city experiences all four seasons with well-distributed precipitation throughout the year. The city is highly urbanized with an average population density of 4491 people per km^2 [36], and a number of universities, parks, and vacant lots for green space. Despite stated efforts to enhance and increase green space in the city [37], from 2008 to 2015, the city increased its urban areas by 11%, while its natural areas decreased by approximately 15% (Table 1).

Table 1. Land use/landcover changes (LULCCs) in Philadelphia from 2008 to 2015 in km^2. *Nature* includes tree canopy, grass, and bare soil. *Urban* includes roads, other paved surfaces, and buildings.

	Area 2008 (km^2)	Area 2015 (km^2)	Difference (km^2)	Percent Change
Nature	155.60	132.71	−22.89	−14.71%
Urban	191.32	212.63	21.31	11.14%

In 2009, the Philadelphia Water Department (PWD) initiated their latest plan to reduce combined sewer overflow events, *Green City, Clean Waters* (GCCW), to meet the regulations of the Pennsylvania Department of Environmental Protection and comply with the federal Clean Water Act [38]. GCCW involves the use of GI as a way to alleviate the amount of runoff that flows into storm drains and, eventually, into the Schuylkill and Delaware Rivers and their tributaries. PWD measures the program's success using the concept of *greened acres* (GAs), or enough GI to manage one inch of stormwater from one acre of drainage area; approximately 27,158 gallons (103 cubic meters) [39]. Greened acres are calculated with the following formula (Equation (1)):

$$GA = IC \bullet Wd, \tag{1}$$

where IC is the impervious cover transformed into GI in acres. This quantity can include the area of the stormwater management feature itself, as well as the area that drains to it. Wd is the depth of water over the impervious surface that can be physically infiltrated into the ground in inches [39].

This program is the first in the United States that prioritizes GI over traditional grey infrastructure to moderate stormwater runoff. Construction of GCCW projects officially began in 2011 and will continue to be implemented until 2036; however, some GI was built before 2011. GI projects are funded through credits to private developers, grants, and public works projects. In an effort to create more GAs, GCCW puts forth eight different best management practices to reduce the amount of impermeable surfaces within the city: 1) Green Streets, 2) Green Schools, 3) Green Public Facilities, 4) Green Parking, 5) Green Open Space, 6) Green Industry, Business, Commerce, and Institutions, 7) Green Alleys, Driveways, and Walkways, and 8) Green Homes [38]. PWD implements a variety of GI practices throughout the city, including downspout planters, green roofs, rain barrels, tree trenches, bump-outs, stormwater planters, pervious pavement, wetlands, and rain gardens [38]. By the year 2036, GCCW will have concluded with at least 9564 GAs, reducing stormwater pollution by 85% from its 2009 levels [38].

3. Materials and Methods

3.1. Data and Preprocessing

To create and validate future landcover distribution, at least three past and current landcover datasets are required. We acquired a 2008 Philadelphia landcover dataset [40], and generated landcover datasets for 2010 and 2015 (Table 2). City boundaries were set using city limits spatial data from the City of Philadelphia [41]. In 2011, the University of Vermont Spatial Laboratory created a 1 m resolution 2008 landcover of Philadelphia for the city government [37]. This landcover is accepted and highly used by the city government. The 2008 landcover was created using object-based image analysis techniques (OBIA) to extract landcover information from a 2008 orthophotography and 2008 LiDAR LAS data [40]. Ancillary data sets were stacked on top of the OBIA data, which included shapefiles of building footprints, roads and railroads, and hydrography provided by the City of Philadelphia [40]. For the purpose of this study, 2008 GI spatial data [42] was stacked onto the 2008 landcover (Figure 1).

To capture the impact of Philadelphia's GI polices, 1 m resolution 2010 and 2015 landcovers of Philadelphia were created by the Kremer lab at Villanova University. Both of the 2010 and 2015 landcovers were created using a supervised classification on 1 m aerial imagery from the United States' National Agriculture Imagery Program (NAIP) for 2010 and 2015 [43,44], respectively, in ESRI ArcGIS 10.5 [45]. Similar to the 2008 landcover, ancillary data sources were stacked on top of each supervised classification, which included shapefiles of building footprints, roads, impervious surfaces, railroads, and hydrography (Table 2). After the landcovers were checked for accuracy, GI spatial data provided by City of Philadelphia [42] were stacked into each landcover (Figure 1). Overall, eight landcover categories were used: tree canopy, grass/shrubs, bare soil, water, buildings, roads/railroads, other paved surfaces, and green infrastructure.

Table 2. Datasets used in the study to develop the future projection of Philadelphia in 2036.

Name	Type	Created by	Spatial Resolution	Reference
Landcover 2008	Raster	University of Vermont Spatial Analysis Laboratory	1 m	[37]
Landcover 2010	Raster	Author	1 m	–
Landcover 2015	Raster	Author	1 m	–
NAIP 2010 & 2015	Aerial Imagery	United States Department of Agriculture	1m	[43,44]
Philadelphia GI	Shapefile	Philadelphia Water Department	1 m *	[42]
2015 Building Footprint	Shapefile	City of Philadelphia	1 m *	[46]
2004 & 2015 Impervious Surfaces	Shapefile	City of Philadelphia	1 m *	[47,48]
2004 Railroads	Shapefile	City of Philadelphia	1 m *	[49]
Hydrology	Shapefile	Philadelphia Water Department	1 m *	[50]
City Limits	Shapefile	City of Philadelphia	–	[41]
08, 10, & 15 DEM	DEM	City of Philadelphia	1 m	[51–53]
08, 10, & 15 Slope	Raster	Author	1 m	–
Distance to roads & rivers	Raster	Author	1 m	–
Evidence Likelihood	Raster	Author	1 m	–

* Shapefiles converted to a raster at 1 m resolution.

Although it was not ideal to use three different landcovers created with two different processes, the OBIA technique used for the 2008 landcover was not replicable within the scope of this project, in part due to over 30,700 manual edits it took to create the landcover [40]; however, the 2010 and 2015 landcovers created with the supervised classification achieved similar results and accuracy (Table 3). The 2010 and 2015 landcover datasets were validated using the accuracy assessment suite in ArcGIS 10.5 [45]. The Accuracy Assessment suite uses ground truthing points and a confusion matrix analysis to calculate user's accuracy, producer's accuracy, overall accuracy, and Kappa coefficient scores. Kappa coefficient is a measure of the proportional improvement by the classifier over a purely random assignment to classes [20]. The user's accuracy measures the proportion of each landcover class that is correct, whereas the producer's accuracy measures the proportion of the land base that is correctly classified [20]. The classified images were assessed for accuracy based on a stratified sample of 105 reference points for each time period, which were visually evaluated using the NAIP imagery. The overall accuracies of the 2010 and 2015 classifications were, respectively, found to be 80% and 90.5%, with Kappa coefficients of 0.767 and 0.889 (Table 3).

Additional datasets used in this study are listed in Table 2. Yearly slope datasets were derived from their respective digital elevation models (DEM) using the slope tool in TerrSet 18.3 [32]. Distance from roads and rivers were calculated using the Euclidean distance tool in ArcGIS 10.5. All of the data was projected into the UTM 18N projection.

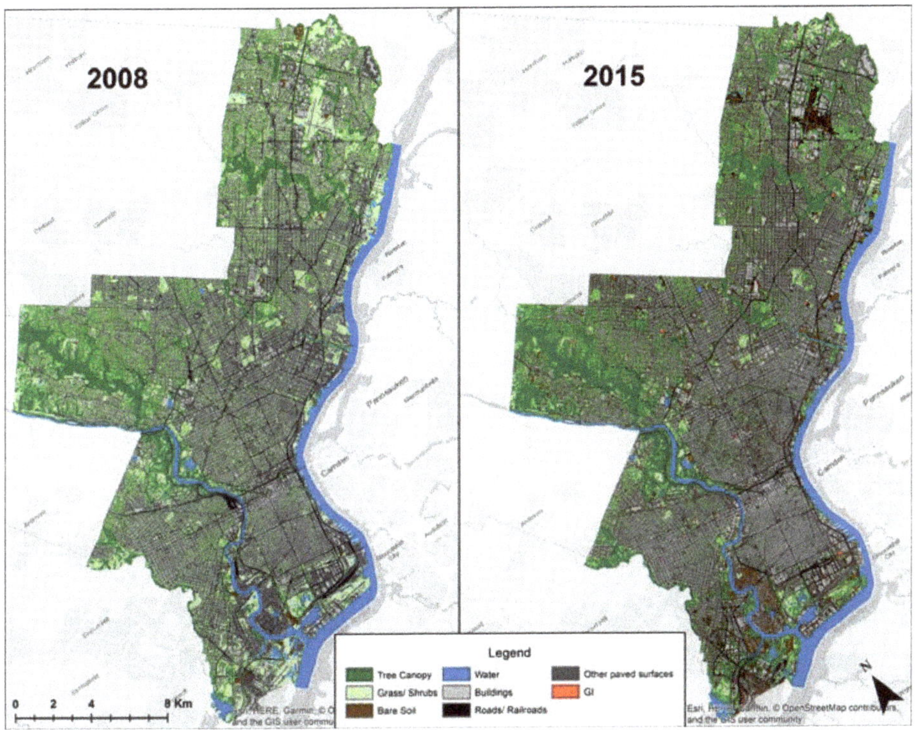

Figure 1. Landcovers of Philadelphia in 2008 [40] and 2015 featuring green infrastructure.

Table 3. Accuracy scores of the 2010 and 2015 supervised classifications of Philadelphia.

Year	Accuracy *	Tree Canopy	Grass/ Shrubs	Bare Soil	Water	Buildings	Roads/ Railroads	Paved Surfaces	Overall Accuracy	Kappa
2010	U Acc.	100%	80.0%	66.7%	86.7%	66.7%	73.3%	86.7%	80%	0.767
	P Acc.	75%	66.7%	83.3%	100%	83.3%	100%	68.4%		
2015	U Acc.	93.3%	93.3%	93.3%	100%	80.0%	93.3%	80.0%	90.5%	0.889
	P Acc.	87.5%	87.5%	77.8%	100%	100%	100%	85.7%		

* U Acc. = User's Accuracy; P Acc. = Producer's Accuracy.

3.2. LULCC Model and Validation

The 2036 LULCC model was created using the Land Change Modeler (LCM) in TerrSet 18.3 [32]. The LCM was used to model continued urbanization based on spatial patterns from 2008 to 2015; specifically, landcover transitions to buildings, roads/railroads, and other paved surfaces. The LCM procedure involves change analysis, determining drivers of change, applying rules and restrictions, Markov chain-based transition predictions, and validation of the model (Figure 2).

3.2.1. Change Analysis

The landcover maps of Philadelphia in 2008 and 2015 (Figure 1) were analyzed for patterns of change, and exact quantities of transition between different landcover classes were calculated. Only transitions representing continued urbanization (i.e., change from trees, grass, and/or soil to buildings, roads/railroads, and/or other paved surfaces) were analyzed, as they were the transitions of interest.

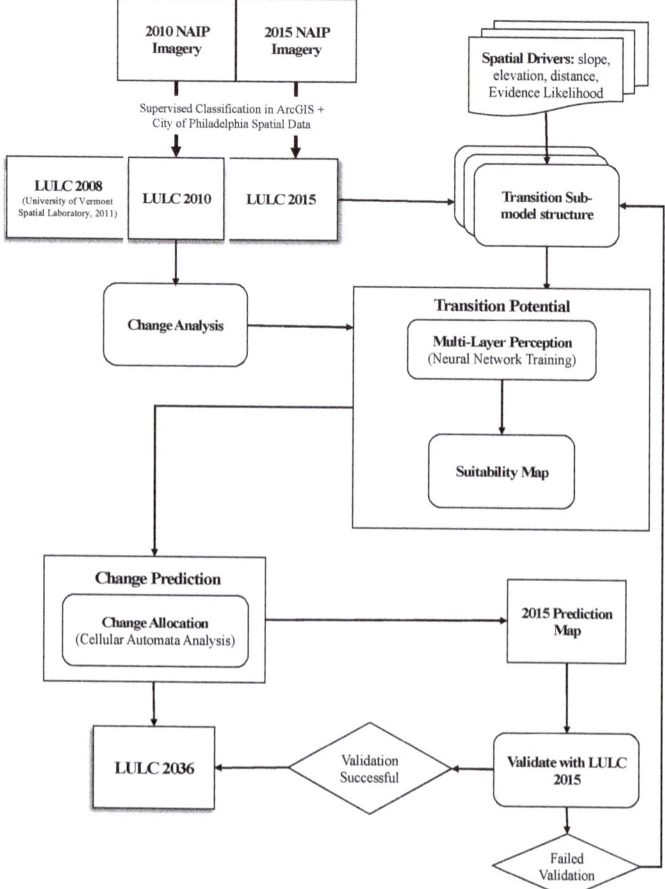

Figure 2. Processes involved in the Land Change Modeler (LCM) to create and validate the 2036 LULCC of Philadelphia.

3.2.2. Drivers of Change

Driver variables in this analysis included DEMs, slope, distance to existing roads and rivers, and evidence likelihood rasters. Distance to roads and distance to rivers were set as dynamic factors to be recalculated over time because as roadways and rivers change over time, so do the distances to these features. Elevation and slope have been documented to influence the location of urban growth [2,17,54]. Evidence likelihood rasters were created for each LULCC that occurred. The evidence likelihood tool in the LCM transforms categorical variables, such as change from one landcover class to another, into numerical values so that they can be used in the modeling procedure [35].

Urban transitions were grouped into respective submodels by what they transitioned into and then used to compute transition potentials. A transition submodel consists of a group of transitions that share the same underlying driver variables [10], and so they can be modeled at once. The modeling of transition potentials is necessary for determining spatial change [55]. The output of this step generates a series of transition potential maps that each correspond to a landcover transition based on the previous change analysis [55]. The transition maps consider the suitability of pixels that have transformed into urban pixels based on a number of driving factors used for modeling processes of historical change.

The transition potentials are created using a multilayer perceptron (MLP) neural network, which allows for transition submodels to be modeled at once [55]. The MLP neural network is a feedforward neural network in which data flows in one direction from an input layer to an output layer through a number of hidden layers in between, as set by the user [56]. A small number of hidden layers were used in this study, which expressed the common underlying themes in the variables [35] and resulted in higher accuracy levels. The computing elements (nodes) are grouped into layers, and each node receives an input signal from other nodes. After processing the signals locally through a transfer function, it outputs a transformed signal to other nodes for the final result [57]. Each signal feeding into a node in a subsequent layer has the original input multiplied by a weight with a threshold added, and is then passed through an activation function [57] that, in this study, was non-linear. The weights are determined in the automatic training process before the network can be used for prediction purposes, aiming at changing the weights as to minimize the error between the observed and the predicted outcomes [57]. Due to the non-linearity of the data, a sigmoid factor of 0.5 was applied to the weighted sum of inputs before the signal passed to the next layer.

3.2.3. Rules and Restrictions

Rules and restrictions were also applied to the model. A Boolean layer expressing areas of conservation and restricted development were added to the model so that urbanization would not occur in areas under conservation and the local Philadelphia airports. This spatial layer includes local urban parks, federal and state conservation areas, and the two airports in Philadelphia.

3.2.4. Transition Predictions Using Markov Chain Analysis

In the final change prediction, the LCM uses the change rates calculated in the change analysis step, as well as the transition potential maps to predict a future scenario for 2036. This step is responsible for determining the quantity of change to urban landcover in each transition using Markov chain analysis [55]. The Markovian process is a method in which a predicted system can be estimated by finding its previous state and the probability of conversion from one state to another [18]. By utilizing the 2008 and 2015 landcover maps, the Markov chain analysis determines, precisely, how much land would be anticipated to transition from 2015 to 2036, on the basis of projection of the transition potentials. A hard prediction was created in this study which yielded a projected map of 2036, where each pixel is assigned one landcover class—the class that it has been calculated to most likely become.

3.2.5. Validation

Validation was used to ascertain the quality of the predicted 2036 map. In order to validate the methods used to create the 2036 landcover map, the methods were replicated to create a predicted map of 2015 and then validated against the actual 2015 map. The 2008 and 2010 landcovers were used to predict changes in 2015. The validation module in TerrSet was used to statistically assess the quality of the 2015 predicted map against the 2015 reference map. The Kappa variation techniques were used in the validate module as the statistical validation procedure. Three variations of Kappa were calculated: Kappa for no information, Kappa for location, and Kappa standard. The Kappa for no information (K_{no}) signifies overall accuracy obtained in the simulation run, while Kappa for location ($K_{location}$) measures agreement level in a location [58]. Considering the difficulty in interpretation encountered with the Kappa for correctly assigned proportion against the proportion of incorrectly assigned by change ($K_{standard}$) [59], the $K_{standard}$ was not used in this study; however, $K_{location}$ was useful for the validation in the absence of $K_{standard}$ [60].

However, Pontius and Millones (2011) have presented that Kappa scores can be useless, misleading, and/or flawed for practical applications in GIS and remote sensing [60]. Instead, Pontius has developed other summary parameters calculated in TerrSet's validate module, two of which are presented in this study: disagreement at the grid cell level and disagreement due to quantity. Disagreement at the grid cell level is defined as the amount of disagreement associated with the fact that the comparison map fails to specify perfectly the correct locations of categories at grid cells within strata [35]. Disagreement due to quantity is defined as the amount of disagreement associated with the fact that the comparison map fails to specify perfectly the correct quantity of each category according to the reference map [35].

3.3. Modeling Green Infrastructure

In order to assess the future implementation of GI, a cellular automata optimization approach was utilized. The GEOMOD module in TerrSet was used to assess the landcover change between GI and a "non-GI" category, which combined all other landcover categories (Figure 3). For the purpose of this analysis, the 2008 and 2015 landcovers were reclassified to GI and "non-GI," as GEOMOD only allows for the transition from "state one" to "state two." Green infrastructure was modeled separately from all other LULCC in GEOMOD because GEOMOD allows the user to specify the forecasted quantity of change. Philadelphia's GI plan indicates the amount of future GI by 2036, broken down by watershed [39], which was incorporated into the model. Future scenarios of GI are identified using driver variables, suitability maps, and calculating the amount of change in each Philadelphia watershed. Currently, there are no spatial plans for the future distribution of GI, as the city develops GI projects as they are needed with target quantities each year, and so it is challenging to evaluate the accuracy of this GI model, and it is just one possible scenario out of many possible scenarios.

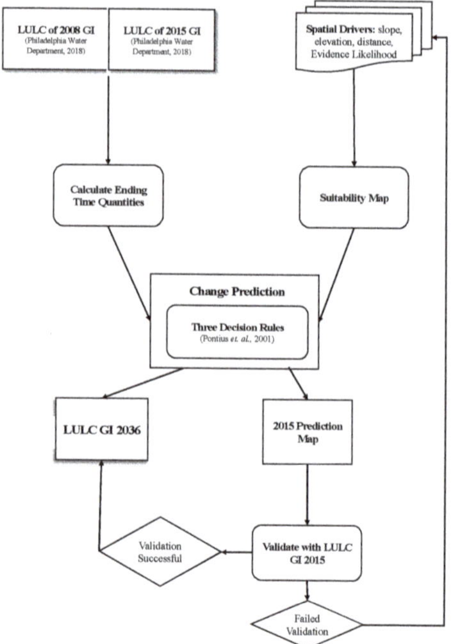

Figure 3. Process to create and validate the 2036 GI prediction using GEOMOD model based on Philadelphia's GI plans outlined in Green City, Clean Waters [38].

3.3.1. Drivers of Change

GEOMOD uses an optimization algorithm that allocates cells to the most suitable locations based on the user-provided site suitability data [27]. Similar to the LCM analysis, DEM, slope, distance, and evidence likelihood rasters were used in GEOMOD as the site suitability data. The suitability image is created by computing a weighted sum of all the driver images for each grid cell [33].

3.3.2. Change Allocation

Locations of land change are selected based on three decision parameters [33]. The first parameter only allows for one directionality of change to occur. Specifically, in this study, all other land categories can change to GI, but GI pixels cannot change to anything else. The second parameter allows for regional stratification, where land change can be simulated differently for a series of user defined boundaries [27]. Finally, neighborhood constraints that restrict growth within any time step to only edge cells between two classes can be set. Unconstrained growth was used in this study as different GI projects are not usually constructed right next to one another.

3.3.3. Amount of Change

The nature of the green infrastructure policy in Philadelphia requires a model restriction that specifies how much change will occur between state 1 (non-GI) and state 2 (GI) over a given amount of time. User-defined quantity of change was necessary for this analysis since it can be predicted how much GI will exist by 2036 according to GCCW. The prediction, in this case, focuses on the most likely location of the GI pixels. Information on 2016 green acreage from PWD's five year review, *Green City, Clean Waters—Evaluation and Adaptation Plan* [39], were used to calculate the amount of GI needed by 2036. PWD calculated the amount of green acreage they had accomplished in each of the four main watersheds in Philadelphia. The percent of GAs of the total 837 GAs was calculated for each watershed in 2016 (Equation (2)). This percentage was assumed to remain the same for 2036 predictions. Using GIS and the GI shapefiles provided by PWD [42], a ratio of physical acres of GI to GAs was then calculated for each watershed (Equation (3)). Predicted physical acres of GI in 2036 were then calculated using the ratio for each watershed (Equations (4) and (5)).

$$\frac{GA_{2016,\,WS}}{837} = GA\%,\ \text{per watershed,} \tag{2}$$

$$\text{Physical acres GI/GA} = \text{GI}_{\text{ratio}},\ \text{per watershed in 2016,} \tag{3}$$

$$9564\ GAs \bullet GA\% = GA_{2036},\ \text{per watershed,} \tag{4}$$

$$GA_{2036} \bullet GI_{ratio} = \textit{Physical acres of GI in 2036, per watershed,} \tag{5}$$

where $GA_{2016,\,WS}$ is the GA in 2016 for each watershed, 837 is the total amount of greened acres in 2016 [39], $GA\%$ is the percentage of greened acres per watershed, GI_{ratio} is the ratio of physical acres of GI to greened acres, and 9564 GAs is the target amount of greened acres in 2036 according to GI policies [39].

Three smaller watersheds were not counted towards PWD's GAs goal. To account for the growth of GI in each of the smaller watersheds over time, the rate of change for each region was calculated and then extrapolated to 2036. The calculations for each watershed are displayed in Table 4.

3.3.4. Validation

The GI predictions were validated using the same approach as the LCM validation process. An intermediate GI prediction of 2015 was created in GEOMOD based on the 2008 landcover using the aforementioned methods. Again, the validate module in TerrSet was used to statistically assess the 2015 GI prediction against the 2015 GI reference image using Kappa statistics and disagreement scores.

3.4. Final 2036 LULC

The results of the GI-specific 2036 prediction were integrated into the 2036 predicted urbanization dataset. The end result is a predicted landcover dataset that includes the spatial distribution of predicted GI locations and continued intra-urbanization (Figure 4).

Table 4. Resulting calculations for the increase in GI by 2036 (1 acre = 4046.86 m^2).

Main Watersheds	2016				2036	
	Gas *	% GA	Acres of GI [†]	Ratio-GI Acres:GAs	GAs	Acres of GI
Darby-Cobb	36	4.3%	7.571	0.210	411.355	86.515
Delaware Direct	334	39.9%	123.701	0.370	3816.459	1413.467
Schuylkill River	306	36.6%	130.426	0.426	3496.516	1490.314
Tacony-Frankford	162	19.4%	41.374	0.255	1851.097	472.755
Total	837	100%	303.071	0.362	9564.000 *	3463.051
Smaller Watersheds				Rate of Change (acres/yr) [†]		
Pennypack Creek			33.396	1.411		53.710
Poquessing Creek			18.881	0.787		30.536
Wissahickon Creek			22.846	0.931		35.202

* Data from Green City, Clean Waters—Evaluation and Adaptation Plan [45]. [†] Data calculated using GIS and the PWD GI polygons [48].

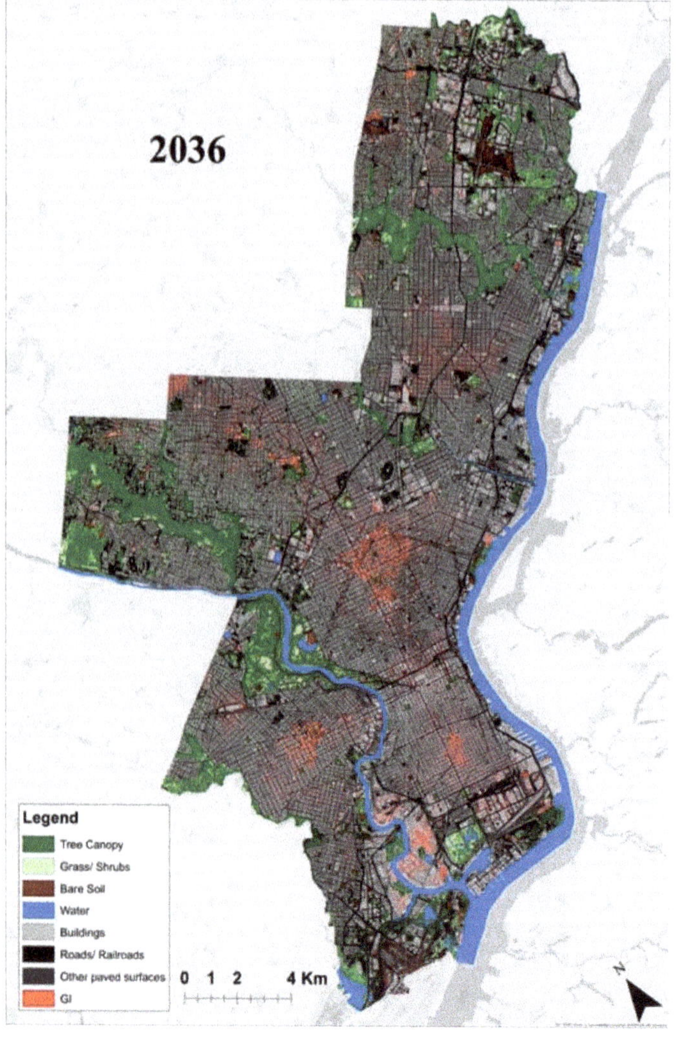

Figure 4. The predicted landcover of Philadelphia in 2036 under green infrastructure policies.

3.5. Landscape Metrics

To understand the changes in the connectivity of green space and urban areas, each landcover was recategorized into two classes: green space (tree canopy, grass/shrubs, bare soil, and GI) and urban (buildings, roads/railroads, and other paved surfaces). To quantify the changes in the landscape over time, landscape metrics were calculated using Patch Analyst 5.0 [61], which integrates the spatial metrics found in FRAGSTATS [62] into ArcGIS. The number of patches, mean patch size, largest patch index (LPI), and patch cohesion index (PCI) were calculated at the class level. Patch cohesion index (PCI) and Shannon's diversity index (SHDI) were calculated for each landcover at the landscape level. FRAGSTATS 4.2 [62] was used to calculate the LPI and PCI for each landcover. Due to restrictions in FRAGSTATS 4.2, the landcover datasets had to be resampled to a 5 m resolution. A description of each metric can be found in Table 5.

Table 5. Description of spatial metrics.

Metric	Description	Unit	Range
Largest Patch Index (LPI)	The area of the largest patch of the corresponding patch type divided by total area of the measured class.	%	$0 < \text{LPI} \leq 100$
Mean Patch Size (MPS)	Average patch size.	m^2	$\text{MPS} > 0$, no limit
Number of Patches (NP)	Number of patches in the landscape.	N/A	$\text{NP} \geq 0$, no limit
Patch Cohesion Index (PCI)	The physical connectedness of the corresponding patch type. PCI approaches 0 as the proportion of the landscape comprised of the focal class decreases and becomes increasingly subdivided and less physically connected, and vice versa.	Dimensionless	$0 < \text{PCI} < 100$
Shannon's Diversity Index (SHDI)	Relationship between the number of classes, the total number of patches, and the relative abundance of patches in each class. It has a value of 0 when no diversity is present and increases as the landscape becomes more fragmented.	Dimensionless	$\text{SHDI} \geq 0$, without limit

4. Results

4.1. Validation Results

The $K_{location}$ score (Table 6) was used for assessing accuracy, as it is difficult to interpret the Kappa for correctly assigned proportion against the proportion of incorrectly assigned by change ($K_{standard}$) [59]. The LULCC prediction created with the LCM received a $K_{location}$ score of 78% (Table 6). The GI prediction created with GEOMOD received a $K_{location}$ score of almost 100% (Table 6). Most of the cells in the landcovers used in the GEOMOD analysis were the "non-GI" class, contributing to the high accuracy of the GI prediction. Additionally, in both the urbanization and the GI predictions, the disagreement scores represent very little disagreement between the models and their respective reference maps (Table 6).

Table 6. The Kappa scores and disagreement scores used to validate the predicted 2015 landcover against the reference 2015 landcover, and the 2015 GI against the reference 2015 GI.

	2015 LULC	**2015 GI**
K_{no}	80.1%	99.8%
$K_{location}$	78.3%	99.7%
$K_{standard}$	69.4%	99.7%
Disagree Grid Cell	1.11×10^{-1}	1.30×10^{-3}
Disagree Quantity	6.58×10^{-2}	0.00

4.2. Landcover Changes

The final 2036 landcover is featured in Figure 4. Overall, buildings are predicted to exhibit the largest net increase, followed by roads/railroads by 2036 (Figure 5). Grass/shrubs are predicted to exhibit the largest net decrease, followed by other paved surfaces by 2036 (Figure 5).

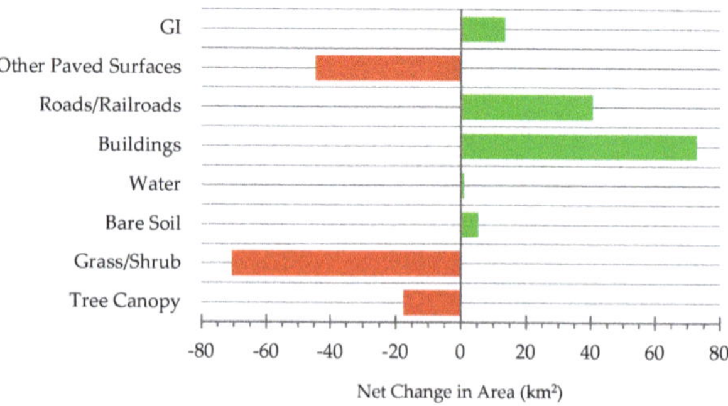

Figure 5. Net change for each land use category from 2008 to 2036.

The model was able to successfully meet Philadelphia's GI goal of at least 9564 GAs by 2036. The 2036 prediction allocated enough GI for 9945 GAs in 2036. A large portion of the new GI is predicted to be located in the middle and south Philadelphia (Figure 6). The new GI is predicted to mostly replace other paved surfaces, followed by grass/shrubs (Figure 7).

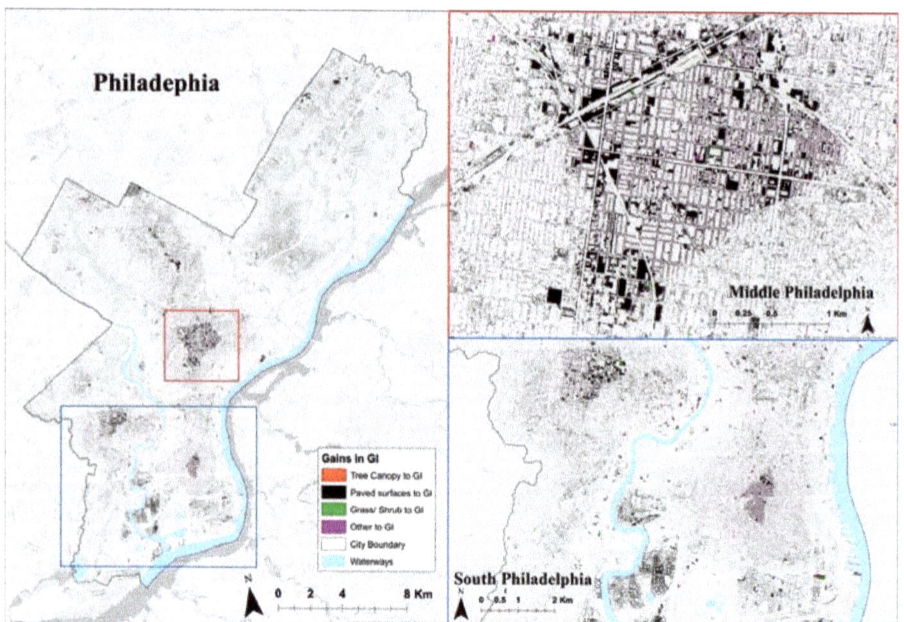

Figure 6. Map of the increase in GI from 2008 to 2036 broken down by category. The middle and southern portions of Philadelphia are highlighted ("Other to GI" includes buildings, soil, and roads/railroads transitioning to GI).

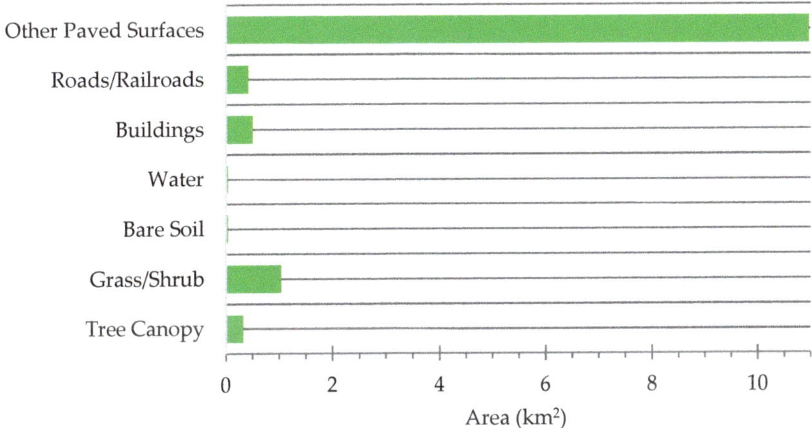

Figure 7. Contributions to the increase in GI by area (km²) from 2008 to 2036.

Buildings demonstrate a large net increase of almost 73 km² (Figure 5). A small reduction in buildings is due to new GI (Figure 7). Similar to increases in GI, most of the gains in building area are predicted to result from a loss in other paved surfaces, followed by grass/shrubs and tree canopy (Figure 8 right). It is forecasted that the new buildings will be well distributed throughout the city, with some concentration in south Philadelphia (Figure 8 left).

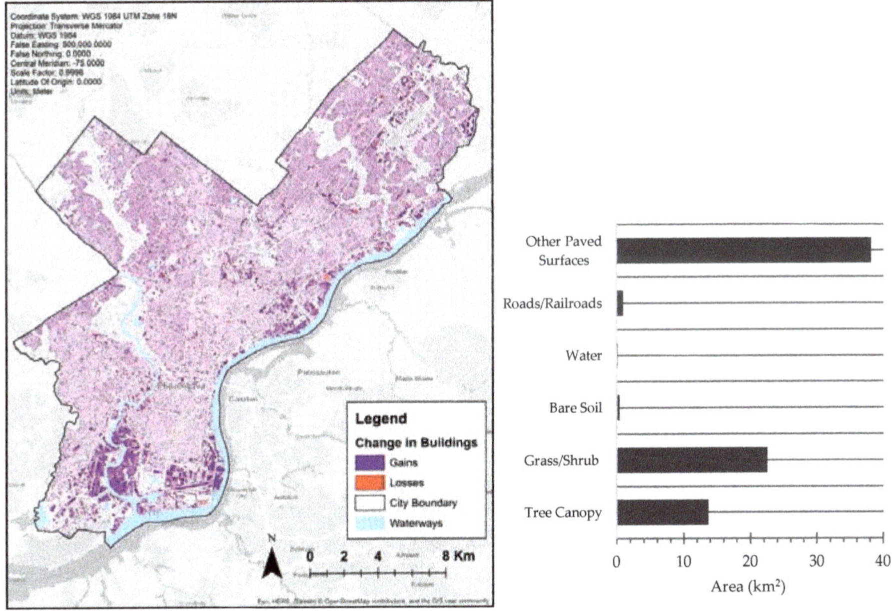

Figure 8. Changes in buildings from 2008 to 2036. **Left:** Map of change in buildings. **Right:** Contributions to increases in buildings.

4.3. Spatial Metrics

SHDI is forecasted to reach its lowest levels of fragmentation throughout the entire landscape by 2036. The number of patches decreased in both categories from 2015. Mean patch size increased for both classes from 2015. However, the largest patch index decreased for green space over time but increased for urban areas. The patch cohesion index decreased over time for green space, increased slightly over time for urban areas, and increased to its highest levels for the entire Philadelphia landscape (Table 7).

Table 7. Spatial metrics of the Philadelphia landscape, urban areas, and green space: Shannon's diversity index (SHDI), number of patches (NP), mean patch size (MPS), largest patch index (LPI), and patch cohesion index (PCI).

	SHDI	NP	MPS	LPI	PCI
2008					
Landscape	1.736				99.947
Green Space		265,562	585.994	2.039	99.246
Urban		168,661	1135.090	18.383	99.985
2015					
Landscape	1.820				99.945
Green Space		2,373,040	55.795	0.612	97.803
Urban		514,426	19.739	20.051	99.986
2036					
Landscape	1.709				99.957
Green Space		1,452,070	56.819	0.433	95.223
Urban		324,990	802.67	24.537	99.992

5. Discussion

5.1. Data Resolution

This research aimed to model fine-scale, multiclass prediction of intra-urban landcover change under GI policies. This study is unique in that it utilized one-meter resolution aerial imagery from the United States' NAIP to create the base landcover datasets. Most LULCC modeling uses Landsat data at a 30 m resolution or larger; however, GI can easily be less than 30 m in size and so higher resolution data is needed to map and model GI.

Remote sensing techniques and the availability of free to low cost satellite imagery and their temporal frequency has greatly enhanced the monitoring of urban growth and land use dynamics around the world [9]. As technology continues to improve, the resolution of satellite images will also be enhanced. Developing and testing LULCC models with fine-scale data, as we do here, allows for more detailed models of the dynamics of change within urban environments. Advanced models can aid policymakers and planners in analyzing the effects of smaller landscape features, such as GI.

5.2. Green Infrastructure

This study spatially modeled the increase of GI in Philadelphia under the city's GI program for the year 2036. Many studies model future urban growth in areas around the world; however, we found only one other study that also modeled LULCC under GI policies [28]. Contrasting to Mitsova et al. (2011), Philadelphia's 2036 model predicts an overall increase in cohesion throughout the entire landscape under GI policies (Table 6), which can be attributed to the decreased number of patches from 2015. Green space patches decreased by 920,970 patches from 2015 to 2036, with the mean patch size also increasing slightly (Table 6).

Although the number of patches decreased and mean patch size increased, the patch cohesion metric for green space still decreased over time (Table 6). GI serves as a nature-based solution to stormwater management in Philadelphia; nevertheless, to maximize the potential of GI's ecosystem services, additional attention is needed regarding the type of GI implemented and to increase its overall connectivity with other green spaces. GI and green spaces enhance ecosystem function in fewer, larger areas compared to numerous, small patches [5], but connectivity cannot be achieved by solely enlarging the total area of GI.

The model predicts that the new GI to be added by 2036 will be fairly well dispersed throughout the city, with the middle of the city and the southern portion of Philadelphia gaining some of the largest increases (Figure 6 top). A majority of the new GI will replace other paved surfaces (Figure 6), which includes urban features such as parking lots, alleys, and sidewalks. Replacing a majority of other paved surfaces with GI is a plausible prediction, as many of the GI strategies that the city currently uses does replace paved surfaces. These GI strategies include trenches, bump-outs, planters, and pervious pavement [63]. A small portion of the expected GI is predicted to replace tree canopy, which is unlikely as the cities would probably not cut down trees for GI development. This prediction could be possible if tree trenches are added to these areas, so the tree could be preserved but, also, the trench underneath it would collect water for later use.

This study only used geophysical driver variables, whereas, in reality, other socioeconomic drivers have an additional influence on the spatial distribution of GI projects. To improve the GI model, GI experts at the City of Philadelphia should be consulted to identify other variables and constraints that may influence GI location.

5.3. Urbanization Land Use/Landcover Changes

Unlike the results presented by Mitsova et al. (2011), where urban growth is forecasted to stall, urban areas in Philadelphia are still predicted to increase, even under GI policy. Continued urbanization in the form of new buildings can be seen in south Philadelphia around the Delaware and Schuylkill rivers (Figure 8). Philadelphia's population is newly re-growing, as it has increased steadily over the past seven years, and is predicted to continue to grow [64], driven by urban regeneration and new employment opportunities in the city. By 2036, urban spaces are predicted to decrease by 189,436 patches across Philadelphia, but the mean patch size and largest patch index still increase from the 2015 measurements (Table 6), indicating further urban development in areas that are already urbanized. Specifically, buildings will experience the largest growth of the urban classes (Figure 5). With the addition of new urban areas throughout the city, the patch cohesion index for urban spaces continues to increase by 2036.

Road/railroad area also increases in 2036, approximately doubling from the 2008 area (Figure 5). New roads and railroads will be needed to meet the transportation needs of the new buildings, residents, and businesses. Other paved surfaces, such as sidewalks and parking lots, decrease in area over time as they are lost to new GI and buildings (Figure 5).

The amount of urban growth forecasted by the model should be analyzed further. The Delaware Valley Regional Planning Commission predicts that the population of Philadelphia will increase by only 6.37% from 2015 to 2035 [65]. Under this population growth, the model most likely overestimates urban growth, as buildings are projected to have a net increase of almost 25%, and roads/railroads to have a net increase of approximately 23%. The demand for new buildings with only a 6.73% increase in population will not meet the rate projected by the model. Many socioeconomic factors affect urban and population growth rates, and they should be added to future versions of this model to accurately assess urban land change.

5.4. Future Research

There is little research on landcover changes in Philadelphia [66], and only one study that applies GI policies to LULCC using similar methods [28]. This study aims to fill those gaps, as Philadelphia is a complex urban system and one of the few cities in the United States that prioritizes GI over traditional grey infrastructure. As the city continues to develop green infrastructure, it may act as a model for other cities in need of implementing nature-based solutions.

The methodology and the resulting landcover models used in this study can serve as a base for further exploration into how GI will function in the future. As a result of urban expansion, cities around the world are finding alterations of energy budgets with modifications to climatic, hydrologic, and biogeochemical cycles, and habitat fragmentation leading to a reduction in biodiversity [67]. In Philadelphia, GI can serve as a solution to these problems. The degree to which GI can resolve these issues can be assessed by using future landcover models with predicted GI distribution as the basis of ecosystem service analyses. Additionally, further analysis of GI distribution based on socioeconomic equity should be studied in the future, but it was beyond the scope of this paper. This is integral to assessing future risk and vulnerability to environmental phenomena, such as climate change [17].

6. Conclusions

This study finds that LULCCs from GI policies can successfully be modeled in a heterogeneous intra-urban environment using fine-scale data. As Philadelphia continues to grow, GI will be implemented throughout the city to meet its GI program goals. Philadelphia city planners should consider GI cohesion to expand the many ecosystem services that GI contributes to the city.

Additionally, this study highlights the usefulness of LULCC modeling as a method for planning for the future. Availability of data and imagery will only improve in resolution and temporality in the future, allowing for more accurate landcover models that can capture smaller features, such as GI. As policymakers and city planners begin to plan for increased populations and the aggravated consequences of climate change, nature-based solutions will receive greater implementation as they allow for inexpensive and creative methods for city adaptation.

Author Contributions: Conceptualization, C.S. and P.K.; Data curation, C.S.; Formal analysis, C.S.; Funding acquisition, P.K.; Investigation, C.S.; Methodology, C.S. and P.K.; Project administration, P.K.; Supervision, P.K.; Writing—original draft, C.S.; Writing—review & editing, C.S. and P.K.

Funding: The Department of Geography and the Environment at Villanova University provided summer research funding for Charlotte Shade for this project.

Acknowledgments: Thank you to Eric Wagner, Jon Graziola, and Danielle Martinez at Villanova University for technical help throughout this project. Additionally, thank you to committee members Keith Henderson at Villanova University and Julia Rockwell at PWD for their advice throughout this project.

Conflicts of Interest: The authors declare no conflict of interest.

Abbreviations

Nature-based solutions (NBS); Land use/landcover change (LULCC); Cellular automata (CA); Green infrastructure (GI); Multilayer perception (MLP); Land Change Modeler (LCM); Green City, Clean Water (GCCW); Greened acre (GA); Philadelphia Water Department (PWD); object-based image analysis (OBIA); Digital Elevation Model (DEM); Shannon's diversity index (SHDI), number of patches (NP), mean patch size (MPS), largest patch index (LPI), and patch cohesion index (PCI).

References

1. Millennium Ecosystem Assessment. *Ecosystems and Human Well-Being: Synthesis*; Millennium Ecosystem Assessment: Washington, DC, USA, 2005.
2. Megahed, Y.; Cabral, P.; Silva, J.; Caetano, M. Land Cover Mapping Analysis and Urban Growth Modelling Using Remote Sensing Techniques in Greater Cairo Region—Egypt. *ISPRS Int. J. Geo-Inf.* **2015**, *4*, 1750–1769. [CrossRef]
3. Bauduceau, N.; Berry, P.; Cecchi, C.; Elmqvist, T.; Fernandez, M.; Hartig, T.; Krull, W.; Mayerhofer, E.; Sandra, N.; Noring, L. *Towards an EU Research and Innovation Policy Agenda for Nature-Based Solutions & Re-Naturing Cities: Final Report of the Horizon 2020 Expert Group on 'Nature-Based Solutions and Re-Naturing Cities'*; Publications Office of the European Union: Brussels, Belgium, 2015.
4. Nesshöver, C.; Assmuth, T.; Irvine, K.N.; Rusch, G.M.; Waylen, K.A.; Delbaere, B.; Haase, D.; Jones-walters, L.; Keune, H.; Kovacs, E.; et al. The science, policy and practice of nature-based solutions: An interdisciplinary perspective. *Sci. Total Environ.* **2017**, *579*, 1215–1227. [CrossRef] [PubMed]
5. Benedict, M.A.; McMahon, E.T. *Green Infrastructure: Linking Landscapes and Communities*, 2nd ed.; Island Press: Washington, DC, USA, 2006; ISBN 978-1559635585.

6. Nakicenovic, N.; Davidson, O.E. *IPCC Special Report on Emission Scenarios*; Cambridge University Press: Cambridge, UK, 2000.

7. Swart, R.J.; Raskin, P.; Robinson, J. The problem of the future: Sustainability science and scenario analysis. *Glob. Environ. Chang.* **2004**, *14*, 137–146. [CrossRef]

8. Peterson, G.D.; Cumming, G.S.; Carpenter, S.R. Scenario Planning: A Tool for Conservation in an Uncertain World. *Conserv. Biol.* **2003**, *17*, 358–366. [CrossRef]

9. Tewolde, M.G.; Cabral, P. Urban sprawl analysis and modeling in Asmara, Eritrea. *Remote Sens.* **2011**, *3*, 2148–2165. [CrossRef]

10. Abuelaish, B.; Olmedo, M.T.C. Scenario of land use and land cover change in the Gaza Strip using remote sensing and GIS models. *Arab. J. Geosci.* **2016**, *9*, 274. [CrossRef]

11. Pontius, R.G.; Boersma, W.; Castella, J.C.; Clarke, K.; Nijs, T.; Dietzel, C.; Duan, Z.; Fotsing, E.; Goldstein, N.; Kok, K.; et al. Comparing the input, output, and validation maps for several models of land change. *Ann. Reg. Sci.* **2008**, *42*, 11–37. [CrossRef]

12. Prestele, R.; Alexander, P.; Rounsevell, M.D.A.; Arneth, A.; Calvin, K.; Doelman, J.; Eitelberg, D.A.; Engström, K.; Fujimori, S.; Hasegawa, T. Hotspots of uncertainty in land-use and land-cover change projections: A global-scale model comparison. *Glob. Chang. Biol.* **2016**, *22*, 3967–3983. [CrossRef] [PubMed]

13. Sakieh, Y.; Salmanmahiny, A. Performance assessment of geospatial simulation models of land-use change—A Landscape metric-based approach. *Environ. Monit. Assess.* **2016**, *188*, 1–16. [CrossRef]

14. Chen, H.; Pontius, R.G. Diagnostic tools to evaluate a spatial land change projection along a gradient of an explanatory variable. *Landsc. Ecol.* **2010**, *25*, 1319–1331. [CrossRef]

15. Poelmans, L.; Van Rompaey, A. Detecting and modelling spatial patterns of urban sprawl in highly fragmented areas: A case study in the Flanders-Brussels region. *Landsc. Urban Plan.* **2009**, *93*, 10–19. [CrossRef]

16. Martins, V.N.; Cabral, P.; Sousa e Silva, D. Urban modelling for seismic prone areas: The case study of Vila Franca do Campo (Azores Archipelago, Portugal). *Nat. Hazards Earth Syst. Sci.* **2012**, *12*, 2731–2741. [CrossRef]

17. Pinto, A.P.; Cabral, P.; Caetano, M.; Alves, M.F.; Pintot, P.; Cabralf, P.; Caetanoi, M.; Alvest, M.F. Urban Growth on Coastal Erosion Vulnerable Stretches. *J. Coast. Res.* **2018**, *II*, 1567–1571.

18. Nelson, E.; Sander, H.; Hawthorne, P.; Conte, M.; Ennaanay, D.; Wolny, S.; Manson, S.; Polasky, S. Projecting global land-use change and its effect on ecosystem service provision and biodiversity with simple models. *PLoS ONE* **2010**, *5*, e14327. [CrossRef] [PubMed]

19. Pickard, B.; Gray, J.; Meentemeyer, R. Comparing Quantity, Allocation and Configuration Accuracy of Multiple Land Change Models. *Land* **2017**, *6*, 52. [CrossRef]

20. Mishra, V.; Rai, P.; Mohan, K. Prediction of land use changes based on land change modeler (LCM) using remote sensing: A case study of Muzaffarpur (Bihar), India. *J. Geogr. Inst. Jovan CvijicSasa* **2014**, *64*, 111–127. [CrossRef]

21. Liu, J.; Li, J.; Qin, K.; Zhou, Z.; Yang, X.; Li, T. Changes in land-uses and ecosystem services under multi-scenarios simulation. *Sci. Total Environ.* **2017**, *586*, 522–526. [CrossRef] [PubMed]

22. Gibson, L.; Münch, Z.; Palmer, A.; Mantel, S. Future land cover change scenarios in South African grasslands—Implications of altered biophysical drivers on land management. *Heliyon* **2018**, *4*, e00693. [CrossRef]

23. Andriamasimanana, R.H.; Cameron, A. Spatio-Temporal Change in Crowned (*Propithecus coronatus*) and Decken's Sifaka (*Propithecus deckenii*) Habitat in the Mahavavy-Kinkony Wetland Complex, Madagascar. *Primate Conserv.* **2014**, *28*, 65–71. [CrossRef]

24. Mahmoud, M.I.; Duker, A.; Conrad, C.; Thiel, M.; Ahmad, H.S. Analysis of settlement expansion and urban growth modelling using geoinformation for assessing potential impacts of urbanization on climate in Abuja City, Nigeria. *Remote Sens.* **2016**, *8*, 220. [CrossRef]

25. Anand, J.; Gosain, A.K.; Khosa, R. Prediction of land use changes based on Land Change Modeler and attribution of changes in the water balance of Ganga basin to land use change using the SWAT model. *Sci. Total Environ.* **2018**, *644*, 503–519. [CrossRef] [PubMed]

26. Reddy, C.S.; Singh, S.; Dadhwal, V.K.; Jha, C.S.; Rao, N.R.; Diwakar, P.G. Predictive modelling of the spatial pattern of past and future forest cover changes in India. *J. Earth Syst. Sci.* **2017**, *126*, 8. [CrossRef]

27. Ahmed, B.; Kamruzzaman, M.D.; Zhu, X.; Shahinoor Rahman, M.D.; Choi, K. Simulating land cover changes and their impacts on land surface temperature in dhaka, bangladesh. *Remote Sens.* **2013**, *5*, 5969–5998. [CrossRef]

28. Mitsova, D.; Shuster, W.; Wang, X. A cellular automata model of land cover change to integrate urban growth with open space conservation. *Landsc. Urban Plan.* **2011**, *99*, 141–153. [CrossRef]

29. Paegelow, M.; Camacho Olmedo, M.T.; Mas, J.-F.; Houet, T.; Pontius, R.G., Jr. Land change modelling: Moving beyond projections. *Int. J. Geogr. Inf. Sci.* **2013**, *27*, 1691–1695. [CrossRef]

30. Eigenbrod, F.; Bell, V.A.; Davies, H.N.; Heinemeyer, A.; Armsworth, P.R.; Gaston, K.J. The impact of projected increases in urbanization on ecosystem services. *Proc. Biol. Sci.* **2011**, *278*, 3201–3208. [CrossRef]

31. Clarke, K.C. *SLEUTH [Software]*; Gigalopolis: Santa Barbara, CA, USA, 1993.

32. Clark Labs. *TerrSet (18.3) [Software]*; Clark University: Worcester, MA, USA, 2017.

33. Pontius, R.G., Jr.; Chen, H. GEOMOD Modeling. In *Idrisi 15: The Andes Edition*; Clark University: Worcester, MA, USA, 2006; pp. 1–44.

34. Carbonell, J.G.; Michalski, R.S.; Mitchell, T.M. An Overview of Machine Learning. In *Machine Learning: An Artificial Intelligence Approach (Volume I)*; Michalski, R.S., Carbonell, J.G., Mitchell, T.M., Eds.; TIOGA Publishing Co.: Palo Alto, CA, USA, 1983; pp. 3–23. ISBN 978-1493303489.

35. Clark Labs. *TerrSet (18.3) [Help System]*; Clark University: Worcester, MA, USA, 2017.

36. U.S. Census Bureau. *2016 American Community Survey 1-Year Estimates*; American Fact Finder: Washington, DC, USA, 2016.

37. Office of Sustainability. *Greenworks: A Vision for a Sustainable Philadelphia*; Office of Sustainability: Philadelphia, PA, USA, 2016.

38. Philadelphia Water Department. *Green City, Clean Waters: The City of Philadelphia's Program for Combined Sewer Overflow Control*; Philadelphia Water Department: Philadelphia, PA, USA, 2011.

39. The City of Philadelphia Water Department. *Green City, Clean Waters—Evaluation and Adaptation Plan*; Philadelphia Water Department: Philadelphia, PA, USA, 2016.

40. University of Vermont Spatial Analysis Laboratory. *High Resolution Landcover*; University of Vermont Spatial Analysis Laboratory: Philadelphia, PA, USA, 2008, 2011.

41. Patel, D. *City Limits*; Philadelphia Department of Planning and Development: Philadelphia, PA, USA, 2016.

42. Philadelphia Water Department. *Green Stormwater Infrastructure Projects*; Philadelphia Water Department: Philadelphia, PA, USA, 2018.

43. USDA-FSA-APFO Aerial Photography Field Office. *NAIP Digital Ortho Photo Image*; USDA-FSA-APFO Aerial Photography Field Office: Philadelphia, PA, USA; Salt Lake City, UT, USA, 2010.

44. USDA-FSA-APFO Aerial Photography Field Office. *NAIP Digital Ortho Photo Image*; USDA-FSA-APFO Aerial Photography Field Office: Philadelphia, PA, USA; Salt Lake City, UT, USA, 2015.

45. ESRI. *ArcMaps (10.5) [Software]*; ESRI: Redlands, CA, USA, 2016.

46. Department of Licenses and Inspection; Office of Innovation and Technology. *Building Footprints*; Department of Licenses and Inspection, Office of Innovation and Technology: Philadelphia, PA, USA, 2015.

47. Office of Innovation and Technology. *Impervious Surfaces*; Office of Innovation and Technology: Philadelphia, PA, USA, 2004.

48. Division of Technology GIS Service Group. *Philadelphia Impervious Surfaces 2015*; Division of Technology GIS Service Group: Philadelphia, PA, USA, 2015.

49. Office of Innovation and Technology. *Railroad Lines*; Office of Innovation and Technology: Philadelphia, PA, USA, 2004.

50. Philadelphia Water Department. *Hydrology Polygons*; Philadelphia Water Department: Philadelphia, PA, USA, 2015.

51. Office of Innovation and Technology. *LiDAR 2008*; Office of Innovation and Technology: Philadelphia, PA, USA, 2008.

52. Office of Innovation and Technology. *LiDAR 2010*; Office of Innovation and Technology: Philadelphia, PA, USA, 2010.

53. Office of Innovation and Technology. *LiDAR 2015*; Office of Innovation and Technology: Philadelphia, PA, USA, 2015.

54. De Noronha Vaz, E.; Caetano, M.; Nijkamp, P. Trapped between antiquity and urbanism—A multi-criteria assessment model of the greater Cairo Metropolitan area. *J. Land Use Sci.* **2011**, *6*, 283–299. [CrossRef]

55. Eastman, J.R. *TerrSet Manual: Geospatial Monitoring and Modeling System*; Clark Labs Clark University: Worcester, MA, USA, 2016; p. 470.

56. Nazzal, J.M.; El-emary, I.M.; Najim, S.A.; Ahliyya, A.; Box, P.O.; Arabia, K.S. Multilayer Perceptron Neural Network (MLPs) For Analyzing the Propoerties of Jordan Oil Shale. *World Appl. Sci. J.* **2008**, *5*, 546–552.

57. Sibanda, W.; Pretorius, P. Novel Application of Multi-Layer Perceptrons (MLP) Neural Networks to Model HIV in South Africa using Seroprevalence Data from Antenatal Clinics. *Int. J. Comput. Appl.* **2011**, *35*, 26–31. [CrossRef]

58. Nadoushan, M.A.; Soffianian, A.; Alebrahim, A. Predicting Urban Expansion in Arak Metropolitan Area Using Two Land Change Models. *World Appl. Sci. J.* **2012**, *18*, 1124–1132. [CrossRef]

59. Kim, I.; Jeong, G.; Park, S.; Tenhunen, J. Predicted Land Use Change in the Soyang River Basin, South Korea. In Proceedings of the 2011 TERRECO Science Conference, Karlsruhe, Germany, 2–7 October 2011; pp. 17–24.

60. Pontius, R.G.; Millones, M. Death to Kappa: Birth of quantity disagreement and allocation disagreement for accuracy assessment. *Int. J. Remote Sens.* **2011**, *32*, 4407–4429. [CrossRef]

61. Rempel, R.S.; Kaukinen, D.; Carr, A.P. *Patch Analyst and Patch Grid*; Ontario Ministry of Natural Resources, Centre for Northern Forest Ecosystem Research: Thunder Bay, ON, USA, 2012.

62. McGarigal, K.; Marks, B.J. *FRAGSTATS: Spatial Pattern Analysis Program for Quantifying Landscape Structure*, 2nd ed.; U.S. Department of Agriculture, Forest Service, Pacific Northwest Research Station: Portland, OR, USA, 1995.

63. Philadelphia Water Department. *Green City Clean Waters*; Philadelphia Stormwater Dialogue: Philadelphia, PA, USA, 2011; p. 29.

64. U.S. Census Bureau Population Division. *Annual Estimates of the Resident Population: April 1, 2010 to July 1, 2017*; U.S. Census Bureau Population Division: Suitland, MD, USA, 2018.

65. Bell, M. County and Municipal-Level Population Forecasts, 2015–2045. Available online: https://www.dvrpc.org/webmaps/PopForecast/ (accessed on 24 January 2019).

66. Kremer, P.; DeLiberty, T.L. Local food practices and growing potential: Mapping the case of Philadelphia. *Appl. Geogr.* **2011**, *31*, 1252–1261. [CrossRef]

67. Seto, K.C.; Fragkias, M.; Guneralp, B.; Reilly, M. A Meta-Analysis of Global Urban Land Expansion. *PLoS ONE* **2011**, *6*, e23777. [CrossRef] [PubMed]

 *land*

Opinion

Lawns in Cities: From a Globalised Urban Green Space Phenomenon to Sustainable Nature-Based Solutions

Maria Ignatieva [1],*, Dagmar Haase [2,3], Diana Dushkova [2] and Annegret Haase [4]

[1] School of Design, The University of Western Australia, 35 Stirling Highway, Perth, WA 6001, Australia
[2] Department of Geography, Humboldt University Berlin, Unter den Linden 6, 10099 Berlin, Germany; dagmar.haase@ufz.de (D.H.); diana.dushkova@geo.hu-berlin.de (D.D.)
[3] Department of Comp. Landscape Ecology, Helmholtz Centre for Environmental Research—UFZ, Permoserstr. 15, 04318 Leipzig, Germany
[4] Department of Urban and Environmental Sociology, Helmholtz Centre for Environmental Research—UFZ, Permoserstr. 15, 04318 Leipzig, Germany; annegret.haase@ufz.de
* Correspondence: maria.ignatieva@uwa.edu.au; Tel.: +61-8-6488-6000

Received: 29 January 2020; Accepted: 27 February 2020; Published: 2 March 2020

Abstract: This opinion paper discusses urban lawns, the most common part of open green spaces and urban green infrastructures. It highlights both the ecosystem services and also disservices provided by urban lawns based on the authors' experience of working within interdisciplinary research projects on lawns in different cities of Europe (Germany, Sweden and Russia), New Zealand (Christchurch), USA (Syracuse, NY) and Australia (Perth). It complements this experience with a detailed literature review based on the most recent studies of different biophysical, social, planning and design aspects of lawns. We also used an international workshop as an important part of the research methodology. We argue that although lawns of Europe and the United States of America are now relatively well studied, other parts of the world still underestimate the importance of researching lawns as a complex ecological and social phenomenon. One of the core objectives of this paper is to share a paradigm of nature-based solutions in the context of lawns, which can be an important step towards finding resilient sustainable alternatives for urban green spaces in the time of growing urbanisation, increased urban land use competition, various user demands and related societal challenges of the urban environment. We hypothesise that these solutions may be found in urban ecosystems and various local native plant communities that are rich in species and able to withstand harsh conditions such as heavy trampling and droughts. To support the theoretical hypothesis of the relevance of nature-based solutions for lawns we also suggest and discuss the concept of two natures—different approaches to the vision of urban nature, including the understanding and appreciation of lawns. This will help to increase the awareness of existing local ecological approaches as well as an importance of introducing innovative landscape architecture practices. This article suggests that there is a potential for future transdisciplinary international research that might aid our understanding of lawns in different climatic and socio-cultural conditions as well as develop locally adapted (to environmental conditions, social needs and management policies) and accepted nature-based solutions.

Keywords: lawns; ecosystem services and disservices; nature-based solutions for lawns; alternative to lawns; sustainable lawns; two natures

---

1. Introduction

The recent worldwide changes in climate, including heatwaves and long drought periods, has resulted in the degradation of urban green spaces. Regardless of climatic conditions, water availability or cultural traditions, lawns are the most common elements of green city spaces across the globe, covering up to 50–70% of urban green areas [1]. In German cities like Leipzig, public park lawns cover at least 50% [2]. Similarly, in Sweden, lawns make up 50% of urban green areas [3]. Interestingly, Chinese cities are currently one of the largest users of lawns [4] and in the US, lawn (often called turf or grass) surfaces dominate urban and suburban landscapes and cover almost 2% of the country's terrestrial area [5]. The total area under turf in Australia is around 4400 hectares making up an average of 11% of the total areas of cities [6].

Lawns are highly recognised and massively prefabricated landscape design element. In many cases, turf is used as the easiest and most cost-effective short-term solution to covering "leftover places" after the demolition of buildings or for the "beautification" of abandoned places [1]. For shrinking cities, lawns act as an interim successional stage after the abandonment or demolition of built structures [7–10]. While urbanisation has led to a dramatic increase in lawn surfaces, these surfaces require significant input of energy and resources and the use of seed mixtures from global lawn nurseries. This has resulted in biological and visual homogenisation of urban environments [11–13].

The recent hot and dry summers in Europe (2017–2019) and severe drought conditions in many other countries around the world—California, Arizona, and Mid-West of USA, Cape Town in South Africa and across Australia—revealed particular issues related to the restrictions of water use [14–16]. Many urban grassy surfaces degrade from trampling and extreme sun or shade exposure, and thus quickly become brownscapes while also losing the ecosystem services that lawns typically perform [7,8].

One of the reactions to lawn degradation is the use of synthetic lawns instead of living grassed surfaces. Along with the growing contamination of aquatic habitats from plastic particles [17], the use of artificial lawns is contributing to the pollution of urban environments. A significant volume of polymer granules and synthetic grass fragments are introduced by water and wind into the environment each year and need to be better recognised as a form of microplastic pollution affecting soil, waterways, and ultimately the ocean [18,19]. These problems have highlighted the need to investigate and develop alternative, more resistant sustainable solutions for lawns that withstand impending climate change conditions and, at the same time, create environmentally friendly and aesthetically acceptable urban green open spaces.

In this article, we analyse the ecosystem services and disservices created by lawns. We further discuss existing alternative visions for urban lawns from different countries in both the northern and southern hemispheres. In this study, we use the concept of nature-based solution as an important foundation for searching of sustainable lawns. We accept the definition of nature-based solutions as proposed by European Commission [20] and Raymond et al. [21], that they are " ... actions and solutions to societal challenges which are inspired and supported by or copied from nature and provide at the same time multiple environmental, social, economic co-benefits such as the improvement of place attractiveness, of health and quality of life, creation of green jobs, etc." Our vision of nature-based solutions is grounded in the acceptance and respect of local peculiarities from country to country and is founded on a complex approach which includes biological, planning and design elements, provide social and economic benefits (such as the improvement of place attractiveness, of health and quality of life, creation of green jobs, etc.) as well as sustainable management and stewardship that is driven by municipalities.

Despite the universal adoption of lawns, there are a variety of lawn types and differences in technological peculiarities of construction and management regimes. These types are rooted in the history of the introduction of lawns and are connected to climatic, economic, and cultural conditions, as well as to specific land use and landscape design traditions [1]. There is an urgent need to explore nature-based solutions in order to better adapt lawns to current changing climatic conditions within particular geographical zones, local cultural perceptions and expectations, social wants and needs, and economic opportunities.

2. Conceptual Analytical Framework

This opinion paper discusses the phenomenon of lawns based on the authors' long-term project experience of working within the interdisciplinary research projects on lawns in different cities of Europe (Leipzig and Berlin in Germany, Uppsala, Malmo and Gothenburg in Sweden, Moscow, St. Petersburg, Kirovsk and Apatity in Russia), New Zealand (Christchurch), USA (Syracuse, NY) and Australia (Perth). These projects provided the opportunity to obtain and analyse large amounts of qualitative and quantitative data as well as to test some alternative nature-based and locally adapted solutions related to lawns.

The detailed conceptual analytical framework of this article is presented in Figure 1. To identify key questions related to existing lawn research and discover a research gap, we used a literature review based on SCOPUS, ISI Web of Sciences and Google Scholar (Figure 2). We specifically targeted key terms related to the particular ecosystem services and disservices of lawns (analysed in detail in Section 3). The search keywords were "lawn as a habitat", "use of lawns", "lawn as a symbol", "plastic lawn", "heat island mitigation by lawn". To support the theoretical hypothesis of the relevance of nature-based solutions for lawns we also suggest and discuss the concept of two natures—different approaches to the vision of urban nature, including understanding and appreciation of lawns (Section 4). This concept helped us to explain the directions for lawn alternatives and their correlation with local natural and social conditions (Section 5). For lawn alternatives, the choice of case studies and references was based on the analysis of existing projects in which authors had participated (urban meadows, pictorial meadows, woody meadows, use of appropriate native groundcovers in private gardens etc.) or related literature review [22,23]. We prioritised the most recent publications (2011–2019) but also included earlier peer-reviewed works. There were 139 publications related to the key words; 92 publications were included in the final list of references and 47 publications were excluded from the final list. There were two main criteria for exclusion: non-urban areas and if the article did not address one of the key searching aspects (key words). The structural analysis of the literature review used in the paper is presented in Table 1.

PROCESS / RESEARCH STEP	DESCRIPTION / ACTIVITIES	
RESEARCH DESIGN	• Identify the key issues related to lawn research • Identify the hypothesis • Reveal the goals, methods (including research tools) used in the related research and a research gap	Research goal and research questions ⬇ Literature review in SCOPUS, ISI Web of Sciences, Google Scholar ⬇
PREMISES FOR CONDUCTING RESEARCH	• Preliminary list for research gaps and suggestions how they can be covered by the current research • Developing concept, thinking models and framework	Key issues identified as the main sections of the research ⬇
ANALYSIS	• Social, cultural, historical and political embedding • Natural system and environmental impact (ecosystem services and disservices) • Planning and management issues • Two natures concept related to lawns • Searching for nature-based alternatives for urban lawns	Analysis of the case study evidence (own experience in lawn research) ⬇ Analysis of the relevant publications and documents ⬇
DISCUSSION	• Comparing findings with the existing concepts and results • Reflect (test) the research hypothesis • Suggest a new concept • Future directions for research on sustainable lawns	Lawn experts' workshop: information sharing and discussion ⬇ Contribution to the knowledge base
CONCLUSION	• Developing future interdisciplinary research framework	

Figure 1. Conceptual framework of this paper.

To support the theoretical hypothesis of the relevance of nature-based solutions for lawns we also suggest and discuss the concept of two natures—different approaches to the vision of urban nature, including understanding and appreciation of lawns.

We also used a workshop as part of the research methodology. The international lawn experts' workshop "Urban Biodiversity and Nature-Based Design methodology and practical applications for interdisciplinary research" in Berlin on 28–29 November 2019 was organised by the Geography Department of Humboldt University, Berlin. Participants from different scientific background and different part of the globe, who have been dealing with the different aspects of urban lawns and nature-based solutions gathered together, discussed methodology and the future perspectives of lawn research. Thus, this article also provides the theoretical and conceptual foundation for future international and interdisciplinary research project on lawns (Figure 14).

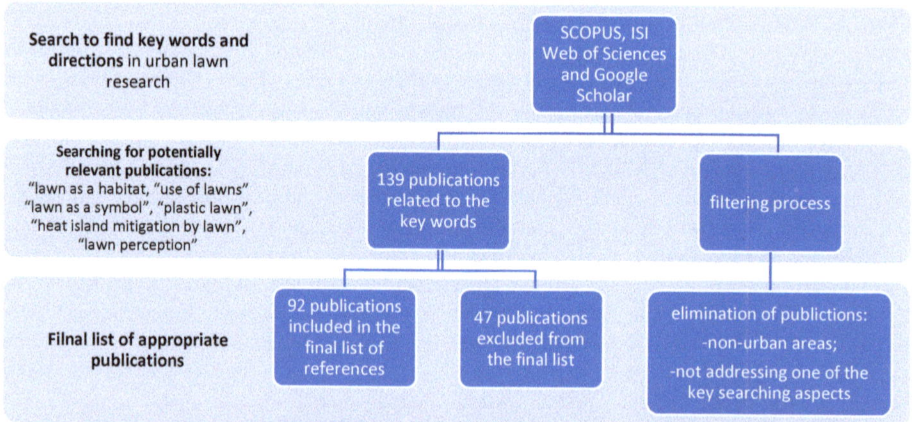

Figure 2. Literature review process.

Table 1. Structural analysis of the literature review used in this paper.

Category of Research on Lawn	Aspects	References
Environmental aspects of lawns	Estimation of lawn cover using remote sensing and earth observation methods	Hedblom et al., 2017 [3]; Milesi et al., 2005 [24]; Robbins and Birkenholz, 2003 [25]
	Biodiversity and vegetation aspects	Gaston et al., 2005 [26]; Hahs and McDonnell, 2007 [27]; Lindenmayer et al., 2008 [28]; Müller, 1990 [29]; Stewart et al., 2009 [30]; Sukopp and Kowarik, 1990 [31]; Threlfall et.al., 2015 [32]
	Ecosystem services provided by lawns	Amani-Beni et al., 2018 [33]; Armson et al., 2013 [34]; Beard and Green, 1994 [35]; Brunton et al., 2010 [36]; Burgin, 2016 [37]; Cumming, 2018 [6]; Fischer et al., 2013 [38]; Fischer et al., 2016 [39]; Haase et al., 2014a [7]; Haase et al., 2014b [8]; Johnson, 2013 [40]; Lele et al., 2013 [41]; Monteiro, 2017 [42]; Stirling et al., 2013 [43]; Thompson and Kao-Kniffin, 2017 [5]; Trigger and Mulcock, 2005 [44]; Wang et al., 2016 [45]; Wastian et al., 2004 [46]
	Ecosystem disservices provided by lawns:	Brunton et al., 2010 [36]; Burgin, 2016 [37]; Campagne et al., 2018 [47]; Cumming, 2018 [6]; Döhren and Haase, 2015 [48]; Dunn, 2010 [49]; Ignatieva and Hedblom, 2018 [1]; Lyytimäki, 2013 [50]; McKinney, 2006 [12]; Milesi et al., 2005 [24]; Müller and Sukopp, 2016 [51]; Priest et al., 2000 [52]; Runola et al., 2013 [53]; Shackleton et al., 2016 [54]; Schapel et al., 2018 [55]; Sharma et al., 1996 [56]; Stirling et al., 2013 [43]; Trigger and Mulcock, 2005 [44]; Wheeler et al., 2017 [13]

Table 1. *Cont.*

Category of Research on Lawn	Aspects	References
Social aspects of lawns	Public perception, attitude and preferences	Elgizawy, 2016 [57]; Han et al., 2013 [58]; Ignatieva, 2017 [22]; Jenkins, 1994 [59]; Pisa, 2019 [60]; Poškus and Poškienė, 2015 [61]; Rall et al., 2017 [62]; Ramer et al., 2019 [63]; Robbins, 2007 [64]; Sewel et al., 2017 [65]; Teysott, 1999 [66]; Müller, 1990 [29]; Trigger and Mulcock, 2005 [44]; Yang et al., 2019 [4,67]
	Urbanisation and homogenisation	Antrop, 2004 [68]; Groffman et al., 2014 [11]; Ignatieva and Hedblom, 2018 [1]; Pondichie, 2012 [69]
	Health and well-being aspects	Elgizawy, 2016 [57]; Payne and Bruce, 2019 [70]; Stolz et al., 2018 [71]
History of lawns	History of development	Ignatieva, 2017 [22]; Ignatieva, 2018 [72]; Ignatieva and Hedblom, 2018 [1]; Fischer et al., 2013 [38]; Gaynor, 2017 [73]; Hipple, 1957 [74]; Jenkins, 1994 [59]; Robins, 2007 [64]; Robinson, 1991 [75]; Yang et al., 2019 [4,67]
	Lawn as a site of conflicts	Greenbaum, 2000 [76]; Harari, 2016 [77]; Trudgill et al., 2010 [78]
Lawn alternatives	Searching for lawn's alternatives and sustainable lawn management	Alumai, 2008 [79]; Burgin, 2016 [37]; Chawla et al., 2018 [80]; Cumming, 2018 [6]; Gaynor, 2017 [73]; Hogue and Pinceti, 2015 [16]; Ignatieva et. al., 2008 [23]; Ignatieva, 2010 [81]; Ignatieva and Ahrné, 2013 [82]; Johnson, 2013 [40]; Pineo and Barton, 2010 [83]; Schapel et al., 2018 [55]; Steinberg, 2006 [84]; Teysott, 1999 [66]; Wasowski and Wasowski, 2002 [85]; Wasowski and Wasowski, 2004 [86]; Wastian et al., 2004 [46]; Wilson and Feuch, 2018 [87]; Zollner, 2018 [88]
	Alternative to lawns	Bormann et al., 2001 [89]; Daniels, 1995 [90]; Hitchmough, 2004 [91]; Ignatieva, 2017 [22]; Ignatieva, 2018 [72]; Ignatieva and Hedblom, 2018 [1]; Robinson, 1991 [75]; Sprajcar, 2017 [92]
	Artificial lawn related aspects	Brooks and Francis, 2019 [93]; Chawla et al., 2018 [80]; Fleming et al., 2013 [18]; Kaminski, 2019 [19]; Loveday et al., 2019 [94]
Two natures	Novel and designed ecosystems	Higgs, 2017 [95]; Hobbs et al., 2006 [96]; Kowarik, 2011 [97]

3. Ecosystem Services and Disservices of Urban Lawns

Lawns are specially designed ecosystems that originated in Europe in Medieval times [22]. We define lawn as a managed, artificially created grass-dominated plant community, designed for fulfilling a range of ecosystem services. This plant community predominantly consists of grass species—cultivars, as well as spontaneously occurring and unwanted herbaceous species known as "lawn weeds" [22,29]. One crucial aspect of lawns is the uniform phenomenon of a turf (sod), which is the upper level of soil that is covered by closely knit grasses and forbs intertwined with their roots or/and stolons, and which are in symbiosis with soil and fauna. Turf, in particular, is responsible for creating the uniform and "durable" surface commonly used by people for recreation and sport.

Lawns provide a full range of ecosystem services such as regulating the water cycle by promoting infiltration, thus facilitating regeneration of ground-water stocks and evapotranspiration [42]. In addition, lawns mitigate the heat-island effect through transpiration and evaporation and provide cooler microclimates [98]. Another important ecosystem service of lawns is habitat provision for some urban fauna species [37]. Lawns also support soil organisms. Since their introduction, the most recognised ecosystem service of lawns has undoubtedly been the cultural aspect, i.e., the creation of the specially designed leisure spaces (Figure 3).

Along with these positive contributions to human life and well-being, there are a number of ecosystem disservices created by urban lawns, such as those presented in Figure 4. According to Shackleton et al. [54], ecosystem disservices are commonly understood as "the ecosystem-generated functions, processes and attributes that result in perceived or actual negative impacts on human well-being". These negative effects arise from ecosystem characteristics that are economically or socially harmful or that endanger health or may even be life-threatening [47–50]. This includes sheltering species such as pathogens and parasites harmful to human health, damaging pests [41] or those that attack humans [49]. In the case of lawns, among the most recognisable disservices are dust pollution and loss of aesthetic qualities during hot and dry summers and surface and ground pollution as a result of using herbicides and pesticides. In the following sub-chapters, we discuss ecosystem services and disservices in details. There is a clear pattern related to human activity, economics (availability of resources for management) and environmental factors beyond human control (draughts, heat waves, floods). At some stage, a definite positive ecosystem service can turn into a definite disservice. For instance, one of the main ecosystem services of lawns that they are a place for recreation, however, when there are too many users and heavy trampling, the lawn's surface becomes degraded, uneven and even dangerous for users. Due to the particular significance of cultural aspects of lawns and a number of studies of this particular phenomenon, the cultural services of lawns received additional scrutiny in this paper (Sections 3.1–3.3).

Figure 3. Lawn as a special leisure space in urban parks worldwide: (**a**) King's Park in Perth, Australia; (**b**) Lene-Voigt Park in Leipzig, Germany; (**c**) one of the public parks in Xian, China. Photos: M. Ignatieva, D. Dushkova.

Ecosystem services provided by lawns	Lawn mediated ecosystem functions	Ecosystem disservices provided by lawns
Cooling/mitigation of heat-island effect by transpiration and evaporation	1. Regulating: 1.1. Air temperature regulation	During dry and hot summers in places with no watering – dust pollution
Carbon sequestration benefit	Regulating: 1.2. Carbon sequestration	Can be negated by using fuel mowers and frequent intensive management
Cooling/mitigation of heat-island effect by transpiration and evaporation	Regulating: 1.3. Air quality	Greenhouse gas emission and dust through intensive management (frequent mowing by fuel operated mowers)
Regulation of the water cycle (infiltration, regeneration of ground-water stocks and evapotranspiration)	Regulating: 1.4. Water quality	Overuse of water; spreading of herbicides and fertilizers - surface and groundwater pollution; pollutions by microplastics from artificial lawns
Irrigated lawns provide habitat for some fauna species (some flowering herbaceous species in "old" lawns available for pollinations, e.g. bees; forage for some birds and mammals)	2. Supporting: biodiversity support and habitat provision	Loss of native habitats, only some specialised 'urban specialist' species benefit. Invasiveness of some grass species
Enhancing private property value, symbol of prestige and power	3. Cultural: 3.1. Aesthetic	Ecological homogenisation, high resource input, unification of design
Recreational activities (including sport)	3. Cultural: 3.2. Recreation	Degradation of ground surface because of overuse (trampling); Dying, unattractive brown surfaces in the absence of irrigation

Figure 4. Ecosystem services and disservices provided by lawns.

The majority of research that analyses lawn ecosystem services (directly or indirectly for example in the research of wildlife in private urban gardens) is based on case studies from the temperate latitudes of the northern hemisphere, namely from Europe (40% of publications) and the US—60 % [3,5,16,22,24,29,42,46,59,65,72,86]. From sources related to the study of lawns (not including alternative lawns), only three were from China directly connected to ecological and cultural aspects of Chinese lawns [4,67,99]. Australia had the most publications related to cultural aspects (history of lawns and their connection to colonial culture [73], positive outcomes of lawns for hot and dry urban environment and how to develop sustainable management of lawns (waterwise irrigation, relevant soil preparation and species selection). There are number of publications on Australian urban private garden wildlife where lawns are mentioned as a new habitat for exotic and native wildlife species [44]. The first direct ecological research on lawns (the biodiversity of lawns) in New Zealand resulted in several publications in the late 2000s [30,81].

3.1. Cultural and Aesthetic Services of Lawns: Historical Roots

From a societal and cultural perspective, lawns are one of the most important and frequently used types of urban green infrastructure. From the very introduction of lawns into Europe as a crucial garden element during the Middle Ages, their most advertised value was primarily cultural and aesthetic function. In actual fact, lawns were introduced purely as a decorative element for human enjoyment, and not associated with any direct economic value. Subsequently, they have been rapidly developed in periods of political stability and technological progress in Western Europe [22].

Since their development, lawns have required both space (land) and labour to provide constant management (especially in the early stages of their use in the 16th and 17th centuries). One important purpose of lawns was intangible—as symbols of power and prosperity. This important symbolism of power, order and control over nature can be found in all countries across the centuries—from French and English gardens designed for the aristocracy, to important contemporary public buildings and private residences of high-income urbanites around the globe [76,77] (Figure 5). From the 18th century onwards, lawns were designed according to the ideas of the picturesque movement (end of the 18th century to the beginning of the 19th century). At that particular time, smooth and gently rolling turf surfaces were revered as the most "beautiful" landscapes [76]. European countries and colonies designed numerous park-like landscapes according to this standard of English "beauty".

(a)   (b)   (c)

Figure 5. Lawns as symbols of beauty, power and prosperity: (**a**) Lawns are a dominant feature of decorative parterre in the formal (French style) garden of Zwinger Palace, Dresden, Germany; (**b**) Lawn in front of a mosque in Doha, Qatar; (**c**) Manicured lawn in a private villa, suburb of Moscow, Russia. Photos: M. Ignatieva.

The English garden with its admiration of the countryside and pastoral landscapes (and using both pastures and lawns in their vocabulary) became a kind of "buffer" between "wildness of nature and the stiffness of art" [74], (p. 241). The majority of English gardening practitioners and scholars of the picturesque movement agreed that the garden area next to the main house should be covered by cut lawns, otherwise it would demonstrate a step away from civilisation [75].

It is very important to note, that English parks used native grassland species for their lawns that were also widely used in pastures. The success of these species on lawn surfaces was due to the mild English climate, high rainfall and appropriate soils.

From the very beginning of the public parks movement in the mid-19th century lawns have served the function of public recreation [59,64,66,75,78,84]. Public parks are based on the model of the British pastoral landscape aesthetic where green grass areas played the essential role. Another highly recognised recreational benefit of lawns is their provision of surfaces for sports like football, cricket and golf. All these games are rooted in the British Isles.

With technological progress and the invention of mowers for the common man, lawns became more widespread in private suburban gardens in the UK, Australia, New Zealand and to a significant extent in the USA (Figure 6). The majority of existing literature on colonial lawns is dedicated to researching the social perception and analysis of the American attachment to perfect home lawns [59,66,79]. American authors believed that, despite the lawn being primarily an English feature, the US "front lawn" became the most powerful sociological manifestation and an obligatory element of the American lifestyle. Lawns in America stand for personal respect and being a good citizen, and are associated with public health and even safety. For example, in short cut lawns dangerous creatures such as snakes or ticks would have no chance to appear [66]. Today, the United States is one of the largest producers and consumers of turf, and turf represents the largest irrigated non-food crop in the country [24,79,80,83].

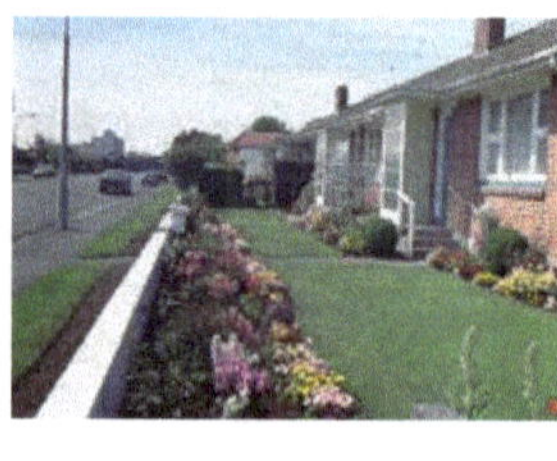

(a)   (b)   (c)

Figure 6. Lawns in private gardens: (**a**) USA (Syracuse, NY), (**b**) Australia (Perth) and (**c**) New Zealand (Christchurch). Lawns are obligatory elements of suburbia. Photos: M. Ignatieva.

3.2. Cultural Services of Colonial Lawns: Australia and New Zealand

Colonial lawns were introduced to Australia and New Zealand as the important aesthetic heritage of England, together with other garden, planning and architectural archetypes. In New Zealand, lawns became part of the newly established public parks, private estates, and suburban gardens and they used the picturesque-gardenesque aesthetics. European grasses were well adapted to New Zealand's temperate climate. Christchurch, for example, is called "the most English city outside of England" [100].

In Australia, establishing turf was more difficult due to the heat, frequent droughts and unsuitable soils for European turf species. In other words, these were unusual and hostile environmental conditions for lawns. Whereas most lawn grasses in Europe have their origins in the native or secondary grasslands of that region, in the southern hemisphere, all suitable turf species were non-native plants. It was therefore a long and painful process to find appropriate non-native species and lawn management regimes for lawns in the southern hemisphere. Particularly in the dry climates of Western and South Australia, the turf industry faced many challenges. Grasses in poor soils would not grow without constant irrigation and fertiliser application.

Concerning the early days of settlers in Perth, Gaynor [73] (p. 4) states that "grass and other garden plants were allies in a war against the heat and dirt that perpetually invaded settlers' homes and their dreams of creating a 'civilised' city, and the alliance was forged and maintained with water". For early Australian settlers, lawns, as part of a cultivated and irrigated garden, were a powerful symbol of the supposed superiority of European civilisation in contrast to the Indigenous (Aboriginal) wilderness. These lawnscapes were in opposition to the wildness of "the bush" and the bush referred to the" wild" places, the shrublands, forests, mountains, deserts and sometimes even the rural countryside [101].

To the first settlers in Australia and New Zealand, native plants were unattractive and appeared "alienating". Plants were never as green as those in England. A similar attitude towards the surrounding "messy" wildness has persisted in the modern landscape. The colonial symbolism of the well-kept private garden has not changed in the 20th and the 21st century. A neat garden indicates neighbourhood status and high property values. This separation of urban landscapes where lawn and exotic decorative plants are delineated from native bushland is the foundation for the existence of two different "natures" in Australian and New Zealand cities (Part 4).

3.3. Cultural Services of Lawns: Recreation

From an urban sociological standpoint, there are various ways that lawns are utilised. One of the main values of conventional lawns is the space they provide for social activities such as picnicking, resting, sunbathing, walking dogs, games and sports [102]. Another important function, which was connected to the 19th century gardenesque style, was the use of lawns as aesthetic backgrounds for architecture and art elements [22].

Recent studies from Europe and USA have revealed that people's love of lawns is connected to the integral role lawns play in the everyday landscape [1,22,60–63,71,72]. Short cut lawnscapes are associated with improved quality of life and personal safety. Lawns, with their openness and good visibility, are opposed to dense shrubs and woodlands which can hide dangerous people. In arid developing countries where water shortage is paramount, lawns are nevertheless used to green workplaces and are seen as primary vehicles for enhancing the quality of human life [57].

In Scandinavian countries, due to the cold climate and subsequent lifestyles, lawns can only be used from late May to October [22]. In parts of Central and Western Europe with milder climates, lawns can be used for longer periods of time. In Europe, due to the changing climate and with warmer winters and extended summer temperatures, lawns are used throughout the whole year. There is evidence that some German city lawns in urban parks are being used from February to November, which is far longer than in previous decades. This prolongation of the growing season has led to lawns being overused and not being afforded necessary recovery periods.

In China, due to the very recent introduction of lawns to urban public spaces and to certain specific socio-cultural practices (overuse of green spaces), lawns are not accessible for general recreation but only play a decorative (aesthetic) role in urban landscapes [4]. In warm, humid or arid climate countries, lawns are used all year round, however, their condition is dependent on irrigation which raises concerns of overuse of water, particularly in arid cities.

These days, lawns are used for a broad range of activities—for quiet recreation (reading, talking and walking) to sports, plays, parties, barbecues and picnics (Figure 7). Due to the "mediteranianisation" of European lifestyles, people would like to spend more time outside [62].

(a)   (b)   (c)

Figure 7. Lawns provide the main arena of human activities in cities today: (**a**) Park in Tokyo (Japan), (**b**) Rabet Park in Leipzig (Germany), (**c**) Gujiazhai Park in Shanghai (China). Photos: M. Ignatieva, D. Dushkova.

Today, the lawn is idealised as a universal cultural norm and is considered the most "beautiful" aesthetic function of urban landscapes, which in turn, helps to create positive human psychological and physical health [22,78,103].

In recent years, the main cultural and aesthetic disservices of lawn are caused by the increasing recreational pressure put on publicly accessible parks resulting in large compacted, trampled areas. There are simply more people who want to use lawns. This leads to degradation of the lawn surface, to a greater input of resources (watering, aeration and fertilisers) and to constant repairs of damaged areas. One of the examples is Görlitzer Park in Berlin which is widely used by local people (parties and festivals) and by tourists. Another example is public parks in Leipzig where recreational pressure significantly increased due to the growing population (Figure 8a). Across Europe, in many cities that have become the destination of youth immigration, park lawns are especially in demand and under pressure [104,105].

In Europe and some countries where watering is restricted or prohibited during dry summers, lawns are turning brown and becoming a significant source of dust [69] (Figure 8).

(a)　　　　　　　　　**(b)**　　　　　　　　　**(c)**

Figure 8. Degradation of lawns from overuse (trampling) and drought. Lawns were established on former post-industrial site (**a**) or wasteland (**c**): (**a**) Lene-Voigt Park, Leipzig (Germany), (**b**) Zaryadye Park, Moscow (Russia), (**c**) public green space near a hospital in Kirovsk, Russia. Photos: D. Dushkova.

In shrinking European post-industrial cities, urban lawns are also frequently used after the demolition of housing or industrial structures. This is a fast method for reviving open space and making it accessible for recreation [106,107]. In such areas, the soil can often be very thin and grass species struggle to survive. In contrast, planted trees and shrubs in such areas receive more attention and better maintained. This is particularly acute in urban parks that are created on former brownfields. Those parks suffer much more during hot and dry summers. Thus, lawns become almost unusable and cannot fulfil their recreational function.

3.4. Mitigation of the Heat Island Effect, Carbon Sequestration and Regulation of the Water Cycle

The cooling effect of lawns is well recognised and is always used as an argument for the importance of grass-covered areas (part of green infrastructure) versus hard urban surfaces (grey infrastructure not covered by vegetation). The cooling capacity of lawns is directly related to the evapotranspiration process and very much depends on water availability. In temperate climates, lawns have shown their capacity to decrease the temperature peaks of hot summer days by approximately 1 °C [42].

Proper irrigation regime enhanced the cooling effect of grasses [33]. Irrigated turf has become an important factor for the mitigation of the urban heat island effect in hot arid climates, such as Australian cities [98]. For example, in Adelaide, where a warming trend is occurring as a result of climate change, by 2070 the maximum temperatures during January and February are expected to exceed 45 °C, which is higher than the average maximum temperatures between 1980 and 1999 of 43 °C and between 2000 and 2012 of 44 °C. Heat mapping of urban areas as well as high resolution thermal infrared imagery of 285 km^2 region of Adelaide's southern suburbs showed that the coolest sites were golf courses, water bodies, dense woody vegetation and irrigated turf, while the hottest areas were generally comprised of buildings, dry agricultural fields, dry/dead grass and vegetation, exposed soil and unshaded hard surfaces [108]. Research into surface temperatures of hard and soft urban landscape elements in Perth, Western Australia, found that areas with grey pavers were the hottest, whilst areas with ground-cover plants were the coolest. In the evenings, grey pavers remained the hottest, whilst decking, soil, and turf grass were the coolest [94].

Another recognised ecosystem service of lawn is carbon sequestration. In temperate zones of Europe and the USA, carbon sequestration has been positively associated with carbon accumulating in the soil [42]. However, other recent studies of the northern hemisphere temperate zones have shown that the positive effects of soil carbon sequestration in intensively managed lawns can be negated by greenhouse gas emissions generated by the routine management operations of mowing, fertiliser application and irrigation [22]. In the Newcastle region in Australia domestic lawn-mowers contributed 5.2% and 11.6% of carbon monoxide (CO) and non-methane hydrocarbons emissions (NMHC), respectively [52].

Park and garden soils in western and eastern European countries have experienced several centuries of enriched soil fertility and, accordingly, increased humus amount in soils. In cities where lawns are created on sandy soils, such as Perth, these soils require a lot of input from the outset to grow turf grasses because of their limited water and nutrient capacity. Recent research has shown that "excess nitrogen and phosphorus leaching beneath urban lawns on sandy soils in metropolitan Perth may pose a serious threat not only to the quality of the underlying groundwater but also to many surface-water bodies" [56] (p. 1).

Northern hemisphere case studies from USA and Europe have outlined the importance of turf for reduction of water runoff and increased water infiltration, with resulting in flooding problems and increase in water recharging [42]. However, the data is limited and, in most scenarios, it is based on cases from temperate climates that are not directly obtained from researching turf grass urban ecosystems. For lawns in arid and semi-arid regions, water-related issues are typically considered disservices rather than services. To maintain living and green turf grass, substantial irrigation is required. Studies from arid zones of the United States have revealed that lawn used up to 75% of the total annual household water consumption [24]. In southern hemisphere cities, for example in Perth, gardens accounted for over half of the city's water use in 1970's. Perth's total water usage, accounting for scheme water usage, regulated bore water abstraction and estimated private garden bore usage between 2017 and 2018 was 629,390 mega litres. Approximately 258,403 mega litres (41%) of this amount was used for the irrigation of lawns and gardens, of which 79% was drawn from groundwater [109].

Without irrigation in such dry conditions, lawns are becoming dry, brown, dusty and unappealing to people. In Germany, Sweden and England, watering is not allowed for public green areas during hot summers, thus lawns and street trees are rapidly degrading. Some grasses recover after late summer rainfalls, however, the damage is visible and turf surfaces often need to be repaired. In many Australian cities (for example in Melbourne and Sydney) and semi-arid states of USA (in California and Arizona) there are a strict water conservation efforts and restriction policies against using water for lawn irrigation [16]. In Australian cities, only some species of turf grasses are capable of reviving after summer droughts. Others just die and a whole new lawn needs to be reinstalled. As they offer a quick solution to keeping an urban yard or playground "clean" and "green", the synthetic lawn industry has boomed all over the globe. (Figure 9). However, ecological and sociological research into such substitute for nature is limited. There is concern that material used for plastic non-living lawn reduces urban habitat, suppresses soil fauna, pollutes runoff via plastic and synthetic particles and other unknown impacts on the environment [18,19,93,110]. Loveday et al. [94] revealed that artificial turf grass can be particularly hot, often more than 30 °C above turf grass.

In many cities where herbicides and pesticides are used to keep lawn uniform and tidy, there are concerns of contamination of groundwater and runoff water. For example, in 2012, US houses applied up to 57.6 million kilograms (12.7 million pounds) of pesticide to lawns [111]. The most recent and widely discussed example is Roundup™—one of the most widely used herbicides on the planet. The International Agency for Research on Cancer (IARC), classified glyphosate (the active ingredient in Roundup™) as "probably carcinogenic to humans" [112].

(a)   (b)

Figure 9. Synthetic (plastic) lawns in a playground in Malaga (Spain) (**a**), and front yard of a residential building in Leipzig (Germany) (**b**). Photos: D. Dushkova.

3.5. Habitat (Biotope) Provision

Since lawns consist of sod, a combination of grass roots and soil, they support particular type of wildlife, for example, insects (ants and some species of beetles), nematodes, earthworms and spiders. Older lawns in the northern hemisphere temperate climate usually include some broadleaf herbaceous species (*Trifolium repens, Potentilla anserina, Prunella vulgaris*), that are capable of adapting to the mowing height. The life habits of these plants adapt to a frequent mowing and allowed them to go through their life cycle and produce flowers, thus attracting pollinators such as bees and bumblebees [22].

Some domestic lawns and moderately visited park lawns in Europe are attractive to small herbivorous animals such as rabbits and hares. Lengthening the mowing interval and creating a timed schedule for mowing to allow for the flowering of broadleaf herbaceous species such as clover, would increase the diversity of plants in the lawn and pollination and grazing opportunities for wildlife.

In the southern hemisphere, especially in Australia and New Zealand, the aim of maintenance of domestic and public lawns is to achieve a homogenous green carpet-like appearance (Figure 10).

(a)   (b)   (c)

Figure 10. Ecological homogenization of urban environment in (**a**) Australia (Perth), (**b**) Singapore (Singapore Botanic Garden) and (**c**) New Zealand (Lincoln University Campus). Photos: M. Ignatieva.

Compared to European and USA temperate zones, Australian and New Zealand lawns consist of far fewer native plants, which are unable to grow amongst the dense exotic grasses due to their very different life strategies. Since native plant communities are either destroyed or replaced by irrigated lawns, native fauna has to adapt and use lawn grasses as a food source. For example, the native bird, little corella (*Cacatua sanguinea*) is regularly seen browsing on irrigated lawns (in parks and sports fields) in Canberra, Sydney and Brisbane. The Australian magpie, ibis and wagtail birds are also a very common forager of Australian urban lawns [44]. Among non-avian taxon who prefers urban lawns in the Pacific coast of Australia is the can toad (*Rhinella marinus*) introduced from South America and becoming a problem. Can Toad's toxin kill household pets and any native species that will attempt to prey on them [37]. Many invasive urban bird species also feed on urban turf grasses. In Australia's increasingly dry environment, especially over the last decade, irrigated urban lawns have become desirable food sources for large marsupials such as kangaroos (Figure 11) [113]. The Eastern and Western grey kangaroo often forages on golf course turfs and urban lawns. For local turf producers and golf courses greenskeepers, kangaroos are seen as a nuisance. They can also ruin fences and cause hazards on the roads. In Canberra, urban lawns are one of the main habitats for the rabbit-an introduced animal that is now considered a pest in Australia. In these cases, the positive ecosystem service of "providing wildlife habitat" has turned to a disservice.

Figure 11. Urban lawns have become a desirable food source for Kangaroos in Western Australia. Photos: M. Ignatieva.

Australian lawns also provide habitat for other harmful pests such as the stinging nematode (*Ibipora lolii*). Infestations of this accidentally introduced parasitic nematode (possibly originating in South America or the Caribbean) have resulted in grass with shallow root systems, sparse turf cover and bare patches in many sports fields and recreational areas [43].

Over the last decade, the most noticeable and widely discussed ecosystem disservice of lawns is aesthetic uniformity resulting in the ecological homogenisation of urban areas, with lawn plant communities becoming similar in composition and structure across numerous biogeographical zones [1]. The demand for these monotonous green surfaces can only be met by using monocultures of one or two species. In temperate climates, four European species, *Poa pratensis, Festuca rubra, Lolium perenne* and *Agrostis spp.* are widely grown in turf grass nurseries. In warm climate countries, the most common are *Cyonodon dactylon* (native to Africa), *Stenotaphrum secundatum* (originally from Central and South America), *Paspalum vaginatum* (from the Americas), *Pennisetum clandestinum* (East Africa) and *Zoysia japonica* (from southeast Asia and Indonesia). The main method for maintaining the homogeneous composition is establishing lawns by seeding or by vegetative planting and eliminating any other species (weeds) by applying herbicides and frequent mowing.

Another ecosystem disservice of lawns, especially in non-European countries, is the invasive capability of some lawn grass species, with many spreading into native biomes. One classic example is the most famous lawn species, *Cyonodon dactylon*. This species was listed by the Global Compendium of Weeds as one of the top 12 cited invasive weeds in the world [114]. Increasing the biodiversity of lawns can be achieved by leaving some native or spontaneously appearing broadleaf flowering plants to attract pollinating insects. However, this is controversial in light of the attitudes to the conventional lawn—that it should be a highly manicured and controlled plant community where other plants are undesirable.

4. Two Natures

Lawnscapes dominate urban landscapes and people perceive urban "nature" through the prism of lawns [22,59,64,66,78,84]. When a person steps on the grass beside of a road or outside a building, it is often their only daily contact with nature. In cities with no or very limited access to wild vegetation or other pristine nature, urban dwellers have an even stronger connection and association with turf grass as nature. Turf grasses together with other "natural features" consisting of living organisms such as trees, flowerbeds, shrubberies and water bodies form this vision of nature [57].

Historically, European preferences for grassy surfaces were transferred and adopted in other countries and communities. For many centuries, European green areas consisted of native species and included natural or semi-natural vegetation. By the 19th century, the English vision of urban green spaces dominated the USA, Australia, New Zealand and European colonies [81]. European settlers literally transferred their values of turf grasses to their colonies and created a new version of urban nature. It was a new "civilised" European nature based on exotic species that were opposed to wild nature. Mowed lawn was often used as a demarcated line between these two natures. We argue that in Europe there is only one urban nature and in Australia and New Zealand there are two urban natures.

4.1. European Urban Nature

European researchers in their post-World War II studies of urban ecosystems saw urban nature as a heterogeneous and complex phenomenon. Their vision of urban nature included all types of urban biotopes-remnants of "pristine" forest, semi-natural modified groves, designed urban parks, small community gardens, abandoned wastelands or ditches or cracks in walls or pavements [31,81,82]. Most Western and Central European landscapes were modified during the long history of human settlements [68]. Some introduced decorative and crop species that had escaped from cultivation and, with time, became integral parts of urban ecosystems. Studies of European urban ecology consider the naturalisation stages of urban flora and vegetation [31]. The degree of naturalisation and invasiveness in Northern, Central, Eastern and Western Europe is still not as severe as urban environments in the New World. For example, in Central Europe the original flora consisted of 2,400 vascular plants. Since 4000 BC. more than 12,000 taxa have been introduced and only 279 (2.3%) have naturalised in natural plant communities [51]. In New Zealand, flora comprises about 2500 indigenous plants (80% of them are endemic). However, since European arrived in the 1840s, over 25,000 exotic plants have been introduced and one tenth of them have already become naturalised, with four more entering the wild each year [81,115].

In Europe, urban nature is still dominated by native flora including urban lawns. Due to the ecology of European native biomes which have undergone numerous disturbances, there are effective recovery mechanisms for disturbed ecosystems. A large number of native pioneer species in the soil seedbank allows urban biotopes to quickly regenerate. The typology of European lawns and their composition and structure is regulated by management and, first of all, by the frequency of mowing. For example, in Sweden there are conventional lawns that are frequently mown and meadow-like lawns (high grass and meadows), that are cut one to two times per year. High grass areas have a greater potential for biodiversity when properly maintained (collecting clippings after cutting to restrict soil fertility) and with a proper mowing schedule (at the right time of the season) [22].

The majority of urban ecology research in North America is based on temperate cities and also regard "urban nature" as an entity that is not separated from wild nature and manmade (designed) nature. In the US, the urban-rural gradient approach is the most popular method of studying urban ecosystems. American cities have the dominant urban planning model: central-business, district-sprawl and suburbia-rural ecosystems [116]. North American urban ecologists have also focused more on remnant indigenous vegetation for example urban forest [117]. However, the USA has quite substantial input on socio-cultural research of urban lawns [64,118].

4.2. Australian Urban Nature

In Australia and New Zealand, "nature" typically refers to natural indigenous ecosystems. While Europe has a long tradition from the 19th century of studying urban flora and vegetation with the most advanced classification of plant naturalisation, urban vegetation and urban biotope mapping,

Australian urban ecology is much younger. Urbanisation patterns were different from Europe and urban development took place in relatively intact native vegetation. Many high-quality remnants of native woodlands, scrublands, grasslands or wetlands survived and could be found scattered through cities and its suburbs [27]. Some of these valuable patches were severely transformed during urbanisation, but some still contain a large proportion of their original vegetation. That is why research of flora and wildlife in urban remnants of indigenous vegetation (forests, woodlands, grasslands, rivers, creeks and wetlands) and principles of their protection and restoration or study of native wildlife species in urban areas are prioritized among Australian urban ecologists [27,28,32]. For Australians, "nature" equates to "the bush". This is a very Australian word for wilderness used by the general public and by governmental and public organisations (https://www.bushlandperth.org.au/bush-forever/).

European and North American ecologists introduced and widely used terms such as "urban ecosystems", "urban plant communities", "urban biotopes", "urban habitats". These terms all include urban plants and their assemblages without divisions such as "cultivated", "spontaneously natural" or "natural". In comparison, Australian ecologists use the concept of "novel ecosystems", meaning ecosystems that differ in composition and/or function from present and past historical (meaning original native) systems [96]. Originally, the term "novel ecosystems" was introduced by USA ecologists Chapin and Starfield [119] to recognise "the response of the boreal forest to current and anticipated climatic changes".

The primary goal of accepting and reinterpreting the concept of novel ecosystems by Australian ecologists is to understand invasive species behaviour in native remnants and provide mechanisms for saving and restoring native vegetation. Australia's rapid urbanisation and use of European landscape models has resulted in a dramatic loss of unique and fragile native ecosystems, which existed for thousands of years in isolation with relatively minor human disturbance. Invasiveness in Australian ecosystems is severe, with many introduced decorative and crop species and associated weeds escaping from cultivation into the wild environment [28].

Recently the new term "designed or engineered ecosystems" has complemented "novel ecosystems". Designed ecosystems are described as "requiring intensive interventions to create them and ongoing management to sustain them" [95]. Novel ecosystem as a term is now recognised by European and USA urban ecologists [97,120] and used to explain the character of biodiversity, its level of "naturalness" and capacity for urban biodiversity conservation and protection [121]. Ingo Kowarik [97] has even suggested the concept of "four natures": 1 nature-pristine (forest, wetlands); 2 nature-agricultural (grasslands, fields); 3 nature-horticultural (parks, gardens); and 4 nature-urban-industrial, vacant lots, industrial sites and transport corridors. This typology of urban nature is also reflected in the European understanding of urban landscapes.

Study and practical application of Australian urban ecology in landscape design typically deal with human modification of "wild" or "natural" systems within urban and agricultural lands and uses this knowledge as a tool for the conservation of "native" nature. As for the other "nature" that dominates urban areas, there are still quite a few gaps in urban ecological research especially at the urban biotope level, such as research of lawns (except some historical history literature), public parks or wastelands and abandoned industrial areas. We suggest this urban man-made nature be called "designed and managed" nature (Table 2). Dominated by introduced exotic tropical and subtropical species, lawns and gardens in the hot and dry Australian climate are completely dependent on irrigation, supplementary nutrients, and management. "Designed and managed" nature in Australian cities is based on global landscape design patterns and similar exotic plant material available in nurseries. There is a sharp boundary between "wild" nature and the "designed" urban nature under total human control.

Residents of Australian cities share similar attitudes to the US where lawns in suburban private gardens form a unique middle ground between nature and the built environment [76]. Actually, the majority of urban ecology research of lawns in Australia concerns wildlife in urban private gardens in suburbia. There is a lack of Australian ecological research into "pure" designed and managed nature".

A better understanding of the structure, composition, flow, succession, resilience, and resistance of urban ecological systems could help bridge the chasm between "designed nature" and "native nature".

In New Zealand cities, which as in Australia, originated as colonial settlements, some ecological research of suburban gardens and urban lawns has been conducted which outlines an important strategy and the potential to return indigenous vegetation into the urbanised environment, thus creating more harmony between "wild" and "designed" nature [23,30,122].

Table 2. Vision of Two Natures.

Europe	Australia and New Zealand
Urban nature: no separation of native and non-native components	*Urban nature means native ecosystems*
Urban biodiversity: all components including remnants of native vegetation (if any), semi-natural, spontaneously appearing and planted exotic plant species	Separation of man-made (designed) nature from native ecosystems (native nature). New Zealand even introduced a separate term: native biodiversity
Small percentage of naturalised and invasive species	Large percentage of invasive and naturalised species as well as introduced species
Europe is the birthplace of the urban nature vision (landscape architecture styles)	Receiver of European nature vision: "beautiful" green nature
More relaxed attitude towards native/exotic approach	Urban green infrastructure, connectivity of green corridors, water sensitive design, protection means conservation, restoration and connectivity of remnants of native vegetation
Europe as the "cradle" of urban ecology science	Very few studies of "designed nature" lawns, private gardens, post-industrial zones, wastelands, road vegetation, etc., and their ecosystem services and potential for sustainable design principles
Elaborated methodology of urban ecology research (flora, vegetation and their related urban ecology aspects, social perceptions), including ecosystem service flows of benefits	Very little research into urban soil characteristics, ecosystem services provision, trample resistance and stability, social acceptance and preferences and constraints among different users

5. Nature-based Solutions—Existing Alternatives to Lawns

Our understanding of the nature-based solutions concept is that it hinges on three main criteria: "actions and solutions to societal challenges" (where landscape design and planning of lawns can help to achieve sustainable solutions); inspiration from nature (inspiration for lawns in local native ecosystems or in self sustained urban plant communities); provision of environmental, social and economic benefits for people [20,21,123,124]. There are several types of alternatives to lawns in Europe, USA, New Zealand and Australia (Table 3).

Alternatives to lawns are usually inspired by different grassy ecosystems or from biomes with low growing vegetation that can withstand heat and drought. Most existing alternatives, nevertheless, are not equivalent to conventional turf—durable sod that withstands recreational pressure (trampling). The purpose of such solutions is to decrease the number of unused lawns surfaces (urban planning) and to avoid homogeneity (visual and ecological) by employing different landscape design patterns (colour and texture) as well as providing more biodiversity, and thus ecologically friendly, wildlife habitats and a healthier environment (decreased mowing and fewer greenhouse gas emissions).

Table 3. Alternatives to lawns.

Alternatives to Lawns	Germany	Sweden	UK	[a] USA	[b] USA	Australia	New Zealand
Go Spontaneous	+						
Meadows	+	+	+				
Grass-free (tapestry lawns)		+	+				
Pictorial meadows	+	+	+	+	+ (road plantings)		
Naturalistic plantings			+				
Prairie gardens				+			
Swale and rain gardens plantings	+	+	+	+	+	+	+
Xeriscape gardens/rock gardens					+	+	+
Verge gardens and woody meadows						+	
Use of appropriate native groundcovers in private gardens	+	+	+	+	+	+	+

Notes: [a] USA—temperate climate states, [b] USA—arid climate states.

In Europe, all ideas about alternative lawns are connected to native grasslands, pasture land, or the open margins of temperate forests (which support some grasses and low growing vegetation). In Sweden and Germany, for example, there are several nurseries that specialise in producing multispecies native meadow mats made up of 70–80% grass and 20–30% native herbaceous wildflowers [22]. English landscape architects such as James Hitchmough and Nigel Dunnett introduced naturalistic plantings that combine native herbaceous and grass species with attractive non-native, flowering prairie plants from North America. These aimed at increasing biodiversity and facilitating low level of management [91].

Pictorial meadows created from flowering annual plants are increasingly popular all over Europe and the US. Often used for road (highway) plantings, such meadows are inspired by the margins of agricultural fields or by natural blossoms in dry Mediterranean ecosystems or semi-desert areas. The use of prairie plants in private gardens and public parks in the suburban Midwest of the USA (e.g., Millennium Park in Chicago) is gaining popularity in the face of increasing temperatures and lengthening drought periods (Figure 8). Similar factors are the driving force behind xeriscape gardening in California, Arizona, New Mexico, Texas, Colorado and Florida as well as in Australian cities [87,125,126]. Xeriscaping is a process of landscaping that reduces/eliminates/minimises the need for supplemental water from irrigation. Local plant species are being promoted as the most tolerant to such harsh conditions [1,35,42,46,89,90].

The German-born "Go spontaneous" approach is connected to the essence of German urban ecology and particularly with the Berlin School. It has been developed in special abundant wasteland sites and regenerating vegetation. Studies into the capacity of urban nature to regenerate to a certain successional stage, to exhibit certain plant strategies and to provide ecosystem services, have showed success and have also fostered acceptance of the "go wild" approach to designing public spaces (Park am Gleisdreieck). Industrial habitats were "reinforced" by seeding "weeds" or leaving nature "alone" [72] (Figure 12).

(a)   (b)

Figure 12. Spontaneous lawn in Gleisdreieck Park, Berlin (**a**), and the latest inspiration in the US: Lurie Garden with prairie plants in Millennium Park, Chicago (**b**). Photos: M. Ignatieva.

Among the tested alternatives to lawns, grass-free (tapestry) surfaces are closest to the idea of conventional turf where roots and stolons produce strong sod that can tolerate human traffic pressure. A few low-growing European native herbaceous plants (*Potentilla, Prunella, Veronica, Trifoluim, Lotus, Hieracium and Polygonum aveculare*), that are already present in actively visited lawns, may be used for a new generation of grass-free lawns where planting is based on a mixture low growing ground covers and forbs (Figure 13). However, such lawns require experimental trials and further research into their resistance to human traffic. Even in non-European lawns, for example in China, there are several native herbaceous species in conventional lawns that can be considered as potential candidates for creating sustainable future alternatives [4].

(a)   (b)   (c)

Figure 13. Nature-based solutions for urban lawns: (**a**) Grass-free lawn in the Ultuna campus of Swedish University of Agricultural Sciences in Uppsala (Sweden); (**b**) Steppe Garden in Zaryadye Park in Moscow, Russia; (**c**) native plants for traffic islands in Wellington (New Zealand). Photos: M. Ignatieva, Dushkova.

In Australia, one particular alternative to lawns, the verge garden, uses native plants in the strip of council land between the street and the footpath in suburban areas. Recently, Perth City Council in Western Australia has encouraged people to transform their verge into native low maintenance gardens using a waterwise approach and planting low-growing native plants instead of lawns. Ongoing interdisciplinary research in Perth is studying the social motives of suburban homeowners who are willing to transform their front verges into native gardens [88].

Another approach, introduced only a few years ago and inspired by the pictorial and naturalistic meadow movement in the UK, is "woody meadows". The idea behind the "woody meadow" is to plant low-growing native plants (herbaceous plants and lower shrubs) within urban "designed" landscapes. "Modelled on natural heathland plant communities across southern Australia, the aim of the project is to create visually interesting landscapes that require little ongoing maintenance, such as irrigation and labour, to sustain them" [127]. This project is a research collaboration between the

University of Sheffield in England, the University of Melbourne and the City of Melbourne. Now the "woody meadow" will be implemented in Perth, using unique, Western Australian plants. Directly influenced by alternative thinking and designing of lawns in England, the idea is to create "a beautiful, meadow-like appearance", similar to what has been done in Europe.

6. Discussion

There is a new landscape architecture approach referred to as "biodiversinesque", which promotes a special design style for sustainable landscape design [72]. It is based on multiscale design with particular emphasis on the mesoscale, or the neighbourhood or park level, and is a detailed design where biodiversity and dynamic ecological process-succession can be implemented and monitored. One fundamental difference of this new vision from other approaches is the appreciation of the complexity of biodiversity instead of the narrower native vs. exotic plants debate. This new design language incorporates the dynamic character of urban biotopes and is believed to make a difference that will be understood and appreciated by people.

When promoting a new generation of nature-based lawns, such novel alternatives to lawns should be vastly different from conventional lawns in terms of being more cost-effective, biodiverse, trample-resistant, and stable under extreme weather conditions. At the same time, they should remain connected to the social needs of their users such as certain lawn qualities, amenities of the green space and different recreational activities. Such novel lawns should serve as valuable and resilient parts of urban green infrastructure in growing cities. Each novel nature-based solution should, on the one hand, be based on natural succession processes that occur within lawn plant communities and local indigenous plant communities and on the other hand from surrounding "designed" ecosystems [22]. The idea is to explore nature's dynamic processes and use this knowledge to address specific problems such as lawn management.

These complex approaches require careful study of existing conventional lawns, their structure, composition, soil quality and hydrological capacity, as well as an establishment and management regime (irrigation, fertilisation, pesticide application, aeration, etc.). One essential component of new research into lawn alternatives should be questionnaire surveys and a qualitative analysis of people's attachments to lawns (through interviews and focus groups). The methodology of researching lawns can be quite universal with some interpretation of the local environmental and social parameters.

However, alternative nature-based solutions should be strictly city- and country-specific. For example, with European urban ecosystems, there is much more opportunity to develop drought and trample resistant plants. Some local plants are already established in intensively used parks as a result of native succession. Many of the "old" lawns in Europe which have existed for several decades and where natural successional stages have occurred could be researched and mimicked in experimental sites.

In the southern hemisphere, especially in Australia, the attachment of urban dwellers to lawns cannot be ignored. There are two "natures" that are defined by a sharp boundary and a lack of existing research of "designed and managed" nature. In order to solve the existing lawn problem and improve ecosystem services, different research directions should be taken. Australian flora is one of the most unique in the world. For example, City of Perth is located in one of the world's 35 internationally recognised biodiversity hotspots. We can search for native plants that could be experimentally tested for their capability to withstand high recreational pressure. In the meantime, sustainable lawn management and careful water sensitive design of lawns should be a priority. Education of urban dwellers on how to water private and public gardens and how to select drought-tolerant turf grasses can be an intermediate measure. One important goal of future urban "designed" landscapes should be changing the "green" lawn psychology to a greater appreciation of the Australian native plant colours, i.e., the olive greens, browns, and yellows, and to view these as signs of healthy sustainable urban environments that can adapt better to a changing climate.

Globally, the role of urban planners, geographers and landscape designers is more important than ever before. They need to solve the question of how to move from the current strategy of the "lawnscape" as an element of urban open spaces to more balanced, ecologically-based urban planning and design. One of the most important factors in creating this new generation of sustainable lawns is raising the awareness of the public about the possibilities of different lawn typologies and the necessity to see a place for "wild nature" in lawns. Another global challenge will be instigating interdisciplinary research projects into lawn alternatives and providing practical outcomes.

Our vision of future sustainable lawns is based on a complex hybrid approach (Table 4). Such lawns would retain their essence—their durable surface (the equivalent of turf) but be created by plants (grasses, herbaceous species and/or ground covers) that can withstand recreational pressure. At the same time, alternatives to lawns should also rely on a whole range of sustainable planning, design, and management strategies. Most likely lawn as a phenomenon will have a long life in future urban ecosystems. This is a time for creating a new conceptual framework for researching lawns.

Table 4. Redesigning lawns, a complex approach towards sustainable lawns.

Urban Planning	Landscape Design	Ecological Design	Maintenance Approach
Reduce conventional lawns by sustainable planning of green areas and green infrastructure and new design styles	Rethink spatial composition (avoid the homogenous mono-species approach), choice of appropriate, site-related plants	Mimic spatial structure and composition of existing resistant biodiverse lawns and surrounding native ecosystems that can be used as inspiration	Self-sustaining system, locally driven (climate, culture and economic appropriateness) cutting the regime approach by reinforcing local biodiversity. Sustainable management (appropriate soil preparation, appropriate mowing regime, use of electric or robotic mowers and smart irrigation schemes)

7. Conclusions

To fill the gaps in our understanding of lawn as a global phenomenon, we propose a framework for future interdisciplinary study and nature-based solutions for lawns, which will be based on data from cities in different climatic zones and social, cultural and geographical conditions (Figure 14).

Lawn should be studied as specifically designed urban habitats/biotopes as well as in cultural and social terms, i.e., complex analysis of ecosystem service flows, related social norms and expectations, and uses and behaviours. This should be achieved by applying the most recent methods and theoretical concepts such as resilience, sustainability [128], biocultural diversity [129], nature-based solutions [20,21,124], the "biodiversinesque" landscape architectural style [72] and ecological landscaping [92]. There is a paucity of research on possible nature-based solutions to lawns and the existing urban plant communities—man-made but influenced by natural and anthropogenic factors—and their successional stages. So far, there have been no attempts to understand the functioning mechanisms of urban lawns as a unique dynamic ecological system and to model sustainable landscape design solutions for different types of lawns. We suggest a novel vision of lawn alternatives a local community-driven approach based on existing native and urban plant ecosystems.

The overall aim of future research on lawns should be to investigate, analyse and understand the phenomenon of lawns from different environmental, socio-cultural and design perspectives, as well to empirically explore, suggest and test different locally adapted nature-based solutions. Studies should identify biodiversity characteristics (lawn composition and structure) as well as research plant communities similar to lawns. The aim is to find alternatives, which have high biodiversity and ecosystem service flows, and that are trample resistant, socially acceptable and offer improved overall resilience to climate change and its effects on urban green infrastructure.

Figure 14. Proposed interdisciplinary research framework for lawns as a complex phenomenon (Source: the authors).

Author Contributions: Conceptualization, M.I., D.D. and D.H.; methodology, M.I., D.D. and D.H.; validation, D.H. and D.D.; formal analysis M.I. and D.D.; investigation, M.I., D.D., D.H.; resources M.I., D.D.; data curation and interpretation of results, M.I., D.D.; writing—original draft preparation, M.I., D.D. and D.H.; writing—review and editing, M.I., D.D., D.H. and A.H.; visualization, D.D.; supervision, D.H.; project administration, M.I.; funding acquisition, M.I. and D.H. The concept of "two natures" and the conceptual vision of lawn alternatives was suggested by M.I. All authors have read and agree to the published version of the manuscript.

Funding: This research was funded by several sources: M.I. research—by the University of Western Australia (UWA) FABLE research grant funded from 2018 to 2019; by Swedish Research Foundation FORMAS (225-2012-1369) "Lawns as a cultural and ecological phenomenon" (2013–2017) and by Swedish University of Agricultural Sciences SLU Climate Project: Towards sustainable lawns: searching for alternative cost effective and climate friendly lawns in Ultuna Campus (2016–2017); D.H. research was supported as part of the project ENABLE, funded through the 2015–2016 BiodivERsA COFUND call for research proposals, with the national funders The Swedish Research Council for Environment, Agricultural Sciences, and Spatial Planning, Swedish Environmental Protection Agency, German Aeronautics and Space Research Centre, National Science Centre (Poland), The Research Council of Norway and the Spanish Ministry of Economy and Competitiveness. In addition, Dagmar benefited from the GreenCityLabHue Project (FKZ 01LE1910A) and the CLEARING HOUSE (Collaborative Learning in Research, Information-sharing and Governance on How Urban forest-based solutions support Sino-European urban futures) Horizon 2020 project (No 1290/2013). D.D. and D.H. research was founded by the Horizon 2020 Framework Programme of the European Union, research and innovation project "CONNECTING Nature—COproductioN with NaturE for City Transitioning, Innovation and Governance", Grant Agreement No 730222.

Acknowledgments: We thank participants of the international workshop "Urban Biodiversity and Nature-Based Design: Methodology and Practical Applications for Interdisciplinary Research" orginised by the Geography Department of Humboldt university Berlin on the 28–29 November 2019 in Berlin for a lot of inspiration for the future transdisciplinary research on lawns. We thank Daniel Martin (UWA) for helping with obtaining numbers on water usage in Perth. The authors also extend great gratitude to the anonymous reviewers and editors for their helpful reviews and critical comments.

References

1. Ignatieva, M.; Hedblom, M. An alternative urban green carpet: How can we move to sustainable lawns in a time of climate change? *Science* **2018**, *362*, 148–149. [CrossRef] [PubMed]
2. Haase, D.; Nuissl, H. Does urban sprawl drive changes in the water balance and policy? The case of Leipzig (Germany) 1870–2003. *Landsc. Urban Plan.* **2007**, *80*, 1–13. [CrossRef]

3.   Hedblom, M.; Lindberg, F.; Vogel, E.; Wissman, J.; Ahrné, K. Estimating urban lawn cover in space and time: Case studies in three Swedish cities. *Urban Ecosyst.* **2017**, *20*. [CrossRef]

4.   Yang, F.; Ignatieva, M.; Larsson, A.; Zhang, S.; Ni, N. Public perceptions and preferences regarding lawns and their laternatives in China: A case study of Xi'an. *Urban For. Urban Green.* **2019**, *46*, 126478. [CrossRef]

5.   Thompson, G.L.; Kao-Kniffin, J. Applying Biodiversity and Ecosystem Function Theory to Turfgrass Management. *Crop Sci.* **2017**, *57*, 238–248. [CrossRef]

6.   Cumming, J. Environmental Assessment of the Australian Turf Industry. Hort Innovation Project No. TU16000 Benchmarking Report. 2018. Available online: https://www.horticulture.com.au/globalassets/hort-innovation/resource-assets/tu16000-benchmarking-report.pdf (accessed on 1 November 2019).

7.   Haase, D.; Haase, A.; Rink, D. Conceptualising the nexus between urban shrinkage and ecosystem services. *Landsc. Urban Plan.* **2014**, *132*, 159–169. [CrossRef]

8.   Haase, D.; Larondelle, N.; Andersson, E.; Artmann, M. A quantitative review of urban ecosystem services assessment: Concepts, models and implementation. *AMBIO* **2014**, *43*, 413–433. [CrossRef]

9.   Haase, A.; Wolff, M.; Rink, D. From shrinkage to regrowth. The nexus between urban dynamics, land use change and ecosystem service provision. In *Urban Transformations—Sustainable Urban Development towards Resource Efficiency, Quality of Life and Resilience*; Future City Series; Kabisch, S., Ed.; Springer: Berlin, Germany, 2018; pp. 197–219.

10.   Haase, D.; Haase, A.; Rink, D.; Quanz, J. Shrinking Cities and Ecosystem Services: Opportunities, Planning, Challenges, and Risks. In *Atlas of Ecosystem Services*; Schröter, M., Ed.; Springer: Cham, Switzerland, 2019; pp. 271–277. [CrossRef]

11.   Groffman, P.M.; Cavender-Bares, J.; Bettez, N.D. Ecological homogenization of urban USA. *Front. Ecol. Environ.* **2014**, *12*, 74–81. [CrossRef]

12.   McKinney, M.L. Urbanization as a major cause of biotic homogenization. *Biol. Conserv.* **2006**, *11*, 247–260. [CrossRef]

13.   Wheeler, M.M.; Neil, C.; Groffman, P.M. Continental-scale homogenization of residential lawn plant communities. *Landsc. Urb. Plan.* **2017**, *165*, 54–63. [CrossRef]

14.   NOAA—The US National Oceanic and Atmospheric Administration. *July 2019 Was Hottest Month on Record for the Planet*; NOAA: Silver Spring, MD, USA, 2019; Available online: https://www.clickondetroit.com/weather-center/2019/08/15/noaa-july-2019-was-hottest-month-on-record-for-the-planet/ (accessed on 11 November 2019).

15.   Goosse, H.; Kay, J.E.; Armour, K.C.; Bodas-Salcedo, A. Quantifying climate feedbacks in polar regions. *Nat. Commun.* **2018**, *9*, 1919. [CrossRef] [PubMed]

16.   Hogue, T.; Pinceti, S. Are you watering your lawn? High-resolution data may help to devise effective water conservation strategies in urban areas around the world. *Science* **2015**, *348*, 6241.

17.   Borrelle, S.B.; Rochman, C.M.; Liboiron, M.; Bond, A.L.; Lusher, A.; Bradshaw, H.; Provencher, J.F. Why we need an international agreement on marine plastic pollution. *Proc. Natl. Acad. Sci. USA* **2017**, *114*, 9994–9997. [CrossRef] [PubMed]

18.   Fleming, P.R.; Forrester, S.E.; McLaren, N.J. Understanding the effects of decompaction maintenance on the infill state and play performance of third-generation artificial grass pitches. *Proc. Inst. Mech. Eng. Part P J. Sports Eng. Technol.* **2015**, *229*, 169–182. [CrossRef] [PubMed]

19.   Kaminski, I. Turf It out: Is It Time to Say Goodbye to Artificial Grass? *The Guardian*. 2 August 2019. Available online: https://www.theguardian.com/cities/2019/aug/02/turf-it-out-is-it-time-to-say-goodbye-to-artificial-grass (accessed on 1 November 2019).

20.   EC—European Commission. *Towards an EU Research and Innovation Policy Agenda for Nature-Based Solutions & Re-Naturing Cities. Final Report of the Horizon2020 Expert Group on Nature-Based Solutions and Re-Naturing Cities*; European Commission: Brussels, Belgium, 2015.

21.   Raymond, C.M.; Berry, P.; Breil, M.; Nita, M.R.; Kabisch, N. *An Impact Evaluation Framework to Support Planning and Evaluation of Nature-Based Solutions Projects. Report Prepared by the EKLIPSE Expert Working Group on Nature-Based Solutions to Promote Climate Resilience in Urban Areas*; Centre for Ecology & Hydrology: Wallington, UK, 2017.

22.   Ignatieva, M. *Lawn Alternatives in Sweden: From Theory to Practice*; Manual; Swedish University of Agricultural Sciences: Uppsala, Sweden, 2017.

23. Ignatieva, M.; Meurk, C.; van Roon, M.; Simcock, R.; Stewart, G. *Urban Greening Manual. How to Put Nature in Our Neighbourhoods: Application of Low Impact Urban Design and Development (LIUDD) Principles, with a Biodiversity Focus, for New Zealand Developers and Homeowners*; Landcare Research Sciences Series No.35; Manaaki Whenua Press: Lincoln, NZ, USA, 2008.

24. Milesi, C.; Running, S.W.; Elvidge, C.D.; Dietz, J.B.; Tuttle, B.T.; Nemani, R.R. A strategy for Mapping and Modeling of biochemical turfgrasses in the United States. *Environ. Manag.* **2005**, *36*, 426–438. [CrossRef]

25. Robbins, P.; Birkenholtz, T. Turfgrass revolution: Measuring the expansion of the American lawn. *Land Use Policy* **2003**, *20*, 181–194. [CrossRef]

26. Gaston, K.J.; Warren, P.H.; Thompson, K. Urban Domestic Gardens (IV): The Extent of the Resource and its Associated Features. *Biodivers. Conserv.* **2005**, *14*, 3327. [CrossRef]

27. Hahs, A.; McDonnell, M. Composition of the plant community in remnant patches of grassy woodland along an urban-rural gradient in Melbourne, Australia. *Urban Ecosyst.* **2007**, *10*, 355–377. [CrossRef]

28. Lindenmayer, D.; Fisher, J.; Felton, A.; Crane, M.; Michael, D.; Macgregor, C.; Montague-Drake, R.; Manning, A.; Hobbs, R. Novel ecosystem resulting from landscape transformation create dilemmas for modern conservation practice. *Conserv. Lett.* **2008**, *1*, 129–135. [CrossRef]

29. Müller, N. Lawns in German cities. A phytosociological comparison. In *Urban Ecology*; Sukopp, H., Hejnỳ, S., Kowarik, I., Eds.; SPB Academic Publishing: The Hague, The Netherlands, 1990; pp. 209–222.

30. Stewart, G.H.; Meurk, C.D.; Ignatieva, M.E.; Buckley, H.L.; Magueur, A.; Case, B.S.; Hudson, M.; Parker, M. Urban Biotopes of Aotearoa New Zealand (URBANZ) II: Floristics, biodiversity and conservation values of urban residential and public woodlands, Christchurch. *Urban For. Urban Green.* **2009**, *8*, 149–162. [CrossRef]

31. Sukopp, H.; Kowarik, I. *Urban Ecology*; SPB Academic Publishing: The Hague, The Netherlands, 1990.

32. Threlfall, C.; Walker, K.; Williams, N.; Hahs, A.; Mata, L.; Stork, N.; Livesley, S.J. The conservation value of urban green space habitats for Australian native bee communities. *Biol. Conserv.* **2015**, *187*, 240–248. [CrossRef]

33. Amani-Beni, M.; Zhang, B.; Xie, G.; Xu, J. Impact of urban park's tree, grass and waterbody on microclimate in hot summer days: A case study of Olympic Park in Beijing, China. *Urban For. Urban Green.* **2018**, *32*, 1–6. [CrossRef]

34. Armson, D.; Stringer, P.; Ennos, A.R. The effect of street trees and amenity grass on urban surface water runoff in Manchester, UK. *Urban For. Urban Green.* **2013**, *12*, 282–286. [CrossRef]

35. Beard, J.B.; Green, R.L. The role of turfgrasses in environmental protection and their benefits to humans. *J. Environ. Qual.* **1994**, *23*, 452–460. [CrossRef]

36. Brunton, E.; Srivastava, S.; Burnett, S. Spatial ecology of an urban eastern grey kangaroo (Macropus giganteus) population: Local decline driven by kangaroo-vehicle collisions. *Wildl. Res.* **2018**, *45*. [CrossRef]

37. Burgin, S. What about biodiversity? Redefining urban sustainable management to incorporate endemic fauna with particular reference to Australia. *Urban Ecosyst.* **2016**, *19*, 669–678. [CrossRef]

38. Fischer, L.; Lippe, M.; Kowarik, I. Urban land use types contribute to grassland conservation: The example of Berlin. *Urban For. Urban Green.* **2013**, *12*, 263–272. [CrossRef]

39. Fischer, L.K.; Eichfeld, J.; Kowarik, I.; Buchholz, S. Disentangling urban habitat and matrix effects on wild bee species. *PeerJ* **2016**, *4*, e2729. [CrossRef]

40. Johnson, P.G. Priorities for Turfgrass management and education to enhance urban sustainability worldwide. *J. Dev. Sustain. Agric.* **2013**, *8*, 63–71. [CrossRef]

41. Lele, S.; Springate-Baginski, O.; Lakerveld, R.; Deb, D.; Dash, P. Ecosystem Services: Origins, Contributions, Pitfalls, and Alternatives. *Conserv. Soc.* **2013**, *11*, 343–358. [CrossRef]

42. Monteiro, J.A. Ecosystem services from turfgrass landscapes. *Urban For. Urban Green.* **2017**, *26*, 151–157. [CrossRef]

43. Stirling, G.; Stirling, M.; Giblin-Davis, R.M. Distribution of southern sting nematode, Ibipora lolii (Nematoda: Belonolaimidae), on turfgrass in Australia and its taxonomic relationship to other belonolaimids. *Nematology* **2013**, *15*, 401–415. [CrossRef]

44. Trigger, D.; Mulcock, J. Native versus exotic: Cultural discourses about flora, fauna and belonging in Australia. In *Sustainable Planning and Development: The Sustainable World*; Kungolos, A., Brebbia, C., Beriatos, E., Eds.; Wessex Institute of Technology Press: Southampton, UK, 2005; Volume 6, pp. 1301–1310.

45. Wang, Z.-H.; Zhao, X.; Yang, J.; Song, J. Cooling and energy saving potentials of shade trees and urban lawns in a desert city. *Appl. Energy* **2016**, *161*, 437–444. [CrossRef]

46. Wastian, L.; Unterweger, P.A.; Betz, O. Influence of the reduction of urban lawn mowing on wild bee diversity (Hymenoptera, Apoidea). *J. Hymenopt. Res.* **2016**, *49*, 51–63.

47. Campagne, C.S.; Roche, P.K.; Salles, J.-M. Looking into Pandora's Box: Ecosystem disservices assessment and correlations with ecosystem services. *Ecosyst. Serv.* **2018**, *30*, 126–136. [CrossRef]

48. Döhren, P.; Haase, D. Ecosystem disservices research: A review of the state of the art with a focus on cities. *Ecol. Indic.* **2015**, *52*, 490–497. [CrossRef]

49. Dunn, R.R. Global Mapping of ecosystem disservices: The unspoken reality that nature sometimes kills us. *Biotropica* **2010**, *42*, 555–557. [CrossRef]

50. Lyytimäki, J. Ecosystem disservices: Embrace the catchword. *Ecosyst. Serv.* **2015**, *12*, 136. [CrossRef]

51. Müller, N.; Sukopp, H. Influence of different landscape design styles on plant invasions in Central Europe. *Landsc. Ecol. Eng.* **2016**, *12*, 151. [CrossRef]

52. Priest, M.W.; Williams, D.J.; Bridgman, H.A. Emission from in-use lawn-mowers in Australia. *Atmos. Environ.* **2000**, *34*, 657–664. [CrossRef]

53. Runfola, D.; Polsky, C.; Nicolson, C.; Giner, N.; Pontius, R.; Krahe, J.; Decatur, A. A growing concern? Examining the influence of lawn size on residential water use in suburban Boston, MA, USA. *Landsc. Urban Plan.* **2013**, *119*, 113–123. [CrossRef]

54. Shackleton, C.M.; Ruwanza, S.; Sinasson Sanni, G.K.; Bennett, S.; De Lacy, P.; Modipa, R.; Mtati, N.; Sachikonye, M.; Thondhlana, G. Unpacking Pandora's Box: Understanding and Categorising Ecosystem Disservices for Environmental Management and Human Wellbeing. *Ecosystems* **2016**, *19*, 587–600. [CrossRef]

55. Schapel, A.; Reseigh, J.; Wurst, M.; Mallants, D.; Herrmann, T. *Offsetting Greenhouse Gas Emissions through Increasing Soil Organic Carbon in SA Clay-Modified Soils: Knowledge Gap Analysis*; Technical Report Series No. 18/05; Goyder Institute for Water Research: Adelaide, SA, Australia, 2018.

56. Sharma, M.L.; Herne, D.E.; Byrne, J.D.; Kin, P.G. Nutrient Discharge beneath urban lawns to a sandy coastal aquifer, Perth, Western Australia. *Hydrogeol. J.* **1996**, *4*, 103–117. [CrossRef]

57. Elgizawy, E.M. Expectations towards green lawns to enhance quality of life and workplaces. *Procedia Environ. Sci.* **2016**, *34*, 131–139. [CrossRef]

58. Han, X.; Burton, O.R.; Sternudd, C.; Li, D. Public Attitudes about Urban Lawns: Social Opportunities Provided by Urban Lawns in Lund, Sweden. In Proceedings of the 2013 International Academic Workshop on Social Science, Changsha, China, 18–20 October 2013; Shao, X., Ed.; Atlantis Press: Paris, France, 2013; pp. 1046–1054. [CrossRef]

59. Jenkins, V.S. *The Lawn: A History of an American Obsession*; Smithsonian Institution: Washington, DC, USA, 1994.

60. Pisa, J. The lawn grew too quickly! Perception of rural idyll by Czech amenity migrants. *Geoscape* **2019**, *13*, 55–67. [CrossRef]

61. Poškus, M.; Poškienė, D. The grass is greener: How greenery impacts the perceptions of urban residential property. *Soc. Inq. Into Well-Being* **2015**, *1*, 22–31. [CrossRef]

62. Rall, E.; Bieling, C.; Zytynska, S.; Haase, D. Exploring city-wide patterns of cultural ecosystem service perceptions and use. *Ecol. Indic.* **2017**, *77*, 80–95. [CrossRef]

63. Ramer, H.; Nelson, K.C.; Spivak, M. Exploring park visitor perceptions of "flowering bee lawns" in neighborhood parks in Menapolis, MN, US. *Landsc. Urban Plan.* **2019**, *189*, 117–128. [CrossRef]

64. Robbins, P. *Lawn People: How Grasses, Weeds, and Chemicals Make Us Who We Are*; Temple University Press: Philadelphia, PA, USA, 2007.

65. Sewel, S.; McCallister, D.; Gaussoin, R.; Wortmann, C. Lawn management practices and perceptions of residents in 14 Sandpit Lakes of Nebraska. *J. Ext.* **2010**, *48*, 2Rib4.

66. Teysott, G. *The American Lawn*, 1st ed.; Princeton Architectural Press: New York, NY, USA, 1999.

67. Yang, F.; Ignatieva, M.; Larsson, A.; Xiu, N.; Zhang, S. Historical Development and Practices of Lawns in China. *Environ. Hist.* **2019**, *25*, 23–54. [CrossRef]

68. Antrop, M. Landscape change and the urbanization process in Europe. *Landsc. Urban Plan.* **2004**, *7*, 9–26. [CrossRef]

69. Pondichie, F.A. Globalization Elements in Romanian Cities: Searching for Sustainable Solutions. Master's Thesis, Uppsala University, Uppsala, Sweden, 2012.

70. Payne, S.K.; Bruce, N. Exploring the relationship between urban quiet areas and perceived restorative benefits. *Int. J. Environ. Res. Public Health* **2019**, *16*, 1611. [CrossRef] [PubMed]

71. Stolz, J.; Schaffer, C. Salutogenic affordances and sustainability: Multiple benefits with edible forest gardens in urban green spaces. *Front. Psychol.* **2018**, *9*, 2344. [CrossRef] [PubMed]

72. Ignatieva, M. Biodiversity-friendly designs in cities and towns: Towards a global biodiversinesque style. In *Urban Biodiversity: From Research to Practice*; Ossola, A., Niemelä, J., Eds.; Routledge (Routledge Studies in Urban Ecology): Abingdon, UK, 2018; pp. 216–235.

73. Gaynor, A. Lawnscaping Perth: Water Supply, Gardens, and Scarcity, 1890–1925. *J. Urban Hist.* **2017**. [CrossRef]

74. Hipple, W.J., Jr. *The Beautiful, the Sublime and the Picturesque in Eighteenth-Century British Aesthetics Theory*; Southern Illinois University Press: Carbondale, IL, USA, 1957.

75. Robinson, S.K. *Inquiry into the Picturesque*; The University of Chicago Press: Chicago, IL, USA, 1991.

76. Greenbaum, A. Lawn as a Site of Environmental Conflict. Ph.D. Thesis, Department of Sociology, York University, Toronto, ON, Canada, 2000.

77. Harari, Y.N. *Homo Deus: A Brief History of Tomorrow*; Random House: New York, NY, USA, 2016.

78. Trudgill, S.; Jeffery, A.; Parker, J. Climate Change and the Resilience of the Domestic Lawn. *Appl. Geogr.* **2010**, *30*, 177–190. [CrossRef]

79. Alumai, A. Urban Lawn Management: Addressing the Entomological, Agronomic, Economic, and Social Drivers. Ph.D. Thesis, Ohio State University, Columbus, OH, USA, 2008.

80. Chawla, S.L.; Agnihotri, R.; Patel, M. Turfgrass: A Billion Dollar Industry. In Proceedings of the National Conference on Floriculture for Rural and Urban Prosperity in the Scenario of Climate Change, Gangtok, India, 16–18 February 2018; pp. 30–35.

81. Ignatieva, M. Design and future of urban biodiversity. In *Urban Biodiversity and Design*; Müller, N., Werner, P., Kelcey, J., Eds.; Wiley-Blackwell: Oxford, UK, 2010; pp. 118–144.

82. Ignatieva, M.; Ahrné, K. Biodiverse green infrastructure for the 21st century: From "green desert" of lawns to biophilic cities. *J. Archit. Urban.* **2013**, *37*, 1–9. [CrossRef]

83. Pineo, R.; Barton, S. Turf Grass Madness: Reasons to Reduce the Lawn in Your Landscape. *Sustain. Landsc. Ser.* **2010**, *130*. Available online: https://cdn.canr.udel.edu/wp-content/uploads/sites/16/2018/03/12024155/Turf_Grass_Madness.pdf (accessed on 18 October 2019).

84. Steinberg, T. *American Green: The Obsessive Quest for the Perfect Lawn*; W.W. Norton & Co: New York, NY, USA, 2006.

85. Wasowski, A.; Wasowski, S. *The Landscaping Revolution: Garden with Mother Nature, Not against Her*; McGraw-Hill/The Contemporary Gardener: New York, NY, USA, 2002.

86. Wasowski, S.; Wasowski, A. *Requiem for a Lawnmower: Gardening in a Warmer, Drier World*; Taylor Trade Publications: Lanham, MD, USA, 2004.

87. Wilson, C.; Feuch, J.R. *Xeriscaping: Creative Landscaping*; Colorado State University Extension: Fort Collins, CO, USA, 2018.

88. Zollner, M. Leading the way for urban green spaces: The CAUL Hub in Western Australia. *Urban Beat* **2018**, *8*, 3–4.

89. Bormann, H.F.; Balmori, D.; Geballe, G.T. *Redesigning the American Lawn: A Search for Environmental Harmony*, 2nd ed.; Yale University Press: New Haven, CT, USA, 2001.

90. Daniels, S. *The Wildlawn Handbook: Alternatives to the Traditional Front Lawn*; Macmillan Publishing Company: New York, NY, USA, 1995.

91. Hitchmough, J. Naturalistic herbaceous vegetation for urban landscapes. In *The Dynamic Landscape*; Dunnett, N., Hitchmough, J., Eds.; Taylor & Francis: London, UK, 2004; pp. 130–183.

92. Sprajcar, J. *More than Just a Yard. Ecological Landscaping Tools for Massachusetts Homeowners*; Massachusetts Executive Office of Energy and Environmental Affairs: Boston, MA, USA, 2017.

93. Brooks, A.; Francis, R. Artificial lawn people. *Environ. Plan. E Nat. Space* **2019**, *2*, 548–564. [CrossRef]

94. Loveday, J.; Loveday, G.; Byrne, J. Seasonal and Diurnal Surface Temperatures of Urban Landscape Elements. *Sustainability* **2019**, *11*, 5280. [CrossRef]

95. Higgs, E. Novel and designed ecosystems. *Restor. Ecol.* **2016**, *25*, 8–13. [CrossRef]

96. Hobbs, R.J.; Arico, S.; Aronson, J.; Baron, J.S.; Bridgewater, P. Novel ecosystems: Theoretical and management aspects of the new ecological world order. *Glob. Ecol. Biogeogr.* **2006**, *15*, 1–7. [CrossRef]

97. Kowarik, I. Novel urban ecosystems, biodiversity, and conservation. *Environ. Pollut.* **2011**, *159*, 1974–1983. [CrossRef] [PubMed]

98. Norton, B.; Couts, A.; Livesley, S.; Harris, R.; Hunter, A.; Williams, N. Planning for cooler cities: A framework to prioritise green infrastructure to mitigate high temperatures in urban landscapes. *Landsc. Urban Plan.* **2015**, *134*, 127–138. [CrossRef]

99. Yang, F.; Ignatieva, M.; Wissman, J.; Ahrné, K.; Zhang, S.; Zhu, S. Relationships between multi-scale factors, plant and pollinator diversity, and composition of park lawns and other herbaceous vegetation in a fast growing megacity of China. *Landsc. Urban Plan.* **2019**, *185*, 117–126. [CrossRef]

100. Wilson, J. *Contextual Historical Overview for Christchurch City*; Christchurch City Council: Christchurch, New Zealand, 2005.

101. Aitken, R.; Looker, M. *The Oxford Companion to Australian Gardens*; Oxford University Press in Association with the Australian Garden History Society: South Melbourne, VIC, Australia; Oxford, UK, 2002.

102. Ignatieva, M.; Eriksson, F.; Eriksson, T.; Berg, P.; Hedblom, M. The lawn as a social and cultural phenomenon in Sweden. *Urban For. Urban Green.* **2017**, *21*, 213–223. [CrossRef]

103. Daniels, B.; Zaunbrecher, B.; Paas, B. Assessment of urban green space structures and their quality from a multidimensional perspective. *Sci. Total Environ.* **2018**, *615*, 1364–1378. [CrossRef]

104. Kabisch, N.; Qureshi, S.; Haase, D. Human-environment interactions in urban green spaces—A systematic review of contemporary issues and prospects for future research. *Environ. Impact Assess. Rev.* **2015**, *50*, 25–34. [CrossRef]

105. Kabisch, N.; Haase, D. Green justice or just green? Provision of urban green spaces in Berlin, Germany. *Landsc. Urban Plan.* **2014**, *122*, 129–139. [CrossRef]

106. Rink, D.; Rumpel, P.; Slach, O.; Cortese, C.; Violante, A.; Bini, P.; Haase, A.; Mykhnenko, V.; Nadolu, B.; Couch, C.; et al. Governance of Shrinkage—Lessons Learnt from Analysis for Urban Planning and Policy. EU 7 FP Project Shrink Smart (Contract No. 225193), WP7. 2012. Available online: https://www.ufz.de/export/data/400/39029_WP7_D13_14_15_FINAL_2.pdf (accessed on 16 October 2019).

107. Haase, D.; Dushkova, D.; Haase, A.; Kronenberg, J. Green infrastructure in postsocialist cities: Evidence and experiences from Russia, Poland and Eastern Germany. In *Post-Socialist Urban Infrastructures*; Tuvikene, T., Sgibnev, W., Neugebauer, C.S., Eds.; Routledge: Abingdon, UK, 2019; pp. 105–122.

108. Available online: https://www.resilientsouth.com/urban-heat (accessed on 18 October 2019).

109. Available online: https://www.watercorporation.com.au/-/media/files/residential/about-us/our-performance/annual-report-2018/water%20corporation%202018%20annual%20report.pdf (accessed on 16 December 2019).

110. Francis, R.A. Artificial lawns: Environmental and societal considerations of an ecological simulacrum. *Urban For. Urban Green.* **2018**, *30*, 152–156. [CrossRef]

111. Lawn & Garden Pesticides Facts & Figures. Beyond Pecticides. 2017. Available online: https://www.beyondpesticides.org/assets/media/documents/lawn/resources/bp-fact-lawnpesticides.081417.pdf (accessed on 14 November 2019).

112. WHO—World Health Organization. *International Agency for Research on Cancer (IARC) Monographs Volume 112: Evaluation of Five Organophosphate Insecticides and Herbicides*; IARC: Lyon, France, 2016.

113. Australia's Beloved Kangaroos Are Now Controversial Pests. National Geographic. Available online: https://www.nationalgeographic.com/magazine/2019/02/australia-kangaroo-beloved-symbol-becomes-pest/ (accessed on 10 November 2019).

114. Global Compendium of Weeds. The Top 12 Cited Invasive Weeds in the World. 2018. Available online: http://www.hear.org/gcw/species/cynodon_dactylon (accessed on 16 October 2019).

115. Meurk, C.D. Recombinant ecology of urban areas. In *The Routledge Handbook of Urban Ecology*; Douglas, I., Goode, D., Houck, M., Wang, R., Eds.; Routledge: Abingdon, UK, 2010.

116. McDonnell, M.J.; Hahs, A.K. The use of gradient analysis studies in advancing our understanding of the ecology of urbanizing landscapes: Current status and future directions. *Landsc. Ecol.* **2008**, *23*, 1143–1155. [CrossRef]

117. Zipperer, W.C. Factors Influencing Non-Native Tree Species Distribution in Urban Landscapes. In *Urban Biodiversity and Design, Conservation Sciences and Practice Series*; Müller, N., Werner, P., Kelcey, J.G., Eds.; Wiley-Blackwell Publishing: Hoboken, NJ, USA, 2010; pp. 243–251. [CrossRef]

118. Polsky, C.; Grove, J.M.; Knudson, C.; Groffman, P.M.; Bettez, N.; Cavender-Bares, J. Assessing the Homogenization of Urban Land Management with an Application to US Residential Lawncare. *Proc. Natl. Acad. Sci. USA* **2014**, *111*, 4432–4437. [CrossRef] [PubMed]

119. Chapin, F.S.; Starfield, A.M. Time lags and novel ecosystems in response to transient climatic change in arctic Alaska. *Clim. Chang.* **1997**, *35*, 449–461. [CrossRef]

120. Botzat, A.; Fischer, L.K.; Kowarik, I. Unexploited opportunities in understanding liveable and biodiverse cities. A review on urban biodiversity perception and valuation. *Glob. Environ. Chang.* **2016**, *39*, 220–233. [CrossRef]

121. Heger, T.; Gessler, A.; Bernard-Verdier, M.; Greenwood, A. Towards an Integrative, Eco-Evolutionary Understanding of Ecological Novelty: Studying and Communicating Interlinked Effects of Global Change. *BioScience* **2019**, *69*, 888–899. [CrossRef] [PubMed]

122. Ignatieva, M.; Meurk, C.; Stewart, G. Low Impact Urban Design and Development (LIUDD): Matching urban design and urban ecology. *Landsc. Rev.* **2008**, *12*, 61–73.

123. Kabisch, N.; Korn, H.; Stadler, J.; Bonn, A. *Nature-Based Solutions to Climate Change Adaptation in Urban Areas. Linkages between Science, Policy and Practice*; Springer: Cham, Switzerland, 2017. [CrossRef]

124. Nesshoever, C.; Assmuth, T.; Irvine, K.J.; Rusch, G.M.; Waylen, K.A.; Delbaere, B.; Haase, D. The science, policy and practice of Nature-Based Solutions: An interdisciplinary perspective. *Sci. Total Environ.* **2017**, *579*, 1215–1227. [CrossRef]

125. Mangiafico, S.S.; Obropta, C.C.; Rossi-Griffin, E. Demographic Factors Influence Environmental Values: A Lawn-Care Survey of Homeowners in New Jersey. *J. Ext.* **2012**, *50*, 1Rib6.

126. Xeriscaping with Texas Natives. Available online: https://www.gardensfortexas.com/xeriscaping-with-texas-natives (accessed on 1 November 2019).

127. A Woody Meadow in the Heart of the City. A Unique Research Project by the Universities of Melbourne and Sheffield Aims to Grow Urban Meadows That Are as Tough as They Are Beautiful by Claire Bolge. Available online: https://pursuit.unimelb.edu.au/articles/a-woody-meadow-in-the-heart-of-the-city (accessed on 15 October 2019).

128. Elmqvist, T.; Andersson, E.; Frantzeskaki, N. Sustainability and resilience for transformation in the urban century. *Nat. Sustain.* **2019**, *2*, 267–273. [CrossRef]

129. Elands, E.; Vierikko, K.; Andersson, E.; Fischer, L.K.; Goncalves, P.; Haase, D. Biocultural diversity: A novel concept to assess human-nature interrelations, nature conservation and stewardship in cities. *Urban For. Urban Green.* **2018**, *40*, 29–34. [CrossRef]

 *land*

MDPI

Article

Afforestation of Urban Brownfields as a Nature-Based Solution. Experiences from a Project in Leipzig (Germany)

Dieter Rink [1,*] **and Catrin Schmidt** [2]

1   Department of Urban and Environmental Sociology, Helmholtz Centre for Environmental Research-UFZ, 04318 Leipzig, Germany
2   Institute of Landscape Architecture, University of Dresden (TU Dresden), 01062 Dresden, Germany; catrin.schmidt@tu-dresden.de
*   Correspondence: dieter.rink@ufz.de

Abstract: In Leipzig, despite strong growth, reurbanization and re-densification, in the last decade it has still been possible for the city to green brownfields with a new type of green space: urban forests. The background to this was of course the city's decades of shrinkage and the emergence of numerous brownfields. The city of Leipzig started urban redevelopment in 2001 and pursued the strategy "more green, less density" in its planning. This included the creation of traditional and new green spaces as well as temporary uses. New green space concepts were also experimented with, including pocket forests and urban forests on larger, inner-city brownfields. This pursued several objectives: the forest was meant to contribute to improving the urban climatic and air-hygienic situation, to enhance the value of adjacent areas, create new recreational opportunities and contribute to increasing biodiversity. Another aspect is also the financing, for instance, the afforestation of brownfields is the cheapest way to create greenery. As a result of almost ten years of interdisciplinary monitoring of the project, it can now be stated that urban forests fulfil the objectives and are accepted and used by the population. Urban forests do not constitute an independent or new type of nature-based solutions they create new ecosystems from existing abandoned, brownfields, or neglected area.

Keywords: greening; brownfields; urban forests; green infrastructure; nature-based solution

check for updates

Citation: Rink, D.; Schmidt, C. Afforestation of Urban Brownfields as a Nature-Based Solution. Experiences from a Project in Leipzig (Germany). *Land* **2021**, *10*, 893. https://doi.org/10.3390/land10090893

Academic Editor: Thomas Panagopoulos

Received: 31 May 2021
Accepted: 20 August 2021
Published: 25 August 2021

Publisher's Note: MDPI stays neutral with regard to jurisdictional claims in published maps and institutional affiliations.

1. Introduction

In the last decade, reurbanization and growth have been the dominant developments in German and many major European cities. As the cities have grown and developed, brownfield and open spaces have often been used for new developments, especially residential buildings. This has led to redensification, and in some cases even to the conversion of green spaces. This trend runs counter to demands and efforts to preserve or even increase green and open spaces. These efforts are driven by the need for climate protection and adaptation as well as by the sustainability goals that many cities are pursuing. Last but not least, open and green spaces are seen as an indispensable part of urban quality of life. In this way, "it has become necessary to strengthen the ecosystem services in the city and implement new urban nature-based solution initiatives with the goals of improving the quality of the environment and creating a greener, more liveable city" [1] (p. 4). According to the European Commission, nature-based solutions are "actions inspired by, supported by or copied from nature; both using and enhancing existing solutions to challenges, as well as exploring more novel solutions, for example, mimicking how non-human organisms and communities cope with environmental extremes . . . Nature-Based Solutions aim to help societies address a variety of environmental, social and economic challenges in sustainable ways," [2] (p. 24) [3,4]. Forests represent a specific nature-based solution: they are considered to be of particular importance for climate protection and for adapting to climate change, because they store greenhouse gases and have a positive influence on the urban microclimate. In addition, urban forests increase the quality of life by providing spaces for local recreation and they contribute to the enhancement of neighbourhoods. This paper will examine the extent to which these goals can actually be achieved through afforestation in the inner city. It is based on an example from the city of Leipzig (Germany), where a model project was carried out in the 2010s with the aim of creating new urban forests. Despite reurbanization, growth and densification, the city of Leipzig succeeded in creating three new urban forests. The afforestation initiative is noteworthy because it took place on inner-city brownfield sites that resulted from urban redevelopment or revitalization. The context of origin and the classification of these brownfields must be taken into account, as they represent the framework conditions for the newly created forests.

Otherwise, urban forests are of course not a new phenomenon; they have been around for a very long time and have been the subject of an intensified discussion in the literature since the 2000s. In the English-speaking world, the terms 'urban forest' or 'urban forestry' have even existed since the 1960s; an overview of the development of the term 'urban forest' in the North American context can be found in Johnston [5,6]. The discourse on urban forests was conducted primarily in the Anglo-Saxon-speaking world until the mid-1990s. From the end of the 1990s onwards, there were various projects that initially dealt with questions of urban forestry, but that also included sustainability aspects—in the sense of a social-participatory perspective. These include, for example: "EUFORIC—European Forestry Research and Information Centre", "Neighbourhood Woods", "COST Action E12 Urban Forest and Trees", or "Urban Wood for People—Demonstration of Ways to Increase Recreationable benefits from Urban Woodlands" [7]. In the last decade, the Food and Agriculture Organization of the United Nations (FAO) has also focused intensively on urban forests; they have presented a collection of international examples [8] and published a special issue with analysis [9].

There are no clear definitions of urban forests, because even the term 'forest' depends on specific and individual types of forest [10] (pp. 50–52). The FAO has defined urban and peri-urban forests as "networks or systems comprising all woodlands, groups of trees, and individual trees located in and around urban areas" [11] (p. 2). A similar concept can be found in Kowarik, who classifies forests in terms of their spatial location and their relationship to settlement areas. In addition to peri-urban and non-urban forests [12] (p. 5), there are urban forests, which he describes as "completely surrounded by developed areas" or as "forest island in the city" [12] (p. 4). Here we follow the definition on which the project was based and where the focus was on planning. According to this definition, "urban forests (…) are forest areas in inner-city, often densely built-up areas. They represent a separate category of open space with special significance for urban redevelopment, for urban ecology—especially for the adaptation of cities to climate change—and for recreation" [10] (p. 32). Urban forests are usually found on anthropogenically deformed sites of residential, railway, industrial and commercial brownfields. Moreover, these forests can be very small, although it must be assumed that a minimum area of 0.3 hectares or a minimum diameter of approx. 50 m is required in order to ensure ecological stability [10] (p. 33). Furthermore, urban forests should have a minimum level of equipment and accessibility in order to be used and accepted by the population [10] (p. 33). This paper specifically deals with the urban forests that are created through afforestation on inner-city brownfield sites. The paper refers to three cases to show the diversity and variety of new urban forests. The aim of the scientific research on the new urban forests in Leipzig was to find out to what extent different ecological and social goals could be implemented here and to what extent they can be recommended as a nature-based solution for urban brownfields.

The aim of the interdisciplinary research was was to investigate whether the new urban forests fulfil the ambitions associated with them and to assess the success of three urban forests implemented as nature-based solution in the shrunken city of Leipzig. Improvements in urban microclimate, environmental quality, added value to neighboring areas, recreational opportunities, and biodiversity has been assessed.

2. Background: Urban Brownfields and Afforestation as an Option for Urban Restructuring

Although urban brownfields are a global phenomenon [13,14], they are portrayed very differently according to world region and city. Brownfields are "sites that have been affected by the former uses of the site and the surrounding land; are derelict and underused; may have real or perceived contamination problems; are mainly in developed urban areas; and require intervention to bring them back to beneficial use" [13] (p. 274). In shrinking cities, the brownfield problem is usually more obvious, caused by the process of deindustrialization, suburbanization as well as population decrease. Known examples for this are Detroit with its 'doughnut effect', or old-industrialized regions in Europe, such as the north-west of England, the Ruhr area in Germany or the Upper Silesian industrial area in Poland. In eastern Germany, the brownfield problem has become particularly prevalent as a result of the concurrency, or the combined effect of massive deindustrialization, suburbanization, and demilitarization, the structural change in the transport sector as well as a large decrease in population [15]. This has led to the creation of a huge number of brownfields in a series of affected cities. This problem has developed to such an extent that it has led to a breakup of urban structures. In planning and urbanistic discourses this is also characterised and discussed using the term 'perforation' [16]. Inner-city (older) areas are particularly affected by this, as they are distinguished by an increasing number of industrial, commercial, and residential brownfields and are also characterised by decay and vacancies. Due to the low or non-existent demand for the sites, there is a general uncertainty in such quarters about their further development. It is unclear whether and, if so, when the brownfields will be needed again; many are in an unsafe state and have succumbed to natural succession, which, in the eyes of the residents, reinforces the perception of their neighbourhoods as shrinking or in decline [17,18].

Against the background of this development, the most obvious question is: how can one generally deal with this problem and how does one want to deal with it? Under these conditions, what might be sensible interim uses or subsequent uses? How can one deal with the unplanned continuing spread of brownfields in design and planning? In view of scarce resources and capacity, particularly in shrinking cities, solutions are required that help enhance the residential quarters, but are also inexpensive in their design and maintenance. They must also be flexible and may not completely exclude other future uses, for instance in the form of construction.

Overall, brownfields posed a new urban problem for which there were neither resources nor instruments. Together with other issues, the topic of brownfields reached the agenda of urban redevelopment in eastern Germany in 2000 and became a topic of intensive research, development and planning work. As a result, some cities in eastern Germany have developed or introduced new strategies, instruments and tools for dealing with brownfields, such as different forms of interim use or renaturation [19] (pp. 383–385) [20]. One form that has been developed is the so-called 'urban forests'—this refers to the afforestation of urban brownfields. Normally, this instrument has been used alone and almost always on the urban periphery, for instance to close up gaps formed by the demolition of houses, to complete greenways or to connect the sites to the surrounding countryside. Forests are considered to have a high potential for the reuse of urban brownfields [21] (p. 201). This use category is regarded as multifaceted in terms of function, design and law, which makes it particularly suited to react to uncertain conditions. As a new category of open space, they would be a space-defining element in urban redevelopment and, in contrast to most brownfields, they would not represent a foreign element, but rather "visually and functionally interlock the perforated urban structures" [10] (p. 119). Accordingly, urban forests are said to have a great effect on urban design and urban structure [21] (p. 201). In particular, the use of different forest structure types allows a targeted influence on urban structure and townscapes [10] (p. 119). According to the German Forest Act, any area permanently planted with woody plants is to be regarded as forest land—irrespective of the actual designation of the land and the respective ownership structure or form of management [10]

(p. 8). For the establishment of forest on brownfields or former residential land, this means that it is neither dependent on planning decisions nor on ownership relationships; land readjustment is also unnecessary [10] (p. 119).

In addition, forest allows for a wide—and far from exhausted—range of designs and uses and can thus be adapted very flexibly to local conditions. This refers to its size and shape, its development as a commercial forest, recreational forest or near-natural forest stand. In addition, the existing vegetation—for example on brownfields—can be included in the planting and design and supplemented depending on the objective. However, forest can also be used to create spatially effective structures—such as road edges and boundaries or connections between previously separate areas. One of the advantages of forests is that they are accessible to the public and have fewer security requirements than, for example, public green spaces. It is also much cheaper to maintain and care for forests than public green spaces; with reference to figures from Berlin, Giseke speaks of one tenth of the costs [21] (p. 202); Burkhardt et al. also refer to the lower investment and management costs and the associated relief effects for the municipal budget [10] (p. 119). They also point out that urban forests would have positive indirect monetary effects, such as the valorisation of neighbouring properties (ibid.). Burkhardt et al. also argue that urban forests can combine the advantages of green spaces (usability and acceptance) with those of brownfields (low cost, ecological effectiveness): "Although they can neither replace green spaces nor brownfields without restrictions, they offer a broad spectrum of functions and possible uses at comparatively low costs and high acceptance" (ibid., pp. 119–121). Burkhardt et.al. also draw attention to the limits of urban forests: for example, they consider the fact that only a short-term interim use is to be established if a minimum size is not reached (for this they state the size of 0.5 ha) and if "exclusively representative or special design requirements exist for an open space", as is the case, e.g., with ornamental or market squares (ibid., p. 121). In contrast, disadvantages are not stated, e.g., that there are lingering stigmatisations, that the areas are initially fenced in and can therefore only be used to a limited extent, that the population rejects the action and sees it as a cheap greening option, etc. A few years ago, Kil said that afforestation—next to natural succession—was by far the most cost-effective option for brownfields in shrinking cities, but that in the previous consensus for inner-city open spaces it was still considered out of the question [22] (p. 144). He blames this on traditional images of cultivated urban nature: "Apparently our society, shaped by Western cultural concepts, cannot simply let go of the image of the 'controlled landscape'" (ibid.). In the federal programme "Stadtumbau Ost" (Urban Restructuring East) there are several examples of the reforestation of urban brownfields, for example in Eisenhüttenstadt, Halle, Schwedt and Weißwasser. While afforestation was carried out on the respective peripheries of these cities, the distinguishing feature in Leipzig is that urban forests were created on inner-city brownfield sites.

3. Case Study Leipzig: From Shrinkage to Urban Restructuring

The background in Leipzig, the city that will be discussed here, is a long period of shrinkage that increased during the period of post-socialist transformation in the 1990s. As a result of deindustrialization in the 1990s, the abandonment of military sites and structural change in the transport sector after reunification, numerous areas fell into disuse. In 2007, there were just under 2000 brownfields in Leipzig, which together covered about 700 hectares and corresponded to about 2.6% of the settlement area—very high values [23] (p. 19). Although these brownfields are found throughout the whole city, they are concentrated in the former industrial and working-class districts in the east and west of the inner city (ibid.). For many, even large areas of up to 20 or even 30 hectares, there was no demand, and no subsequent use was feasible. The city of Leipzig started urban redevelopment in 2001 and pursued the strategy "more green, less density" in its planning [24]. This included the creation of classic and new green spaces as well as interim uses. It also experimented with new green space concepts, such as planting trees on small inner-city urban redevelopment sites in the mid-2000s. Onn Leipzig's east side, the 'light grove' and the 'dark forest' were created, which, however, provoked criticism and protests. There were complaints about the demolition of valuable Gründerzeit (Wilhelminian-style) houses and the loss of urban qualities. A few years later, the idea arose to create urban forests on larger inner-city brownfield sites.

The Leipzig project: "Urban Forests: Ecological Urban Renewal Through the Creation of Urban Forest Areas on Inner-City Sites Undergoing a Change of Use" is one of the most ambitious renaturation projects. While reforestation in other cities is mainly concentrated on residential brownfields that arose in the course of the deconstruction of peripheral large housing estates, in Leipzig's inner city, brownfields are also being considered for conversion into forest. In addition, the Leipzig project is embedded in a "testing and development project" (E+E project) initiated by the city of Leipzig and the Federal Agency for Nature Conservation (BfN). The BfN's funding programmes pursue various overarching goals. In addition to the integration of nature conservation into urban development, these include the implementation of reliable research results in practice, testing new and improved applications of already tried and tested methods, and processing the experience gained for generally usable recommendations. Since the E+E projects are intended to have a domino effect nationwide, the practical application of the concepts is the focus of the projects. Characteristic for the Leipzig project is the strong heterogeneity of the brownfields, which underlines the model character of the project. The spectrum ranges from residential and commercial brownfields to railway and industrial brownfields, which are located in the inner city but also in peripheral locations and which are partly in municipal and partly in private hands. The wide variety can also be seen in the size of the sites, which range from very large sites (20 ha) to very small sites (<1 ha). In contrast to other urban redevelopment municipalities, the city of Leipzig has the ambition to see renaturation not merely as a reaction to the emergence of brownfields, but to develop differentiated, innovative and transferable solutions. The aim is to develop urban forests as a new land and planning category and to establish them as an instrument for urban redevelopment. However, the task set is by no means trivial; above all, the procurement of land appears to be problematic. Whereas in other municipalities negotiations were held with a few, mostly municipal or semi-municipal landowners, in Leipzig new negotiations have to be held with each landowner, which greatly delays planning and implementation and—if conflicts of interest are too strong—may even prevent them altogether.

The model project pursues several goals: the urban forests are intended to improve the urban microclimate and air quality. They should improve the surrounding neighborhoods, they should create new recreational opportunities for the residents. They should enrich biodiversity in the city by contributing to an increase in species diversity. Furthermore, the aim for planning is that a new category of open space is created that gives urban forests their own status. One aspect is also financing, as afforestation of brownfields is the cheapest form of greening. First, a feasibility study was conducted in 2007–2008, which yielded positive results. This was followed by the testing and development project (E + E project), which was funded by the Federal Agency for Nature Conservation (BfN) and carried out together with the city of Leipzig from 2009 to 2018. It involved the replanting of three urban forests on inner-city brownfield sites.

4. Methods and Materials

4.1. Methods

In the accompanying scientific research of the project, an interdisciplinary approach was chosen in which natural and social science methods were used. In order to actually be able to make sound statements about the contributions of the new urban forests to urban biodiversity, climate protection, improvement of the microclimate, improvement of neighbourhoods with greenery and local recreation, a longer period was chosen. The scientific monitoring started with the establishment of the first urban forest in 2010, after which repeat studies were carried out until 2018—i.e., over a period of almost ten years.

This was to ensure that robust statements could be made on the questions and issues of interest.

With regard to the public perception and use of the new urban forests, a household survey was conducted by the Helmholtz Centre for Environmental Research—UFZ Leipzig at the beginning in 2010 in the neighbourhoods where the new urban forests were to be established, which were four urban areas. A radius of 500 m was made around each of 4 locations of the planned urban forests, in which the questionnaires were then distributed and collected again according to a fixed key [25]. The focus of this survey, which worked with pictures, was the residents' preferences for the design and use of the new urban forests [25]. In order to record the development of recreational use of the three new forests, counts, observations, and surveys of residents were conducted on each site. This was done after the creation of the forest and for comparison two to three years later. In addition, differences in the recreation profile of urban forests and parks were investigated by counting recreationists in five other, also older forests and comparatively six parks in Leipzig and Dresden by the Technical University Dresden [26]. The impact of urban forests on surrounding urban neighbourhoods was analyzed by on-site surveys of the vacancy rate of apartments in the surrounding area of each of five brownfields, parks, and forests in Leipzig. Similarly, the vacancy rate of stores was also investigated in the vicinity of two brownfield sites, parks and forests in Leipzig [27].

In order to document the development of the vegetation in the three newly established forests, forest botanists from the Technical University Dresden examined vegetation records according to Braun-Blanquet once a year over a total period of ten years on representative permanent observation plots with an area of 100 m^2 or 25 m^2 (subplots). The development of the trees was analyzed in detail on the basis of the development of shoot length, stem diameter as well as the annual shoot lengths on firmly marked specimens. In addition, a comparative study was carried out between parks, urban brownfields and urban forests. In this study, model plots were selected for each district of Leipzig (19 in total), in which a vegetation survey was carried out [28]. In parallel, the change in fauna in the three new forests was recorded by the Nature Conservation Institute of the Leipzig Region e.V. by mapping the avifauna and reptiles in 2009, 2014 and 2017/2018. In addition, a mapping of bats and a comparative study of avifauna were carried out in four reference forests of different ages in Leipzig [29]. Soil profiles and soil samples were investigated by the Chair of Landscape Planning of the Technical University (TU) Dresden on six brownfields in Leipzig and, for comparison, in six forests of different ages according to different soil parameters. At the same time, soil from the three newly established urban forests was analyzed in depth [30]. The influence of urban forests on temperature development was investigated by the Chair of Meteorology of the TU Dresden based on measurements of a specially established climate station (Stadtgärtnereiholz) as well as mobile measurement tours through built-up urban quarters of varying densities and differently aged forests in Leipzig. In addition, the microclimatic conditions for two of the three new forests were simulated using the three-dimensional urban climate model ENVI-met. The impact of urban forests on air quality was assessed by measurements on the one hand and model-based simulations on the other hand. In addition, a simulation on the carbon sink capacity of urban greenery was carried out using the SVAT-CN and HIRVAC climate models [31].

4.2. Materials

The first "Stadtgärtnerei-Holz" area was created in 2010 on the site of the former municipal garden in Anger-Crottendorf, a district in the east of Leipzig. The city-designated garden had ceased operations in 2005, after which it was unclear what would happen to the area. The city, which owned the site, tried to sell it, but failed. Afterwards, new concepts or ideas for use were sought. Due to its size—approx. 3.8 ha—it came to the attention of the project and was selected as a possible reforestation area. To do so, the area had to be cleared at relatively high cost, the greenhouses and permanent buildings were demolished, but the main path was left as a passageway through the site. A number of trees and shrubs from the previous use have been preserved and incorporated into the new design. This also applies to some concrete elements that now serve as seating. The reforestation was based on the former use and the areas were divided into parcels. The choice of tree species and shrubs was intensively discussed, and a decision was made early on in favour of a variety of species. As with all new forests, the selection of tree species was based primarily on the local conditions (soil, climate, water), but also on the history of the site, the design and forestry conditions. At the same time, different tree species were to be tested. For the tree species, oaks, lime trees, hackberry, rowan and walnut trees were chosen, while hazel, hawthorn, barberry and elderberry were planted as shrubs. Since the Stadtgärtnerei-Holz and also the other two new forests were to be managed by forestry, mainly 'forest products' were used. In key space-defining areas, larger trees were planted in exceptional cases. The row spacing was 2 m, the spacing of plants within a row 0.5–0.9 m [32].

The afforested plots were initially fenced off for five years. When the fences were dismantled in 2015, the entire area was handed over to the municipal forestry department and is now the responsibility of the foresters. To reinforce the impression of a forest, three high seats were installed in the area. If you ascend them, you can get a good view of the area and the surroundings. Young people had been using the area illegally and spray-painting the old greenhouses; to give them a run for their money, the wall of an adjacent house was officially graffitied. This could not completely prevent individual acts of vandalism, but overall, the area is in good condition, the trees are now about 4 to 5 m high, and the ambience of a forest is visibly developing (see Figure 1).

Figure 1. The urban forest Stadtgärtnerei-Holz, ten years after planting.

The second area, the Schönauer Holz, was afforested three years later, in 2013, in the large Grünau housing estate on the western outskirts of Leipzig. This urban forest is located in the middle of a still densely populated residential area between blocks. One of the city's largest prefabricated slab buildings once stood here, a long eleven-storey building. It was popularly nicknamed the 'Eiger North Wall' by the people of Grünau because of its size and height. Due to the severe shrinkage in Grünau from almost 90,000 inhabitants at the end of the 1980s to about 45,000 at the beginning of the 2010s, Grünau had become the focus of urban redevelopment. The eleven-storey building on Neue Leipziger Straße had come into focus early on and was demolished with public funds in 2007. Afterwards, grass was sown, which only partially grew in due to the poor substrate. The area was hardly used by the population, but served as a pathway between the blocks and the nearby shopping and service facilities. The city bought part of the 5.5 ha area and developed a design concept for the urban forest. Here, the population was involved in the planting and design from the very beginning. During the reforestation, school children participated in a tree planting campaign. Existing elements were incorporated here as well, such as the existing paths and trees or some concrete elements that serve as seating. In addition, high stands were erected again (two), which have become landmarks of the urban forests in the meantime. As far as the tree species are concerned, rowan, wild service tree and European aspen were chosen. This area was also well accepted by the population, although there has been vandalism there as well, such as theft of trees or spraying of the signs. In the meantime, the trees are 2–3 m tall and the fences were removed in 2018, making all areas accessible (see Figure 2).

Figure 2. The urban forest Schönauer Holz seven years after planting.

The third site, Bürgerbahnhof Plagwitz, is located to the west on the edge of Leipzig's Wilhelminian development and is the largest site at around 15 ha. Once one of Germany's largest freight stations, it was completely derelict in the early 1990s. After that, there were only illegal interim uses, but at the end of the 2000s the area came to the attention of local residents. In 2009, the Initiative Bürgerbahnhof Plagwitz (IBBP) was founded, which also gave the area its new name. Since 2010, the initiative has organised 'discovery walks' or 'track breakfasts' and developed its own ideas for transforming the brownfield. Since 2015, a major event, the "Westbesuch" district festival, has also taken place at the Plagwitz civic railway station. The concept of the urban forest on the former railway site is consequently quite different from the two urban forests described above. Here, various other uses were integrated into the "Gleis-Grün-Zug Bahnhof Plagwitz" from the outset, and nature conservation concerns had to be taken into account. The groups and initiatives were involved in the planning and design from the beginning; a series of workshops and participation opportunities served this purpose. It is noteworthy that the existing succession forest in the southern part of the area—approx. 4.3 ha—was included in the overall design and continues to be left alone (see Figure 3). Here there are also sites that are valuable for nature conservation, such as special species living in very dry conditions. Only after long and complicated land negotiations was the city able to buy the area from the German railroad company, Deutsche Bahn. The entire area was gradually developed; in 2013, a bouldering rock and an air swing were inaugurated at the so-called 'Nordkopf', in 2016, a playground and scouts followed, and a year later, the first trees were planted in the fruit grove and orchard, which are tended by neighbourhood residents. An urban forest with oak trees has been created on approximately 1.5 ha and a sport field was opened in 2019. The City of Leipzig, the Free State of Saxony and the European Union also participated in the financing of the various designs and projects at the Bürgerbahnhof Plagwitz. In contrast to the first two examples, the Bürgerbahnhof Plagwitz is used much more intensively; numerous groups, initiatives and associations as well as citizens from the surrounding neighbourhood have 'appropriated' it. There has been no major vandalism here so far.

Figure 3. The urban forest "Karl-Heine-Holz"—natural succession.

All three model areas represent different types of brownfield land (horticultural fallow, residential and railroad brownfields) with different site conditions. At the same time, different types of forest were created (multi-layered dense or rather light, high or rather lower) and different strategies of forest planting were pursued (reforestation, succession). The planning recommendations derived from this were summarized in a 'toolbox' for all interested parties. This is publicly available on the homepage [32]. The location of the three new urban forests in the city area can be seen on the following map (see Figure 4).

Figure 4. Urban forests in the Leipzig urban area.

5. Results

5.1. Acceptance of the New Forests by the Population

The sociological survey conducted in 2010 showed that the design of brownfield sites as urban forests is generally accepted and positively perceived by the population [25]. From the population's point of view, urban forests enhance brownfield sites, make them usable and enrich the cityscape. Forests are generally highly valued by the urban population, but urban forests are not necessarily perceived as forest areas in the first few years after they are planted, and sometimes they cause confusion due to their appearance. Especially in the first years of planting, the area is still associated with a brownfield or 'wilderness' and only secondarily with a usable green space or a forest [25]. As growth progresses, acceptance of the area and its planting increases, but a sense of forest only emerges over time. This is also attached to the relatively small size of the area. Furthermore, the limited usability of urban forests makes acceptance by the population more difficult. By means of a picture survey, the acceptance of succession and afforestation to urban forests was measured in the quantitative survey. A Likert scale was used, from 1 = "like it not at all" to 4 = "like it very much". These two variants were explained to the respondents at the beginning as alternatives and then compared pictorially. The different succession stages are rated significantly worse than afforestation and achieve values between 1.19 (early succession stage) and 2.58 (advanced succession stage) on the four-point scale [25]. Afforestation, on the other hand, reaches values between 2.45 (early afforestation stage) and 3.46 (advanced forest stage) [25]. The fencing off of the afforestation areas limits accessibility and there is a lack of direct experience. The residents expect something similar from an urban forest as they do from classic green spaces, most likely a park-like design and equipment. It has to be said that the afforestation of brownfields is only a second-best solution for the population. The participants in the 2010 household survey wanted paved paths, rubbish bins, benches and seating as well as lighting (ibid.). With regard to use, surveys carried out in 2014 and 2017 revealed that the areas are primarily used by pedestrians and cyclists as passageways [33,34]. The duration of use is accordingly short, usually less than 15 min. Other uses include jogging or walking the dog; families and children in particular use the urban forests more intensively or for longer periods. The uses thus correspond to the character of the areas, because an increased number of visitors or typical park activities such as picnics, barbecues or ball games would be less appropriate for the setting of urban forests. The user groups are mainly residents or people from the immediate vicinity [34]. More intensive forms of use can be found where there are other offers, such as at the Bürgerbahnhof Plagwitz, where there is urban gardening, a playground and even gastronomy.

5.2. Ecological Benefits of Urban Forests

One positive ecological effect of urban forests is the reduction of heat effects. The measurements carried out in Leipzig as part of the research project prove, for example, that the shading effects of old trees in summer can lead to a reduction in the maximum daily global radiation of up to 1/14 of the reference value. This means that the daytime maxima of the air temperature under the trees in summer can be more than five Kelvin degrees lower. This was proven on the one hand by mobile measurement tours through Leipzig, and on the other hand by two measurement stations. One of them was located under old trees, while the reference station is unshaded (see Figure 5).

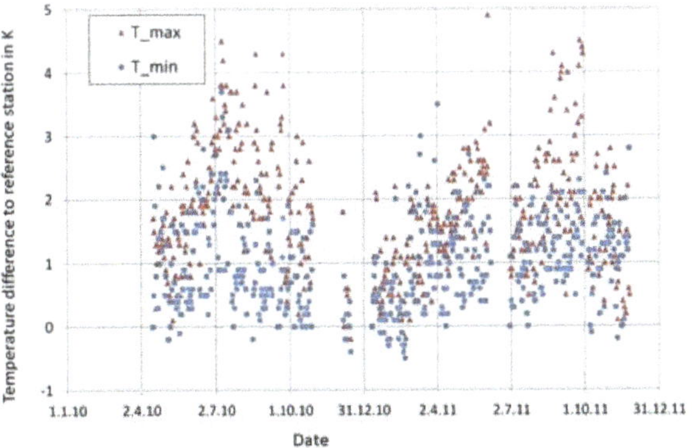

Figure 5. Temperature difference between a measuring station under old trees (KLIMA) and the reference station at the University of Leipzig [27].

In one temperature measurement run, for example, it was about 5–6 K cooler in the Stötteritz forest than in the city center [27]. Depending on the size and location of the area, the cooling effects can extend up to 400 m [27] into the surrounding built-up urban area. As a rule, the larger the forest, the farther its effect extends. Urban forests can thus reduce heat loads even better than parks dominated by open land during the day. In turn, parks dominated by open land contribute to cooling down at night to a greater extent, so that it is a mosaic of both types of green spaces in cities that matters.

Climate change also causes more heavy rainfall events in certain regions, which can lead to problematic scenarios depending on the degree of sealing. Therefore, field studies were conducted on unwooded brownfield sites and reference forests of different ages. Throughout the study, the investigated brownfields could only retain little rainwater [30]. The rainwater tends to run off the surface because the land is mostly composed of typical urban soils with technogenic admixtures of building rubble. The usable field capacity in relation to the effective rooting depth was only low to very low on brownfields without forest, and their water retention capacity was estimated to be low to medium [30]. The comparative studies in reference forests of different ages (10 to one hundred and fifty years) show that the forest vegetation causes clearly visible positive changes in the soil after just ten to twenty years—namely, increased porosity, aeration, water capacity and rooting depth of the soil as well as improved availability of nutrients. The smallest determined water storage capacity of the investigated forests corresponded to the highest levels of the brownfield areas (see Figure 6). Therefore, the planting of urban forests on brownfields does indeed tend to help to counteract flood risks.

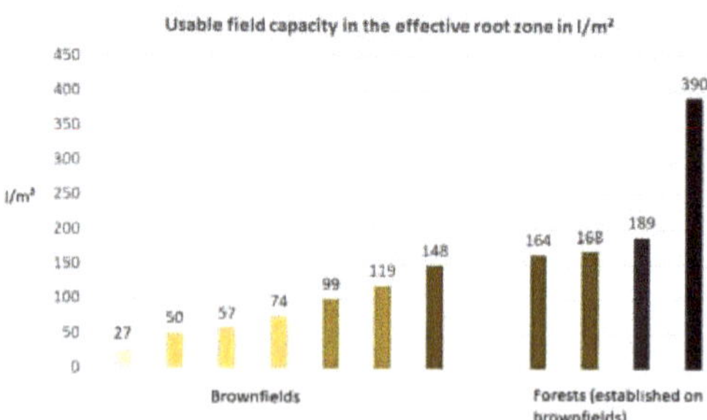

Figure 6. Investigations of the usable field capacity in the effective root depth of fallow land compared to forests [30].

One additional comparative study of different reference forests (according to age and location) with parks dominated by open land and brownfield sites was carried out [32]. The study's results show that brownfields have the highest species diversity in absolute terms, while there were no significant differences between urban forests and parks (see Figure 7). The high species richness of brownfields can be explained by the high variability of site conditions and habitat structures. Forests are the more species-rich, the more diverse their site mosaic is. A study in four reference forests also shows that both the number of breeding bird pairs and the number of breeding bird species tend to increase with the increasing age of the forest [31].

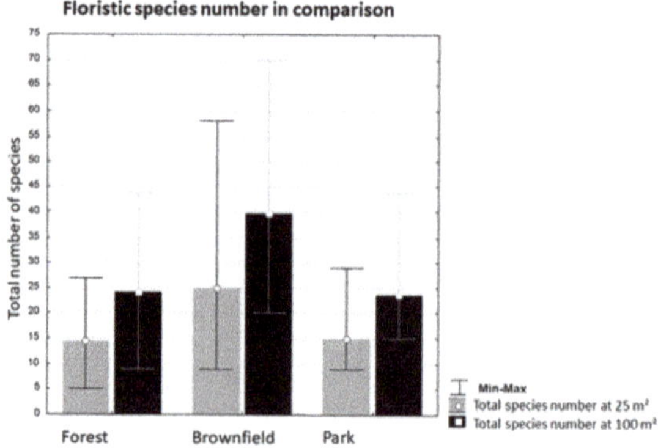

Figure 7. Comparative investigation of forest, brownfields and parks according to the total number of species [28].

The establishment of urban forests on brownfields usually requires soil improvement. This is because the soils of brownfields tend to be dry and often have a high skeletal content. In formerly sealed areas, an application of topsoil of at least 30 cm is recommended after unsealing [32]. Although topsoil was applied to the unsealed sites before afforestation, the required 30 cm was not achieved throughout, which later led to problems with tree growth. Underground infrastructures were not a problem; with the demolition of the aboveground structures, all underground infrastructures were also removed. The afforestation sites did not show any problematic contamination, as systematic soil tests before planting had shown. Brownfields with contamination must of course be treated before planting trees. The tree species that are planted must be very robust and adapted to climate change. On the other hand, they must also fulfil a variety of aesthetic qualities. Thus, an urban forest must take the surrounding urban structure into account [32]. Urban forests should reflect and respond to the type and height of the surrounding buildings or the facade structures. Their creation and maintenance, however, at 2.0–2.7 euros/m^2, is considerably less costly than that of parks [32]. However, at the same time, this means that they are not usually irrigated in the initial phase. Instead of larger trees, small 'forest products' are planted. Accordingly, it takes time before a full forest is created.

5.3. Effects of Urban Forests on Neighbourhoods

The afforestation of urban brownfields can have a positive impact on the attractiveness of a neighbourhood for its residents. This is also reflected in a mapping of the vacancy rate of apartments in 2013, which examined the apartment vacancy rate in the surrounding area of five forests compared to five brownfields and five parks (a total of 36,297 apartments). The apartment vacancy rate in the surrounding area of the brownfields studied was the highest at 21.7%, significantly higher than in the vicinity of the parks (18.1%) and forests at 12.2% [35]. Overall, the apartment vacancy rate was lowest in the urban neighbourhoods considered around forests (see Figure 8).

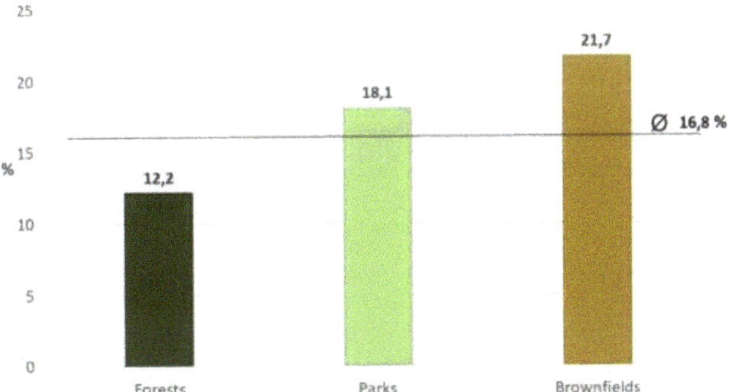

Figure 8. Comparative investigation of forest, brownfields and parks according to apartment vacancy rate in the surrounding area in percent [35].

The effect is even clearer if we focus on the visual range of open spaces. At about 10%, apartment vacancy was lowest within the visual range of urban forests, and highest at nearly 25% within the visual range around brownfields [35]. While the vacancy rate in the visual range of forests was also about 3% lower than outside the visual range, the exact opposite effect was observed for brownfields. Here, the apartment vacancy rate within the visual range actually increased by approx. 4% as compared to the areas outside the visual range. The visual effect of forests and parks is evidently positive compared to that of brownfields. This becomes clear at the same time when compared to the average of the city quarter. In four of the five forests surveyed, the apartment vacancy rate was lower than the average for the urban neighbourhood. In contrast, apartment vacancy rates in the surrounding area of four of the five brownfields were higher than the average for the urban neighbourhood in which they were located [35].

The brownfields that were afforested in the project also showed positive effects. For example, the vacancy rate of apartments around the site Bürgerbahnhof Plagwitz decreased from 21.6% in 2013 to 7.8% in 2018. However, the reason for this was not only the afforestation, but also Leipzig's overall positive population development [35].

In addition, the level of leasing of stores in the vicinity of two forests was compared with that in the vicinity of two brownfields and two parks (701 stores in total, mappings in 2010 and 2016). This shows a similar trend to that of apartments. For example, store vacancy was significantly higher in the surrounding area of brownfields (24.8%) than in the surrounding area of parks (20.1%) or forests (13.3%). However, the correlation between store vacancy and surrounding open space is not as close as for apartments. Overall, though, it has been confirmed in Leipzig that the development of green spaces promotes positive urban development [35].

6. Discussion

The concept of urban forests originates from the phase of reurbanization and urban redevelopment in the 2000s [10]. At that time, the effects of shrinkage were still clearly visible in Leipzig and sensible interim and subsequent uses were being considered for a large number of inner-city brownfield sites. At that time, it was assumed that there would be no demand or building use for them in the foreseeable future. However, practically as soon as the 'urban forests' project began in the early 2010s, the city of Leipzig entered a phase of dynamic growth that lasted until the start of the COVID-19 pandemic. This manifested itself in relatively high population growth rates (2–3% per year), increased demand for and use of land, and redensification. Even a few years ago, it had to be stated that some of the areas that had been shortlisted for an urban forest are now already planned and partly even built on. Thus—as it appears now—there are hardly any possibilities for the realization of further urban forests in Leipzig at present or in the near future. This clearly shows the limitations of the concept of urban forests in (re)growing cities, and thus the limits of nature-based solutions.

Leipzig is now pursuing the concept of 'double internal development' under growth conditions, which includes the establishment and improvement of green spaces, as a growing city also requires (additional) open spaces. Ecological, climatic and recreational effects should be weighed against uses for housing, commerce and social infrastructure [36]. Urban forest can, therefore, definitely be an option for growing cities as well, in terms of sustainable urban development, especially for adaptation to climate change. Thus, so far, urban forests have met with a positive response in Germany, with numerous cities learning about this new green space concept during the course of the project and at the final conference in autumn 2018. It now remains to be seen whether the urban forests will be imitated in other German cities, as was the intention of the project from the beginning.

As a new form of using open space, the urban forest is not comparable with conventional green spaces or existing forests. Its specific usability, its setting, must first be gradually adopted by the population. Planning and administration therefore need 'staying power', as the forest qualities and thus the use only develop over time. However, urban forests contribute to the enhancement of the respective residential environment and to the valorisation of the areas immediately after their creation. They have an enriching effect on the open space supply, and they can compensate for deficits in the respective neighbourhoods. Urban forests as defined here represent a new category of open space within the framework of sustainable urban development. The forest characteristics of urban forests develop only gradually and are only slightly pronounced, especially in the initial years after their establishment. The creation of urban forest should, therefore, mean a long-term and permanent use of the area and be in line with long-term urban development strategies.

The Leipzig urban forest project is a pioneering project. Thus, so far, no other studies have been conducted on other urban forests established on brownfields. As far as is known, urban forests in other German cities such as Eisenhüttenstadt, Halle, Schwedt and Weißwasser experience a similar acceptance as those in Leipzig. Internationally, the tree farms that have been established on brownfields in Detroit should also be mentioned here. They enjoy an extraordinarily high level of acceptance and are seen as a way of upgrading deprived neighborhoods: "The value of neighbourhood homes is increasing" (http://www.hantzfarmsdetroit.com/; accessed on 24 August 2021).

Extensive recommendations for the establishment of new urban forests were developed from the experiences in Leipzig [32]. For example, a toolbox was developed that is freely available to all interested parties via the homepage http://urbane-waelder.de/ (accessed on 24 August 2021). It contains, for example, a search filter for tree species selection, with which one can obtain a selection of suitable tree species and a wealth of information about them according to certain site parameters (https://baumartenauswahl.

urbane-waelder.de/; accessed on 24 August 2021). At the same time, there is a practice-oriented and extensively illustrated handout for planning and designing urban forests (http://urbane-waelder.de/Bilder/Toolbox_B.pdf; accessed on 24 August 2021).

The accompanying research within the project could not sufficiently answer or clarify all questions. The studies took place before the urban forests were established or in the first years after afforestation. It can be assumed that the acceptance and use of the forests will increase as they grow and that a sense of being in the forest will develop among visitors over time. In order to investigate this, further surveys and research are necessary, which should take place in the next few years. Further questions could be: How is the specific setting of the urban forest accepted by the population? At what point or under what conditions does a sense of being in the forest develop in the population? Which uses have become firmly established and what differences can be observed over time? How is biodiversity developing in the three forest areas? What effects do the urban forests have on the surrounding neighbourhoods in the long run? From this, further insights can be gained for future urban forests and recommendations for action can be formulated. It would also make sense to include other existing urban forests in Germany and Europe in such studies. Valuable insights could be gained from this, and impulses given for the dissemination of this new urban forest type.

Dushkova and Haase have already analysed several NBS using Leipzig as a case study and have shown their different ecosystem services [1] (p. 20). The question is whether the new urban forests in Leipzig represent an independent NBS type or whether they can be assigned to one of the NBS types elaborated by Dushkova and Haase. In our opinion, the urban forests do not constitute an independent or new type of NBS. Rather, they can be assigned to Type 3: "NBSs that involve creating new ecosystems from existing abandoned, brownfields, or neglected area" [1] (p. 20). The concept of nature-based solutions (NBS) "has been developed in order to operationalize an ecosystem services approach within spatial planning policies and practices, to fully integrate the ecological dimension, and, at the same time, to address current societal challenges" [1]. As shown in this paper, the new urban forests fulfil these requirements in a particular way and function as nature-based solutions.

7. Conclusions

It can be stated that urban forests are fundamentally suitable and recommended as a new form of revitalizing brownfield sites. Urban forests are obviously a suitable greening concept for shrinking or shrunken cities. Thus, they can be regarded as an instrument or new open space category of urban redevelopment. They are an NBS with which new ecosystems are created on urban brownfields that fulfil a range of ecosystem services. They have several positive ecological effects, for example on the microclimate and biodiversity. They have a positive impact on the surrounding neighbourhoods, for example on the level of housing vacancy.

In the Leipzig model project, urban forests are considered as a separate, new open space category with specific tasks and services. Urban forests are accepted as a second-best solution and are usually used extensively. The model project in Leipzig can be deemed successful in this respect; it was even possible to plant new urban forests on inner-city brownfield sites despite reurbanization, redensification and (re)growth. However, the limitations are also evident here, as no more suitable areas are available for further urban forests in the present or in the future. In Germany, forests enjoy special legal protection, but urban forests should nevertheless be established as an independent planning category in order to firmly anchor them in green concepts, master plans, climate protection and sustainability concepts. In general, it is important to examine what role the urban forest can play as a nature-based solution in (re)growing cities and in the context of redensification.

Author Contributions: Conceptualization, D.R., Data curation, D.R. and C.S., Formal analysis, D.R. and C.S., Funding acquisition, C.S., Investigation, D.R. and C.S., Methodology, D.R. and C.S., Project administration, C.S., Resources, C.S., Supervision, C.S., Validation, D.R. and C.S., Visualization, C.S., Writing—original draft, D.R. and C.S. All authors have read and agreed to the published version of the manuscript.

Funding: The research project was funded as scientific accompanying research of the testing and development project "Ecological urban renewal through the establishment of urban forest areas on inner-city sites under change of use—A contribution to urban development in Leipzig" funded by the Federal Agency for Nature Conservation in the period 2009–2019.

Institutional Review Board Statement: Not applicable.

Informed Consent Statement: Not applicable.

Data Availability Statement: Data sharing not applicable.

Conflicts of Interest: The authors declare no conflict of interest.

References

1. Dushkova, D.; Haase, D. Not Simply Green: Nature-Based Solutions as a Concept and Practical Approach for Sustainability Studies and Planning Agendas in Cities. *Land* **2020**, *9*, 19. [CrossRef]
2. European Commission. Towards an EU Research and Innovation Policy Agenda for Nature-Based Solutions and Re-Naturing Cities. Final Report of the Horizon 2020 Expert Group, Directorate-General for Research and Innovation, Bruxelles 2015. Available online: https://ec.europa.eu/programmes/horizon2020/en/news/towards-eu-research-and-innovation-policy-agenda-nature-based-solutions-re-naturing-cities (accessed on 28 May 2021).
3. Nesshöver, C.; Assmuth, T.; Irvine, K.N.; Rusch, G.M.; Waylen, K.A.; Delbaere, B.; Haase, D.; Jones-Walters, L.; Keune, H.; Kovacs, E.; et al. The science, policy and practice of nature-based solutions: An interdisciplinary perspective. *Sci. Total Environ.* **2017**, *579*, 1215–1227. [CrossRef] [PubMed]
4. Kabisch, N.; Korn, H.; Stadler, J.; Bonn, A. (Eds.) *Nature-Based Solutions to Climate Change Adaptation in Urban Areas: Linkages between Science, Policy and Practice*; Springer: Berlin/Heidelberg, Germany, 2017.
5. Johnston, M. A brief history of urban forestry in the United States. *Aboricult. J.* **2006**, *20*, 257–278. [CrossRef]
6. Konijnendijk, C. New Perspectives for Urban Forests: Introducing Wild Woodlands. In *Wild Urban Woodlands: New Perspectives for Urban Forestry*; Kowarik, I., Körner, S., Eds.; Springer: Berlin/Heidelberg, Germany, 2005; pp. 33–45.
7. Food and Agriculture Organization of the United Nations (FAO). Forests and Sustainable Cities. Inspiring Stories from around the World, Rome, Italy. 2018. Available online: http://www.fao.org/3/I8838EN/i8838en.pdf (accessed on 26 July 2021).
8. Food and Agriculture Organization of the United Nations (FAO). *Forests and Sustainable Cities. Unasylva* **2018**, *69*, 1. Available online: http://www.fao.org/3/I8707EN/i8707en.pdf (accessed on 26 July 2021).
9. Konijnendijk, C.; Nilsson, K.; Randrup, T.B.; Schipperijn, J. (Eds.) *Urban Forests and Trees*; A Reference Book; Springer: Berlin/Heidelberg, Germany, 2005.
10. Burckhardt, I.; Dietrich, R.; Hoffmann, H.; Leschner, J.; Lohmann, K.; Schoder, F.; Schulz, A. *Urbane Wälder-Ökologische Stadterneuerung durch Anlage Urbaner Waldflächen auf Innerstädtischen Flächen im Nutzungswandel—Ein Beitrag zur Stadtentwicklung*; Bundesamt für Naturschutz: Bonn, Germany, 2008.
11. Food and Agriculture Organization of the United Nations (FAO). Guidelines on Urban and Peri-Urban Forestry, Rome, Italy. 2016. Available online: http://www.fao.org/3/i6210e/i6210e.pdf (accessed on 26 July 2021).
12. Kowarik, I. Wild urban woodlands: Towards a conceptual framework. In *Wild Urban Woodlands: New Perspectives for Urban Forestry*; Kowarik, I., Körner, S., Eds.; Springer: Berlin/Heidelberg, Germany, 2005; pp. 1–32.
13. Oliver, L.; Ferber, U.; Grimski, D.; Millar, K.; Nathanail, P. The Scale and Nature of European Brownfields. In Proceedings of the CABERNET 2005—The International Conference on Managing Urban Land, Nottingham, UK, 13–15 April 2005; CABERNET (Concerted Action on Brownfield and Economic Regeneration Network). Land Quality Press: Nottingham, UK, 2005; pp. 274–281.
14. de Sousa, C.A. *Brownfields Redevelopment and the Quest for Sustainability*; Oxford University Press: Oxford, UK, 2008.
15. Rink, D. Urban Shrinkage as a Problem of Post-Socialist Transformation. The Case of Eastern Germany. *Sociol. Rom.* **2011**, *9*, 20–34.
16. Florentin, D. The "perforated City": Leipzig's Model of urban Shrinkage Management. *Berkely Plan. J.* **2011**, *23*, 83–101. [CrossRef]
17. Rink, D. Wilderness: The Nature of Urban Shrinkage? The debate on urban restructuring and renaturation in eastern Germany. *Nat. Cult.* **2009**, *4*, 275–292.
18. Mathey, J.; Arndt, T.; Banse, J.; Rink, D. Public perception of spontaneous vegetation on brownfields in urban areas—Results from surveys in Dresden and Leipzig (Germany). *Urban For. Urban Green.* **2018**, *29*, 384–392. [CrossRef]
19. Rößler, S. *Freiräume in Schrumpfenden Städten: Chancen und Grenzen der Freiraumplanung im Stadtumbau*; Institut für Ökologische Raumentwicklung: Dresden, Germany, 2008.

20. Mathey, J.; Rink, D. Greening Brownfields in Urban Redevelopment. In *Encyclopedia of Sustainability Science and Technology*; Meyers, R.A., Ed.; Springer: Berlin/Heidelberg, Germany, 2020; Available online: https://link.springer.com/content/pdf/10.100 7%2F978-1-4939-2493-6_211-5.pdf (accessed on 18 September 2020).
21. Giseke, U. Und auf einmal ist Platz. Freie Räume und beiläufige Landschaften in der gelichteten Stadt. In *Stadtlichtungen— Irritationen, Perspektiven, Strategien*; Giseke, U., Spiegel, E., Eds.; Birkhäuser: Basel, Switzerland, 2007; pp. 187–217.
22. Kil, W. Luxus der Leere. In *Vom Schwierigen Rückzug aus der Wachstumswelt*; Müller + Busmann: Wuppertal, Germany, 2004.
23. Muschak, C.; Banzhaf, E.; Weiland, U. *Brachflächen in Stadtentwicklung und Kommunalen Planungen am Beispiel der Städte Leipzig und Stuttgart*; Research Report; Umweltforschungszentrum: Leipzig-Halle, Germany, 2009.
24. *Stadt Leipzig: Stadtentwicklungsplan Leipzig 2010*; Dezernat Stadtentwicklung und Bau: Leipzig, Germany, 2002.
25. Rink, D.; Arndt, T. Investigating perception of green structure configuration for afforestation in urban brownfield development by visual methods—A case study in Leipzig (Germany). *Urban For. Urban Green.* **2016**, *15*, 47–53. [CrossRef]
26. Schmidt, C.; Böttner, S.; Schmidt, U.; Modul Erholung. Wissenschaftliche Begleitforschung zum Erprobungs- und Entwicklungsvorhaben der Stadt Leipzig "Urbane Wälder" im Auftrag des Bundesamtes für Naturschutz. TU Dresden: Dresden, Germany. 2018. Available online: http://urbane-waelder.de/ (accessed on 2 July 2021).
27. Moderow, U.; Goldberg, V.; Modul Stadtklima. Wissenschaftliche Begleitforschung zum Erprobungs- und Entwicklungsvorhaben der Stadt Leipzig "Urbane Wälder" im Auftrag des Bundesamtes für Naturschutz. TU Dresden: Dresden, Germany. 2018. Available online: http://www.urbane-waelder.de/ (accessed on 18 January 2021).
28. Roloff, A.; Heemann, S.; Forkert, M.; Modul Flora. Untersuchung der Vegetations- und Gehölzentwicklung. Wissenschaftliche Begleitforschung zum Erprobungs-und Entwicklungsvorhaben der Stadt Leipzig "Urbane Wälder" im Auftrag des Bundesamtes für Naturschutz. TU Dresden: Dresden, Germany. 2019. Available online: http://urbane-waelder.de/ (accessed on 2 July 2021).
29. Mäkert, R.; Klaus, D.; Schmidt, C.; Schmidt, U.; Modul Fauna. Wissenschaftliche Begleitforschung zum Erprobungs- und Entwicklungsvorhaben der Stadt Leipzig "Urbane Wälder" im Auftrag des Bundesamtes für Naturschutz. TU Dresden: Dresden, Germany. 2019. Available online: http://www.urbane-waelder.de/ (accessed on 18 January 2021).
30. Schmidt, C.; Feger, K.-H.; Böttner, S.; Lachor, M.; Schwarz, D.; Hoffmann, D.; Neumann, R.; Schmidt, U. Modul Boden und Wasserhaushalt. Wissenschaftliche Begleitforschung zum Erprobungs- und Entwicklungsvorhaben der Stadt Leipzig "Urbane Wälder" im Auftrag des Bundesamtes für Naturschutz. TU Dresden: Dresden, Germany. 2019. Available online: http://urbane-waelder.de/Bilder/Modul_Boden.pdf (accessed on 2 July 2021).
31. Moderow, U.; Goldberg, V. Modul Stadtklima, 2018. Wissenschaftliche Begleitforschung zum Erprobungs- und Entwicklungsvorhaben der Stadt Leipzig "Urbane Wälder" im Auftrag des Bundesamtes für Naturschutz. TU Dresden: Dresden, Germany. 2018. Available online: http://urbane-waelder.de/Bilder/Modul_Klima.pdf (accessed on 2 July 2021).
32. Schmidt, C.; Edelmann, T.; Haak, J.; Schmidt, U.; Dietrich, R.; Lehmann, L.; Toolbox Urbaner Wald. Handreichung zum Erprobungs- und Entwicklungsvorhaben der Stadt Leipzig "Urbane Wälder" im Auftrag des Bundesamtes für Naturschutz. TU Dresden: Dresden, Germany. 2019. Available online: http://urbane-waelder.de/ (accessed on 2 July 2021).
33. Eube, S. Urban Forests in Leipzig. An Empirical Study on the Perception, Use and Evaluation of a New Way of Sustainable Open Space Development. Master's Thesis, University of Leipzig, Leipzig, Germany, 2015.
34. Untertrifaller, L. Nutzung und Aneignung Neuer Urbaner Grünflächen im Wohngebiet—Am Fallbeispiel Bürgerbahnhof Plagwitz in Leipzig. Master's Thesis, Friedrich-Schiller-University, Jena, Germany, 2018.
35. Schmidt, C.; Böttner, S.; Meier, M.; Modul Stadtumbau. Wissenschaftliche Begleitforschung zum Erprobungs- und Entwicklungsvorhaben der Stadt Leipzig "Urbane Wälder" im Auftrag des Bundesamtes für Naturschutz. TU Dresden: Dresden, Germany. 2018. Available online: http://www.urbane-waelder.de/ (accessed on 18 January 2021).
36. Stadt Leipzig. Integriertes Stadtentwicklungskonzept Leipzig 2030 (INSEK), Leipzig 2018. Available online: https://static.leipzig.de/fileadmin/mediendatenbank/leipzig-de/Stadt/02.6_Dez6_Stadtentwicklung_Bau/61_Stadtplanungsamt/Stadtentwicklung/Stadtentwicklungskonzept/INSEK_2030/INSEK-Leipzig-2030A-Zielbild.pdf (accessed on 6 January 2021).

Article

Vegetation Structure, Species Composition, and Carbon Sink Potential of Urban Green Spaces in Nagpur City, India

Shruti Lahoti [1],*, Ashish Lahoti [2], Rajendra Kumar Joshi [3] and Osamu Saito [4]

[1]   United Nations University Institute for the Advanced Study of Sustainability (UNU-IAS), Tokyo 150-8925, Japan
[2]   Independent Researcher, Tokyo 136-0073, Japan; revatimal@gmail.com
[3]   Jawaharlal Nehru University, New Delhi 110067, India; rajendrakrjo@gmail.com
[4]   Institute for Global Environmental Strategies (IGES), Kanagawa 240-0115, Japan; o-saito@iges.or.jp
*   Correspondence: arshruti.lahoti@gmail.com; Tel.: +81-3-5467-1212

Received: 24 February 2020; Accepted: 30 March 2020; Published: 1 April 2020

Abstract: Nagpur is rapidly urbanizing, and in the process witnessing decline in its green status which is one of the identities of the city. The study aims to understand the current species diversity, composition and structure in different classes of greens prevalent in the city. As urban green spaces (UGS) are also reservoirs of carbon stock, the study estimates their biomass. Through rigorous field work, data were collected from 246 sample plots across various UGS classes as pre-stratification. Then the biomass was estimated using non-destructive method with species-specific equation. The diversity of tree species recorded in UGS varies, with high diversity recorded in avenue plantation and institutional compounds. The overall variation in species composition among UGS classes was 36.8%. While in managed greens the species composition was similar, in institutional greens and forest it was different. Particularly, in forest the evenness was high with low diversity and low species richness. The structural distribution indicate lack of old trees in the city, with high number of tree species between diameter classes of 10–40 cm. Biomass was recorded high in road-side plantations (335 t ha^{-1}) and playgrounds (324 t ha^{-1}), and trees with bigger girth size where the main contributors. The dominant species indicates that high growth rate, tolerance to drought and pollution are the key attributes considered for species selection by local authorities. Though the city holds green image, vegetation along the avenues and institutions are stressed, exposed, and threatened by felling activities for grey infrastructure expansions. In such scenario, protection and preservation of older trees is crucial to maintain the carbon stock of the city. In addition, local authorities need to focus on effective afforestation programs through public participation to achieve high survival rate and reduce the maintenance cost. For species selection in addition to phenology and growth rate, tree biomass and life span needs to be considered to significantly enhance the urban environment and increase the benefits derived from UGS.

Keywords: urban green space; carbon stock; biomass; species diversity; vegetation structure

1. Introduction

Urban green spaces (UGS) act as "lungs of city" and reservoirs of "carbon stock." The vegetation patches in and around the urban areas sequester and store large amount of carbon [1,2]. UGS also contribute toward mitigation of climate change impacts via carbon sequestration and provide various benefits [3]. UGS provide a broad range of ecosystem benefits [4], and through application of nature-based solutions the urban vegetation can generate co-benefits [5,6] by restoring ecological flow in urban areas and strengthen sustainable urbanization with stimulated economic growth as well as improved environment [7,8]. For example, organic farming in urban vacant patches can allow to sequester carbon from the atmosphere through increased organic matter content in soil [9]. Further, though the accumulated carbon of these "non-forested" areas are lower as compared to forested areas, they are important to maintain the local and national carbon balances [10]. With increasing urban areas, even the smaller share of carbon sink from the urban vegetation is playing an important role and is also increasing significantly in size [11]. However, in developing countries the urban carbon reservoirs are significantly affected by fast pace urbanization which leads to alteration in land-cover and change in overall vegetation structure [12,13].

Particularly in India, urbanization is engulfing significant portion of peri-urban arable land, causing substantial loss in green spaces [14]. In the emerging urban centers (1 million population and above) the urban planning efforts are disproportionate as compared to the metropolitan cities, with less priority toward UGS provisions against other infrastructure demands like housing, water and sanitation, energy supply, which accelerates the challenges. In these emerging urban centers, though the benefits of UGS are recognized [15], in general they are undervalued and are facing either destruction or degradation in all major cities of India [16]. The increasing urban sprawl and infrastructure development make the UGS vulnerable [17]. In such scenario, urban planning efforts to save hectares of arable land [18] are urgently needed to save these reservoirs of carbon. Hence, Government of India (GOI) has launched missions like National Mission of Green India and National Mission on Sustainable Habitat under National Action Plan on Climate Change (NAPCC). The aim of the missions is to enhance carbon sink potential of urban areas by undertaking afforestation programs [19]. However, at present, scanty information of carbon sink potential of urban areas and non-availability of local vegetation data of the urban trees hinder afforestation and mass-planting programs. Thus emerges the need of more local level studies to record vegetation structure and carbon sink potential of greens to achieve low carbon scenario envisioned by NAPCC [20] mission [21].

In developed countries, like North America and Europe vegetation data are recorded for many urban forest and urban greens because of freely available assessment and modelling tools like i-Tree Eco and UFORE (urban forest effects model). The tools provide detailed plant inventory and species-specific data, which makes carbon stock assessment process easier and faster, hence widely applied in local areas. However, these tools are not applicable in other areas because of substantial difference in pattern of urbanization, biophysical variable, vegetation type and structure [22]. Thus, lack of local species data and unavailability of modelling tools make the vegetation studies dependent on intensive field work with high resource requirements. Also, most of these studies are limited to national and regional forest reserves. Only a few studies of urban forest carbon assessment are carried out so far; further, a limited tree inventory data and biomass assessment of Indian cities lead to immense gap in this research area [23]. These few and limited studies include, vegetation study of Bangalore, urban forest of Vishakhapatnam, Chandigarh's urban vegetation, Delhi and Gandhinagar's biomass data [12,24–28]. Some local carbon stock studies are also available for Bhopal, Delhi, and Pune [29–31] however, the studies used low-resolution remotely sensed data which fail to capture the finely grained mosaic of land-covers represented by cities [32]. Moreover, most of these studies have focused on vegetation survey with the inventory list, which provides valuable information about local flora. However, this does not account for the variation in vegetation structure and composition associated with different land ownership aspect prevalent in the city landscapes [33].

For the selected study area "Nagpur city," the available vegetation data have been recorded in the form of an inventory list devoid of any compositional, structural, or biomass assessment. The carbon sink potential of the urban forest has also not been recorded so far. The ecosystem assessment of NEERI urban forest [34] was conducted a decade ago and the floral diversity of the city was captured in 2013 by Chaturvedi et al. However, the authors have identified the need of quantification of species diversity and carbon stock in different land use. The earlier cited studies of Indian cities also indicate the necessity of focusing on non-forested but tree-dominated areas of the city like institutional campuses, reserved public greens, and road-side plantation for carbon assessment [23]. Additionally, Tripathi and Bedi have highlighted the importance of measuring carbon sink potential of urban greens along with the detailed inventory of species composition and distribution within different green spaces [21]. Further, within the studied city, the UGS are fragmented and discontinuously dispersed throughout the built-up matrix. The urbanization and urban sprawl studies clearly indicate changing land cover of city and reduction in green cover area of the city [35,36]. Thus, in urban transition scenario it is vital to understand the vegetation composition prevalent in UGS classes and based on their potential understand their relevant importance to act as carbon sink areas. The quantified data on UGS carbon sink potential based on UGS classes could act as one the benefits derived from UGS, thus helping in prioritization of UGS planning in urban policy reforms to a more granular level in the urban mosaic. Taking this and the identified gaps in literature into consideration, the overarching aim of the study is to record the much-needed local vegetation data and evaluate the carbon sink potential of UGS for their effective management. The record of local vegetation data is aimed to develop local greening strategies by understanding the species distribution, diversity, and composition among different UGS classes with their current carbon stock. The main objectives are: (1) Understanding the tree species structure, diversity, and composition differences among UGS classes; (2) estimating the biomass and carbon stock of UGS; (3) establishing linkage between vegetation structure, species diversity, and carbon stock to guide strategic vegetation planting and management for enhancement of urban carbon sink. Thus, through the adopted approach of capturing the details of urban vegetation the research hypothesizes that the diversity and carbon sink potential of UGS varies among UGS classes.

2. Materials and Methods

2.1. Introduction to Study Area

Nagpur city, the 13th largest urban area in India and the third biggest in the state of Maharashtra, is selected for phytosociology and biomass assessment (Figure 1a). The city is situated at a latitude of 21∘9′ N and 79∘6′ E with the average elevation of 303 m above sea level. The city has tropical savannah climate (Aw in Köppen climate classification) with typically hot, dry, and tropical weather with an average annual rainfall of 1162 mm, where summer temperature escalates to 48 °C and the winter temperature dips to 10 to 12 °C. With several identities "zero-mile city," "orange city," and "garden city of Maharashtra," Nagpur is an interesting case for vegetation assessment. The district records rich plant composition of 1136 plant species [37] and 124 tree species at city scale [38]. The public UGS classes present in the city are recreational UGS, open UGS, public institutionalized greens, infrastructure and utility corridor greens, and vacant lands [39].

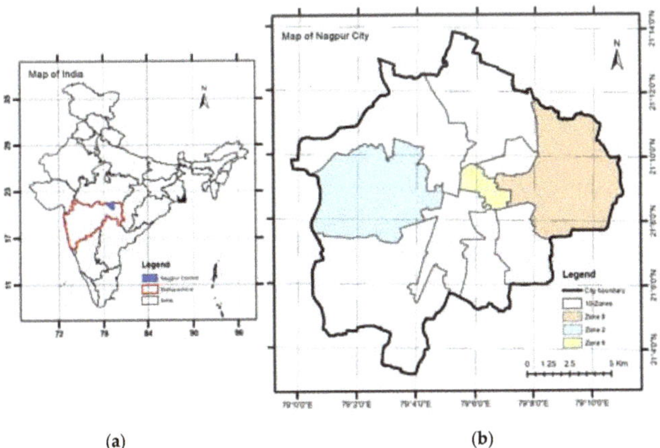

(a)   (b)

Figure 1. (a) Geographic Location of Study Area Nagpur District, Maharashtra, India; (b) Nagpur city administrative boundary with ten administrative zone and the selected typical zones.

2.2. Sampling and Data Collection

UGS are the combination of both public and private types of green spaces. Among these different classes, the vegetation and landscape character differ depending upon the ownership, management, and available resources [40]. Private greens also contribute toward urban environment; however, because of accessibility issue as well as relatively less predictability over their future development, only public UGS are considered for this study. Among the ten administrative zones present in the city, three representative zones were selected from east, west, and center as highlighted in Figure 1b. The identified zones fairly represent the overall vegetation conditions of the city, population density, per-capita green space availability, and public UGS classes [41]. The UGS classes present in each zone are indicted in Table 1.

Table 1. Urban green space classes used as pre-stratification along with the number of samples within each representative zone, the plot size and plot shape used in sampling.

Green Space Classes	Number of Plots	Plot Size (m^2)	Plot Shape
Parks and Garden	30	314	Circular
Playground	18	314	Circular
Lake	20	300	Linear
Forest	50	314	Circular
River	25	300	Linear
Institution	32	314	Circular
Road	53	300	Linear
Vacant land	18	314	Circular

For the field work, stratified random sampling was carried out using UGS classes as pre-stratification. This allowed increased efficiency as variation among the stratum is lower and hence smaller sample data can represent larger parcel of the entire stratum [42]. For pre-stratification, thematic map of UGS with an overall accuracy of 95% and kappa statistic as 0.93 was used [39]. For informal greens, as prior permission was required from Governing bodies, some of the institutes were pre-identified; however, sample plots were randomly selected. The field work was conducted between 10th December 2018 to 25th January 2019, by a group of botanist and landscape planners. Tree species with diameter above 10 cm at breast height (DBH) of 1.3 m within each plot were identified at species level. The height was measured by clinometer and DBH for trees was measured considering multiple stems. The plot locations as per field GPS points for respective zones are shown in Figure 2.

The plot size, plot shape, and number of sampled plots varied as per UGS classes as indicated in Table 1. For circular plots, 10 m radius was considered, while for road network a rectangular plot of 100 m × 1.5 m on both side of the road was used and for edges of lakes and river rectangular plot of 100 m × 3 m was considered. The number of sample plots under each stratum varied considering the vegetation structure and composition [42], based on observations from previous field visit (February-April 2018). For example, in zone 2 of the forest the sampled plots were high (n = 50), while for vacant land and playgrounds the sampled plots were lower. For rough calculation of number of plots to be inventoried i-Tree manual was referred, which mentions a general rule of minimum 20 plots in each stratum to represent the whole city with a standard error of 10%.

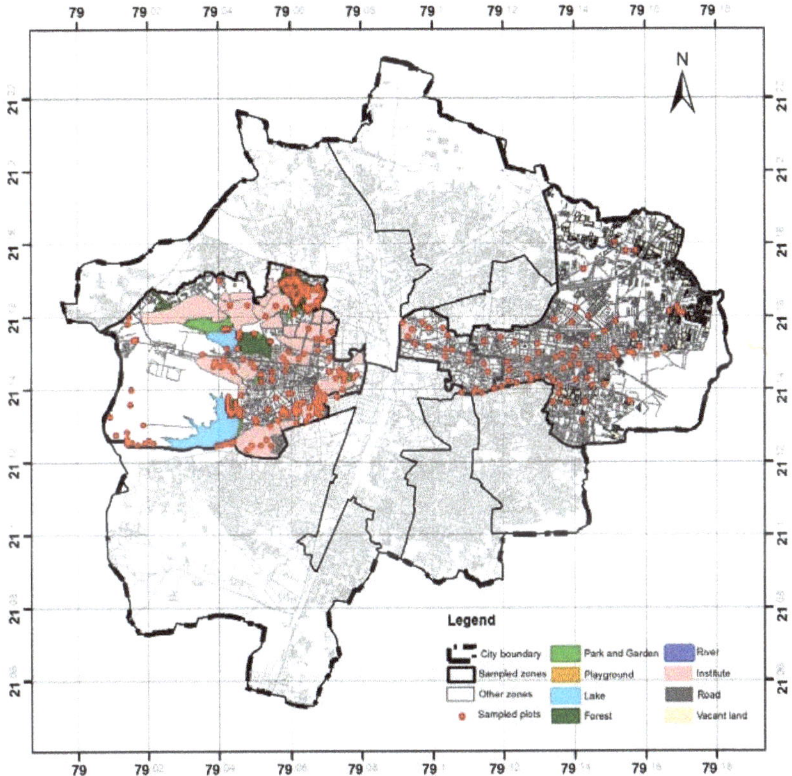

Figure 2. Urban green spaces (UGS) map showing different classes and sampled plots.

2.3. Data Analysis

The basal area (BA) for each tree was calculated using Equation (1) and the aggregate BA (m^2 ha^{-1}) was calculated by multiplying BA with the scaling factor of the UGS class. The importance value index (IVI) of tree species for each UGS class was calculated by summation of total relative abundance, relative density, and relative frequency [43]. IVI is mainly for understanding the share of individual tree in UGS class. Tree species diversity Shannon (H′) and Simpson's (1/D) index were calculated using the below Equations (2) and (3) (Borah et al. 2013). The Pielou's evenness (J) was determined by comparing the diversity (H′) with the maximum diversity (ln of total number of species). For species richness Menhinick's index was used, where number of different species found in a sample is divided by square root of total number of species found in the sample.

$$\text{Basal area (m}^2) = \pi^* \text{ DBH (cm)}^2/40000 \tag{1}$$

$$H' = -\sum_{i=1}^{s} p_i \ln p_i \tag{2}$$

$$D = \frac{1}{\sum_{i=1}^{s} pi^2} \tag{3}$$

In Equation (1), basal area is in m^2 and DBH is in cm^2. In Equations (2) and (3), H′ is Shannon-Wiener diversity index, D is Simpson index, p is the proportion (n/N) of individuals of one particular species found (n) divided by the total number of individuals found (N), ln is the natural log, Σ is the sum of the calculations, and s is the number of species.

The species accumulation curve was plotted to cross-check the sample size, and the composition of the tree species was analyzed using multivariate analysis. Using ordination method, a canonical correspondence analysis (CCA) was performed in PAST software (3.24 version). The two-dimensional diagram of CCA graphically depicts the similarity in vegetation composition among UGS classes. The structural composition of tree species was studied through DBH class distribution.

For biomass assessment, field sample method is preferred over remote sensing method because of the accuracy [44], following non-destructive biomass estimation. By using species-specific volumetric equation with the measured biophysical variables, the biomass of tree was calculated [45–48]. The equations consider measurable parameters like DBH, height of tree, and wood density [49]. All the volumetric equation for the inventoried tree species were derived from literature and are recorded in Appendix A. In case of unavailability of species-specific equation, generalized equation by Chave [50] or equation for same species group was used [1,2]. For unidentified trees regression equation derived by Brown et al. was used [45]. The volume of tree biomass (m^3 ha^{-1}) is calculated by using species-specific volumetric equation by inputting field data (DBH and height) [51]. The above ground biomass (AGB) was calculated by multiplying the tree biomass and wood density of tree species obtained from Forest Survey of India [52]. As urban trees have different surroundings than natural forest trees, to adjust the variation in biomass derived by using forest tree equation, the estimated AGB is multiplied by a factor of 0.8 [48,50,53,54]. For estimation of below ground biomass (BGB), regression equation suggested by Cairns et al. (1997) as in Equation (4) is used. Total biomass was derived by adding AGB and BGB, and to calculate the complete dry weight carbon stock (Cstock), a conversion factor of 0.475 is applied [55–57]. The above multiple equation is combined and used for individual species (with different DBH) for biomass (AGB+BGB) and Cstock estimation.

$$BGB = EXP(-1.059 + 0.884 \times \ln(AGB) + 0.284) \tag{4}$$

3. Results

In all, 2362 individuals belonging to 86 species were recorded among eight UGS classes of Nagpur. The identified 73 species belonged to 58 genus and 22 family. In all, 13 species were unidentified and in 9 plots no trees were recorded. The plots with no vegetation were also considered in biomass assessment for which the woody biomass was considered zero. The largest tree identified was *Mitragyna parvifolia* along the road with DBH of 175 cm, followed by *Ficus religiosa* with 143 cm DBH in institution and *Azadirachta indica* with 127 cm DBH in playground. Among the UGS classes the mean DBH was highest in institutions (36.6 ± 13 cm) and lowest in vacant land (22.4 ± 15 cm). Apart from institutions, parks and garden and playgrounds had relatively higher mean DBH of 32.0 ± 16 cm and 32.5 ± 19 cm. Forest on the contrary had lower mean DBH value of 26.0 ± 9 cm. Analysis of tree species indicates that few species were unique to particular UGS classes only (31 out of 87 which is 36%). In forest, 11 species (13%) were found that were not recorded in another classes. Likely, the uniqueness was high is road and institutions with 9% and 7% respectively.

3.1. Species Richness and Diversity

The diversity was high in roads with 46 species, followed by institutions and parks and garden (as in Table 2). The same was reflected through the dominance index (Simpson) where high diversity was in road followed by parks and garden and river. While as per the Shannon index that considers even rare species, highly diverse stratum was road followed by institutions and parks and garden (Table 2). Thus, road is significantly diverse based on both Shannon and Simpson index, followed by institutions. Forest on the other side is low in diversity with lower evenness and high variation in abundance of species. Considering the sample sizes of both forest and road were same, road is less homogeneous. While vacant land has high degree of evenness where all species are equally common with very low variation in abundance followed by river and playground. Species richness (S) varied among UGS classes. The highest species richness was found in institutions (2.72) and the lowest was in forest (1.2).

Table 2. Phytosociology and diversity attributes of eight UGS classes of Nagpur city.

UGS Classes	No. of Species	No. of Families	S (Species Richness)	Shannon (H)	Simpson (1-D)	Pielou J (Evenness)	Basal Area (m^2 ha^{-1})
Parks and Garden	31	29	1.89	3.06	0.94	0.55	32.06
Playground	24	19	2.08	2.77	0.92	0.57	31.90
Lake	22	20	1.91	2.7	0.92	0.55	31.30
Forest	33	25	1.20	1.84	0.64	0.28	28.99
River	30	25	2.36	2.98	0.94	0.59	17.38
Institutions	46	35	2.72	3.14	0.93	0.55	29.30
Road	49	39	2.11	3.3	0.95	0.53	37.55
Vacant land	23	20	2.52	2.7	0.92	0.67	11.91

The plotted species accumulation curve (Figure 3) indicates an increasing trend in number of individuals. The number of species saturated with increased number of plots for park and garden, playground, forest, river, and vacant land which indicates the sufficiency of sampled plot. However, for lake, road and institutions, as the curve is progressive more sampling would have revealed more species richness. The lowest rate of species accumulation was observed in forest, which was also reflected in lowest species richness, while institutions class showed highest rate of accumulation throughout with highest richness (as in Table 2). Though lake had low rate of species accumulation over first few plots it raised and was progressive indicating more sampling efforts needed.

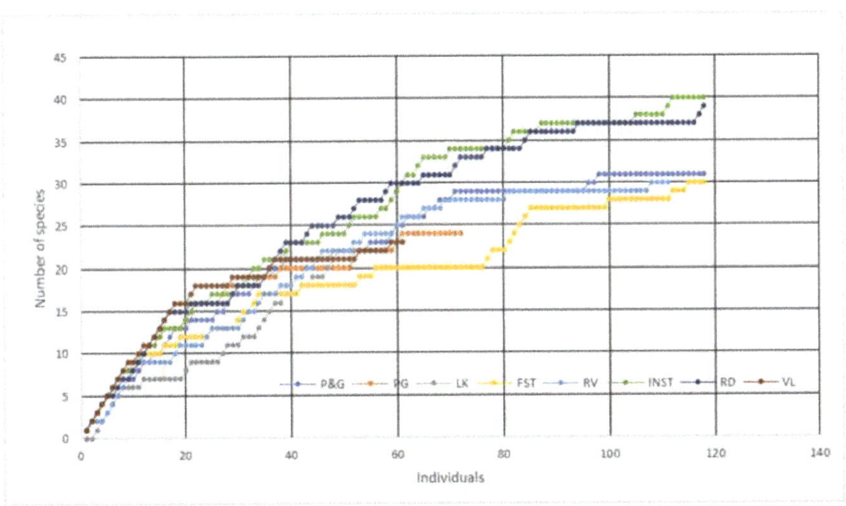

Figure 3. Species accumulation curve for UGS class: Parks and Garden (P&G), Playground (PG), Lake (LK), Forest (FST), River (RV), Institutional greens (INST), Road network (RD), and Vacant land (VL).

3.2. Species Composition and Structure in UGS Class

CCA ordination analysis further identified the similarity and dissimilarity in composition of species among different UGS classes [58]. CCA performed on species IVI collected from 246 sampled plots showed 57% of association along two axes (Figure 4). UGS classes shown along axis 1 explained 36.8% of the variation. The similarity in species composition was observed among the managed greens like park and garden, playground, lake, institutions, and road. The unmanaged greens like river and vacant land; and forest, where composition is indicated by closeness of points and aggregation of species between two points. In axis 2, tree species total variation was found to be 20%. The most significant deviation is recorded around lake with species like *Butea frondosa, Dalbergia sissoo, Ceiba pentandra, Gmelina arborea, Mimusops elengi, Mitragyna parvifolia, Plumeria alba, Sapindus mukorossi,* and *Soymida febrifuga*. The next most prominent deviation is near vacant land with invasive species like *Prosopis juliflora* and *Ziziphus mauritiana*. The last most important deviation is in cluster forest, which is dominated by *Boswellia serrata, Hardwickia binate,* and *Tectona grandis*.

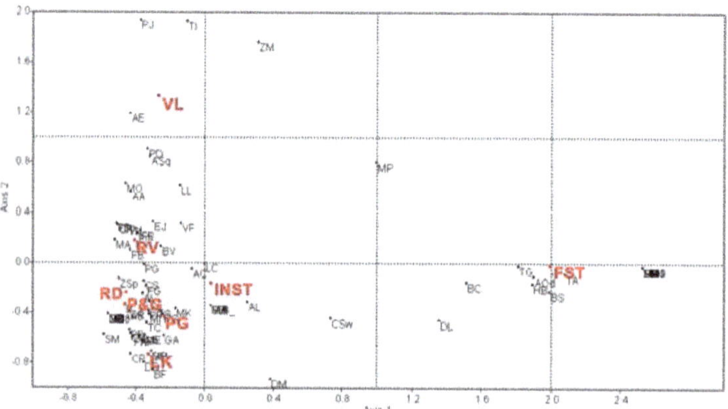

Figure 4. Outcome of canonical correspondence analysis (CCA) for UGS classes in study area, where axis 1 represents green space class and axis 2 represents tree species. Abbreviation used for tree species: AC, *Acacia catechu*; AA, *Acacia arabica*; AL, *Acacia Lecophloea*; AE, *Ailanthus excelsa*; AL, *Albizia Lebbeck*; Aod, *Albizia odoratissima*; AS, *Alstonia scholaris*; Asq, *Annona squamosa*; AI, *Azadirachta indica*; BV, *Bauhinia variegata*; BC, *Bombax ceiba*; BS, *Boswellia serrata*; BF, *Butea frondosa*; CV, *Callistemon viminalis*; CF, *Cassia fistula*; CS, *Cassia siamea*; CP, *Ceiba pentandra*; CSw, *Chloroxylon swietenia*; DL, *Dalbergia latifolia*; DP, *Dalbergia paniculata*; DS, *Dalbergia sissoo*; DR, *Delonix regia*; DM, *Diospyros melanoxylon*; EG, *Eucalyptus globulus*; EJ, *Eugenia jambolana*; FB, *Ficus benghalensis*; FE, *Ficus elastica*; FG, *Ficus glomerata*; FR, *Ficus Religiosa*; FSp, *Ficus sp.*; GR, *Gardenia resinifera*; GP, *Garuga pinnata*; GA, *Gmelina arborea*; HB, *Hardwickia binata*; KP, *Kigelia pinnata*; LP, *Lagerstroemia parvifolia*; LS, *Lagerstroemia speciosa*; LC, *Lannea coromandelica*; LL, *Leucaena leucocephala*; ML, *Madhuca latifolia*; MI, *Mangifera indica*; MH, *Manilkara hexandra*; MA, *Melia azedarach*; MHo, *Millingtonia hortensis*; ME, *Mimusops elengi*; MP, *Mitragyna parvifolia*; MO, *Moringa oleifera*; MAl, *Morus alba*; MK, *Murraya koenigii*; NC, *Neolamarckia cadamba*; NA, *Nyctanthes arbortritis*; PP, *Peltophorum pterocarpum*; PD, *Pithecellobium dulce*; PA, *Plumeria alba*; PL, *Polyalthia longifolia*; PPi, *Pongamia pinnata*; PJ, *Prosopis juliflora*; PG, *Psidium guajava*; PM, *Pterocarpus marsupium*; SM, *Sapindus mukorossi*; SS, *Schrebera swietenioides*; SF, *Soymida febrifuga*; SU, *Sterculia urens*; TI, *Tamarindus indica*; TS, *Tecoma stans*; TG, *Tectona grandis*; TA, *Terminalia arjuna*; TC, *Terminalia catappa*; TD, *Trichilia dregeana*; UI1; UI2; UI3; UI4; UI5; UI6; UI7; UI8; UI9; UI10; UI11; UI12; UI13; VF, *Vachellia farnesiana*; VN, *Vachellia nilotica,* ZM, *Ziziphus mauritiana*; ZSp, *Ziziphus sp.*

The structure of vegetation is the city is studied through tree diameter class distribution among different UGS classes. Overall, the number of individuals decreased with increasing in diameter class as shown in Figure 5. In all the classes, the highest frequency of individuals belonged to >20–30 cm diameter class (33%) followed by >10–20 cm diameter class (30%). Thus, the dominant diameter class is of young trees between >10–30 diameter class (63%). The lowest frequency is of >90–120 (1%) and >60–90 diameter class with 5% share which clearly indicates lack of old trees in the city. The retained old trees were mainly found in park and garden and playground with high number of >60–90 diameter class trees. Along the water bodies (lake and river), the trees with high diameter class are relatively low. The diameter class analysis of forest indicated a reverse J shaped curve, where the number of individuals between diameter class of >10–20 cm is very low, however it picks at 20–30 (44%) and gradually decreases. This vegetation structure indicates that forest is regenerating forest with less old trees, as share of trees with DBH above 40 cm is below 6%. Overall, the DBH class in >60–120 cm has in total BA of 65 m^2 ha^{-1}. Among >30–40 cm class the BA is highest (46 m^2 ha^{-1}) while lowest BA is in >10–20 (14 m^2 ha^{-1}).

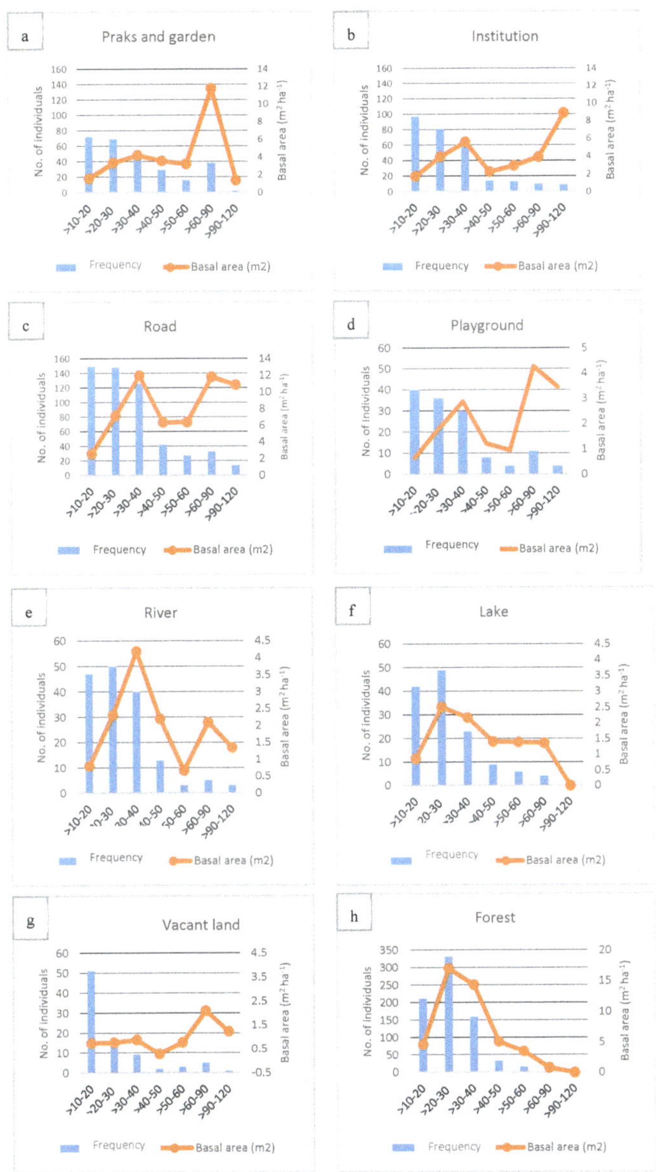

Figure 5. Size class distribution of tree frequency and basal area (BA) (m^2) in different UGS classes (X-axis represent the diameter class of trees and on Y-axis number of trees and basal area is represented. For the UGS classes the number of individuals varied and hence Y axis in the bar chart varied between range of 0–160 for park and garden (**a**), institutions (**b**), road (**c**) and for playground (**d**), river (**e**), lake (**f**) and vacant land (**g**) between 0–60, while for forest (**h**) it is between 0–350).

3.3. Tree Biomass and Carbon Stock

The biomass stored in UGS classes varied significantly between 70.42 t ha^{-1} in river and 334.61 t ha^{-1} in road (Figure 6a). Playground follows road with 323.68 t ha^{-1} of biomass. Institutions though rich in diversity has relatively low biomass and Cstock. Unmanaged greens like river and vacant land has lowest share in cities biomass with 70.42 t ha^{-1} and 110.40 t ha^{-1} respectively. In playground though the tree frequency is lower, biomass is high owing to the presence of trees of high DBH class (>60–90 cm and >90–120 cm). This proves that BA has positive correlation with AGB as found in other studies [59–62]. Though the correlation between BA and AGB varies, in forest and road it is significant ($R^2 = 0.90$ and $R^2 = 0.94$ respective) and for other classes as well it is positive (Supplementary Materials). In Nagpur, the main contributing DBH class toward biomass and Cstock among eight UGS classes is >30–40 cm followed by >60–90 cm and >20–30 cm (Figure 6b). Though the lower DBH class has high tree frequency the share of Cstock is lowest. Overall trees between 20–40 cm hold 40% share and trees above 60 cm DBH hold 32% of Cstock share. Study by Nero et al. indicates similar pattern of exponential increase in carbon frequency with girth size of trees [3].

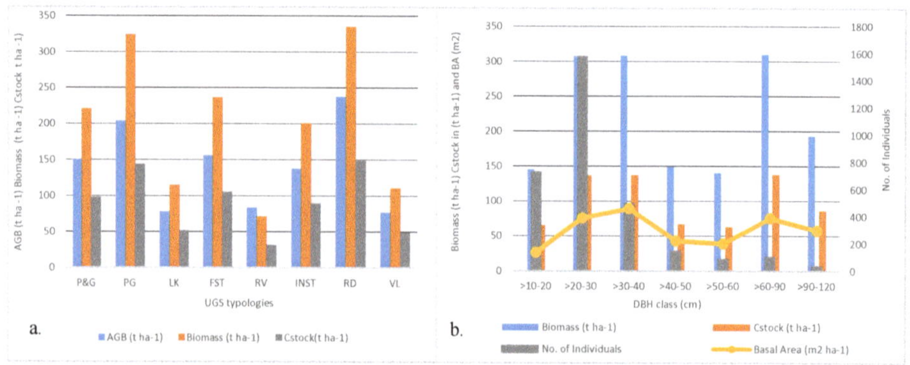

Figure 6. (**a**) Above ground biomass (AGB), biomass (TB), and Cstock distribution in different UGS classes. (**b**) Distribution of TB and Cstock with no. of individuals in tree with different diameter class along with BA (m^2) in different UGS classes.

The high biomass of institutions in Nagpur (137 t ha^{-1}) is comparable to Pune University campus (108 t ha^{-1}) studied by Waran and Patwardhan, while in park and garden (150 t ha^{-1}) have higher biomass as compared to gardens studied in Pune (110 t ha^{-1}) [31]. For road, estimated biomass ranged from the value 56.75–380.11 t ha^{-1} recorded by Rahman et al. in Bangladesh [63]. As for the estimate of biomass (236 t ha^{-1}) in forest, the value is within the identified national range for tropical dry deciduous forest 83–370 t ha^{-1} and 33–315 t ha^{-1} as studied Joshi et al. and Gandhi et al. respectively [64,65]. The Cstock by forest (105. 16 t ha^{-1}) is also comparable to Delhi urban forest biomass range 107–169 t ha^{-1} [66]. However, the Cstock of forest is on the higher side as compared to forest plantation in Italian cities (99 t ha^{-1}) studied by [67] and Shenyang urban forests with 33 t ha^{-1} studied by Liu and authors [68]. Study by Nero et al. on carbon sink potential of different UGS estimated that overall Kumasi has a 228-t ha^{-1} of carbon in different greens [3].

3.4. Characteristics of Dominant Species

The tree species showed positive correlation between IVI and AGB ($R^2 = 0.75$), thus based on IVI dominant tree species among all UGS classes were identified and their AGB is represented in Figure 7. *Azadirachta indica* is the most dominant species with high biomass and IVI. Followed by *Azadirachta indica* is *Tectona grandis* which has high biomass, but density is higher only in forest, owing to its mass plantation. Following this the dominant species are *Mitragyna parvifolia*, *Ficus Religiosa*, *Delonix regia*, *Polyalthia longifolia*, *Albizia Lebbeck*, and *Bauhinia variegate*. For the other identified species, though the AGB value is moderate (Figure 7) they have high IVI with high abundance across the city. *Mitragyna parvifolia*, *Bauhinia variegate*, and *Mimusops elengi* particularly have low IVI but the AGB is very high, as recorded by [69]. Particularly, *Ficus benghalensis*, *Tamarindus indica*, and *Dalbergia sissoo* are high-biomass yielding species. The characteristic of dominant species indicates that tolerance to weather condition and functional attributes are the main criteria considered by local authorities for planting the identified trees. In addition, growth rate and phenology are considered. Further, it is evident that the outstanding old trees in the city have cultural and religious significance, thus being a strong reason for their protection against all odds.

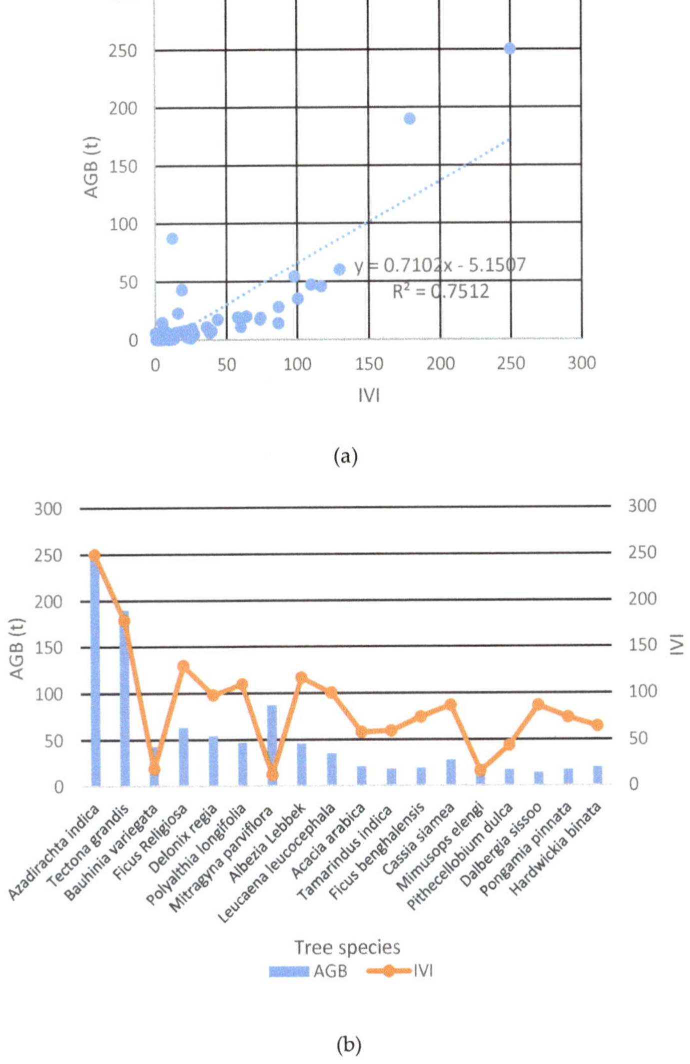

Figure 7. (**a**) Correlation between importance value index (IVI) and AGB (**b**) IVI and AGB of dominant species identified in cities UGS.

4. Discussion

The study records species diversity, composition, structure, and Cstock stored in UGS classes in the rapidly urbanizing city of Nagpur. Though the city was once recognized as one of the greenest city [70], so far local vegetation data and carbon sink data have not been recorded. However, with increasing grey infrastructure expansion the urban trees are declining with reduction in green cover [13], thus the study complies much-needed local vegetation data of urban trees in UGS and in the process explores ways to enhance the carbon sink of the city.

4.1. Vegetation Diversity, Species Composition, and Structure in UGS Classes

The city is rich in floristic diversity, in contrast to other global cities like Oakland and Athens where the density is high but urban tree assemblage is dominated by few species [48,71]. The tree diversity varies among the UGS classes studied, however is found to be lower than the floral diversity previously recorded by Chaturvedi et al. in 2013 [38]. Also as compared to Bangalore's public and private greens tree diversity captured by Sudha and Ravindranath [24], Nagpur showed low diversity. However, Nagpur's tree diversity was found to be close to Delhi's and Bangalore's managed greens tree diversity [12,72]. In comparison to other tier cities like Allahabad, the diversity was high [73]. Further within Nagpur's UGS classes the highest diversity was recorded in avenue plantation. However, when compared with Lutyen's Delhi and Bangalore roadside plantation, Nagpur's avenues showed lower diversity [12,74]. While the institutional compounds of Nagpur hold high diversity and high species richness as observed in the campuses of Pune and Bangalore [24,31]. Among the studied UGS classes, the reserve forest has the lowest species richness and diversity, even lower that the local urban forest of NEERI [34]. The low diversity indicates the mass plantation strategy of certain species adopted by forest department. The data about locally available tree diversity indicate that more species can be introduced to enhance the diversity, particularly along the water bodies, parks and playgrounds, and in the reserve forest area.

The tree diameter class indicates vegetation structure [75], and in Nagpur this structure is dominated by tree species below 40 cm DBH with record of very few old trees. The vegetation structure is more like Mexico [76] and Bangkok [77]. Because of varying level of maintenance, exposure to pollution, disturbance, and stress, the occurrence of old trees varies significantly among UGS classes. In managed greens like parks and playground the old trees are protected, while along the avenues the old trees have been cleared because of expansion of grey infrastructure projects like road widening and metro corridor creation, as witnessed in other Indian cities [78,79]. Further lack of planting effort in newly established road network has further increased the vulnerability of the green corridor of the city. The declining number of old and mature trees in the city is altering the vegetation structure and this effects the carbon sink potential of UGS, as mature and old trees have been identified as the reservoirs of biomass and carbon sink. Hence, felling trees in name of development is adversely impacting the city's environment and hence overall development.

Further, the built-up expansion within institutional compounds is leading to tree felling of old trees as reported by the local media. The negligence on part of authorities, lack of stringent policies, and un-availability of monitoring tools, altogether are leaving the avenue plantation and institutional greens in a highly vulnerable state. Thus, these UGS classes need more attention in terms of preservation action and compensatory planting efforts along with their monitoring. The forest structure is dominated by similar species young trees with occurrence of saplings which indicates the mass planting efforts done by forest department in recent years. However, the lack of old trees because of the severity of fire incidences occurred in past warrants for immediate measures for control over fire occurrences along with enhancement in the diversity and trees structure in reserve forest. The study of girth class distribution with high number of young trees shows recent efforts made by the local authority, however the peripheries and eastern areas of city showcase less planting efforts and need immediate attention for greening interventions.

4.2. Tree Biomass and Carbon Sink Potential of UGS

Tree biomass and Cstock data indicate significance of old trees in carbon accumulation. Although playground contributes less toward the tree diversity, the presence of old trees which have higher BA leads to increased Cstock in playgrounds. Similarly, abundance of mature and old trees along the roads, park, and institutional compounds enhances the carbon sink potential of these UGS and the city. Overall, in comparison to other studies Nagpur has good Cstock among the managed greens. While the unmanaged greens like lake peripheries, nag river corridor, and vacant lands which lack trees are recommended as potential areas for afforestation programs, which has also been acknowledged in the proposed city development plan [70]. The Cstock of urban forest is within the range identified for dry deciduous forest [64], however through effective maintenance and management the carbon sink can also be enhanced. As forests are being regenerated by planting high number of young trees of similar species, maintaining the structural diversity, controlling the disturbance and sporadic unwanted event can lead to increased Cstosk [80].

The evaluated Cstcok values can be used by local authorities to safeguard the existing trees and prioritize planting efforts in the identified UGS classes. The established correlation between BA and biomass as well the role of DBH structure in carbon accumulation can guide the local authorities toward species selection to enhance the overall carbon sink of the city. Hence, the recorded data and the Cstock evaluation acts as a reference data set to develop local greening strategies based on specific UGS classes requirement and guide toward strategic planting and afforestation efforts. Further, as the old trees are reservoirs of high carbon, the study recommends implementation of stringent policies for protection and conservation of old trees.

4.3. Tree Species Selection, Management and Maintainance

Trees in urban areas need to withstand "pollutants, high temperatures as a result of heat island effects, limited rooting space, and less water availability in compacted soil" [81,82]. Thus, different attributes of trees play important role in plant selection. Like dense and broad canopy trees lead to more AGB with enhanced aesthetic values, while the thick foliaged trees allow removal of air pollutants as well as give shade which leads to temperature amelioration [83–85]. Hence, species selection is an important aspect in urban greening, as an appropriate choice of species could significantly enhance the urban environment and increase the benefits derived from UGS. In Nagpur, the vegetation is composed of both native and introduced species with a mix of evergreen and deciduous trees. The most common native species in all classes is *Azadirachta indica* which has high tolerance, fast growth rate, and dense canopy. The highly abundant introduced specie recorded is *Polyalthia longifolia* which is mainly planted in parks and gardens because of its thick foliage to create a screening and fast growth rate. Among the ornamental trees *Delonix regia*, *Cassia siamea*, *Peltophorum petrocarpum*, and *Bahunia variagata* are obvious choices owing to the flowering characteristics of the species, the fast growth rate, and the shady canopy. These species are dominant in managed greens, as also identified in studies of other Indian cities [24,86,87].

In forest area, deciduous tree species with high drought tolerance dominates the tree species characteristics, making *Tectona grandis*, *Hardwickia binate*, and *Boswellia serrata* as the obvious choice for mass plantation. The species recorded with higher DBH class where mainly identified as species with cultural and religious significance. This characteristics has significantly resulted in their preservation and conservation as identified in other cities [88]. Overall, growth rate, phenology, canopy type, tolerance to drought, and resistance to pollution are identified as key attributes considered by local authorities for species selection. Another study in Bangalore also indicated that growth rate is considered as an important criterion for species selection [24], while some researcher have also highlighted the importance of productive trees with longer life span to mitigate the carbon concentrations in cities [11]. Selection of native tree species like *Azadirachta indica*, *Ficus religiose*, *Ficus benghalensis*, *Dalbergi sissoo*, *Alistonia scholaris*, *Bahunia variegate*, and *Mitrangya parvifolia* with high biomass and high efficiency of carbon fixation as well as other ecological benefits are highly recommended. From management and maintenance perspective it is critical to protect the old trees from further felling, thus local authorities need to focus on both preservation of existing tree along with afforestation efforts. The compensatory planting within same areas and same class should be made mandatory, and the same should be monitored and reported periodically. Further learning and adopting ideas from the successive initiatives in other Indian cities like "Green Leap Delhi" and "Tree Ambulance" is recommended [13]. Moreover, public participation and residents' involvement at every stage is recommended as it allows high survival rate of planted saplings and reduced maintenance cost.

The study is novel in terms of using UGS class as a basis to understand the variation in vegetation structure and carbon stock at a granular level in the complex urban mosaic. However, because of classes-based data, comparison of finding and validation was limited because of lack of biomass studies in Indian context [23]. Though for some classes international cases were reviewed, however because of the difference in methodology used to determine the carbon sink, direct comparison was difficult. In the study, only tree biomass has been considered, while ground cover biomass, litter biomass, and soil biomass are recommended for future detailed studies. Further, to understand the dynamic behavior of ecosystem, a thorough understanding of biophysical systems including soil as a carbon pool [89], role of properly functioning soil to enhance ecosystem benefits [90], as well as role played by UGS in addressing the sustainable development goals (SDGs) is recommended for future studies. Also, generating spatial data using high resolution satellite imagery and use of unmanned aerial vehicle is recommended to allow frequent assessment of vegetation carbon data, monitor structural change in urban vegetation over time, keep an account on tree felling and compensatory planting, and guide toward effecting afforestation programs [91]. We also recommend recording vegetation data of private greens for future works.

5. Conclusions

The green city of Nagpur is witnessing urbanization. The increase in grey infrastructure is leading to a decline in existing urban tree cover. Because of scanty information about city's vegetation structure and carbon sink potential, a rigorous field work was conducted to record vegetation data for phytosociology and carbon sink assessment of different UGS classes in the city. Focusing on the UGS classes, the research tries to understand how the vegetation density, diversity, composition, and their structure vary among urban greens. The findings highlight that avenue plantation and institutional greens are highly diverse with high tree density. Cstock is also high in avenue plantation and playgrounds. The managed greens have higher girth trees, which contributes toward increased Cstock, however trees along the road and institutions are subjected to felling in road widening and other infrastructural demands and are under threat. In the forest area, low diversity is recorded with a lack of higher girth tree species; additionally, the sporadic fire events reported by local media highlights the need of more strategic planting, monitoring, and maintenance policies. The city's tree structure is dominated by young and mature trees that indicate that afforestation efforts have been made in recent past. However, lack of saplings along the new road networks and lower tree frequency along the river and playground highlight them as potential areas for future afforestation efforts. In addition to phenology and growth rate, tree biomass and life span are recommended to be consider for species selection to significantly enhance the urban environment and increase the benefits derived from UGS. Large canopy tree with high basal area has the ability to mitigate future climate challenges, thus should also be considered in species selection in addition to functional and tolerance levels. Lastly, local level spatial data of vegetation including private greens as future research are highly recommended.

Supplementary Materials: The following are available online at http://www.mdpi.com/2073-445X/9/4/107/s1, Figure S1. Correlation between BA and AGB of tree species in different UGS typology.

Author Contributions: S.L. has conceptualized the original idea and carried out the field work. The methodology and data analysis were also conceived by S.L., R.K.J. and O.S. supervised for data analysis. S.L. worked on the writing of original draft preparation. A.L. provided critical feedback throughout to finetune the manuscript and proofread the final draft. All authors have read and agree to the published version of the manuscript.

Funding: This research received no external funding.

Acknowledgments: We are thankful to Dhiraj Buradkar, Khusbu Singh, Deepika Sarda, Ashwini Rangari, Harsha Nandurkar, and Pooja Fulwadhani for their support in conducting the field work.

Conflicts of Interest: The authors declare no conflict of interest with respect to the research, authorship, and/or publication of this article.

Appendix A

Volume equations and species specific wood density used in the reaserch, where V stantds for volume in m3, D is diameter at breast height in m, H is height of tree in m, V is square root, Source for * is Chave et al. 2001 and for other FSI and FRI (2001 and 2006).

Species	Volume equation	Specific gravity	Source
Acacia arabica	$AGB = \rho \times exp\,(-0.667 + 1.784\,\ln\,(D) + 0.207\,(\ln\,(D))^2 - 0.0281(\ln\,(D))^3)$	0.7	Chave et al. (2005)
Acacia catechu	$V = -0.02471 + 0.16897D + 1.12083D^2 + 2.9328D^3$	0.88	FSI 1996
Acacia leucophloea	$\sqrt{V} = -0.00142 + 2.61911D - 0.54703\sqrt{D}$	0.76	FSI 1996
Ailanthus excelsa	$V/D^2.0.32056 + 5.16781D - 1.83345\sqrt{D}$	0.81	FSI 1996
Albizia lebbeck	$\sqrt{V} = -0.07109 + 2.99732D - 0.26953\sqrt{D}$	0.55	FSI 1996
Albizia odoratissima	$\sqrt{V} = -0.07109 + 2.99732D - 0.26953\sqrt{D}$	0.76	FSI 1996
Alstonia scholaris *	$AGB = \rho \times exp\,(-0.667 + 1.784\,\ln\,(D) + 0.207\,(\ln\,(D))^2 - 0.0281(\ln\,(D))^2)$	0.36	Chave et al. (2005)
Annona squamosa	$V/D^2 = 0.0697/D^2 \cdot 1.4597/D + 11.79933 - 2.35397D$	0.619	FSI 1996
Azadirachta indica	$V/D = -0.00342/D - 0.0922/D + 2.28178 + 9.46641D$	0.693	FSI 1996
Bauhinia variegata	$V = 0.01475 + 0.2982D^2H$	0.67	FSI 1996
Bombax ceiba	$V/D^2 = 0.18573/D^2 - 2.85418/D + 15.03576$	0.33	FSI 1996
Boswellia serrata	$\sqrt{V} = -0.1503 + 2.79425D$	0.50	FSI 1996
Butea frondosa	$\sqrt{V} = -0.24276 + 2.95525D$	0.48	FSI 1996
Callistemon viminalis *	$AGB = \rho \times exp\,(-0.667 + 1.784\,\ln\,(D) + 0.207\,(\ln\,(D))^2 - 0.0281(\ln\,(D))^3)$	0.69	Chave et al. (2005)
Cassia fistula	$V = 0.066 + 0.287D^2H$	0.71	FSI 1996
Cassia siamea	$V = 0.05159 - 0.53331D + 3.46016D^2 + 10.18473D^3$	0.746	FSI 1996
Ceiba pentandra *	$AGB = \rho \times exp\,(-0.667 + 1.784\,\ln\,(D) + 0.207\,(\ln\,(D))^2 - 0.0281(\ln\,(D))^4)$	0.23	Chave et al. (2005)
Chloroxylon swietenia	$V = -0.003156 + 2.043969\,D^2$	0.458	FSI 1996
Dalbergia latifolia	$V = 0.04422 + 2.328465\,D2 + 0.309150\,D^2H$	0.75	FSI 1996
Dalbergia paniculata	$V = 0.18945 - 2.46215D + 10.54462D^2$	0.64	FSI 1996
Dalbergia sissoo	$V = -0.013703 + 3.943499D^2$	0.692	FSI 1996
Delonix regia *	$AGB = \rho \times exp\,(-0.667 + 1.784\,\ln\,(D) + 0.207\,(\ln\,(D))^2 - 0.0281(\ln\,(D))^3)$	0.8	Chave et al. (2005)
Diospyros melanoxylon	$V = 0.15581 - 2.2075D + 9.175590D^2$	0.68	FSI 1996
Eucalyptus globulus	$V = 0.02894 - 0.89284D + 8.72416D^2$	0.3	FSI 1996
Eugenia jambolana *	$AGB = \rho \times exp\,(-0.667 + 1.784\,\ln\,(D) + 0.207\,(\ln\,(D))^2 - 0.0281(\ln\,(D))^3)$	0.65	Chave et al. (2005)
Ficus benghalensis	$\sqrt{V} = 0.03629 + 3.95389D - 0.84421\sqrt{D}$	0.39	FSI 1996
Ficus elastica	$\sqrt{V} = 0.03629 + 3.95389D - 0.84421\sqrt{D}$	0.39	FSI 1996
Ficus glomerata	$\sqrt{V} = 0.03629 + 3.95389D - 0.84421\sqrt{D}$	0.39	FSI 1996
Ficus religiosa	$\sqrt{V} = 0.03629 + 3.95389D - 0.84421\sqrt{D}$	0.385	FSI 1996
Gardenia resinifera	$V = 0.078 - 1.188D + 6.7511D^2$	0.62	FSI 1996
Garuga pinnata	$V/D = 0.077965/D - 1.481043 + 9.797028D$	0.51	FSI 1996
Gmelina arborea	$V = 0.01156 + 0.21230D + 5.10448\,D^2$	0.41	FSI 1996
Hardwickia binata	$V = 0.063632 + 5.3554860D3$	0.73	FSI 1996
Kigelia pinnata *	$AGB = \rho \times exp\,(-0.667 + 1.784\,\ln\,(D) + 0.207\,(\ln\,(D))^2 - 0.0281(\ln\,(D))^3)$	0.81	Chave et al. (2005)
Lagerstroemia parviflora *	$AGB = \rho \times exp\,(-0.667 + 1.784\,\ln\,(D) + 0.207\,(\ln\,(D))^2 - 0.0281(\ln\,(D))^3)$	0.81	Chave et al. (2005)
Lagerstroemia speciosa	$V = -0.000001 + 0.35751D^2H$	0.53	FSI 1996
Lannea coromandelica *	$AGB = \rho \times exp\,(-0.667 + 1.784\,\ln\,(D) + 0.207\,(\ln\,(D))^2 - 0.0281(\ln\,(D))^3)$	0.54	Chave et al. (2005)
Leucaena leucocephala	$V/D = -0.00342/D - 0.0922/D + 2.28178 + 9.46641D$	0.55	FSI 1996
Madhuca latifolia	$V = 0.063632 + 5.3554860D3$	0.74	FSI 1996
Mangifera indica	$V = 0.288 - 2.913D + 13.869D2$	0.74	FSI 1996
Manilkara hexandra	$V = 0.0245 - 0.00497D + 0.000719D^2$	0.89	FSI 1996

Figure A1. *Cont.*

Species	Equation		Reference
Melia azedarach	$V = -0.03510 + 5.32981D^2$	0.619	FSI 1996
Millingtonia hortensis *	$AGB = p \times exp(-0.667 + 1.784 \ln(D) + 0.207\{\ln(D)\}^2 - 0.0281\{\ln(D)\}^3)$	0.72	Chave et al. (2005)
Mimusops elengi *	$AGB = p \times exp(-0.667 + 1.784 \ln(D) + 0.207\{\ln(D)\}^2 - 0.0281\{\ln(D)\}^3)$	0.72	Chave et al. (2005)
Mitragyna parviflora	$V/D^2 = 0.099768/D^2 - 1.744274/D + 10.086934$	0.56	Chave et al. (2005)
Moringa oleifera *	$AGB = p \times exp(-0.667 + 1.784 \ln(D) + 0.207\{\ln(D)\}^2 - 0.0281\{\ln(D)\}^3)$	0.5	Chave et al. (2005)
Morus alba *	$AGB = p \times exp(-0.667 + 1.784 \ln(D) + 0.207\{\ln(D)\}^2 - 0.0281\{\ln(D)\}^3)$	0.53	Chave et al. (2005)
Murraya koenigii *	$AGB = p \times exp(-0.667 + 1.784 \ln(D) + 0.207\{\ln(D)\}^2 - 0.0281\{\ln(D)\}^3)$	0.636	Chave et al. (2005)
Neolamarckia cadamba *	$AGB = p \times exp(-0.667 + 1.784 \ln(D) + 0.207\{\ln(D)\}^2 - 0.0281\{\ln(D)\}^3)$	0.636	Chave et al. (2005)
Nyctanthes arbor-tritis *	$AGB = p \times exp(-0.667 + 1.784 \ln(D) + 0.207\{\ln(D)\}^2 - 0.0281\{\ln(D)\}^3)$	0.88	Chave et al. (2005)
Peltophorum pterocarpum	$VV = -0.08150 + 2.48467D$	0.62	FSI 1996
Pithecellobium dulce *	$AGB = p \times exp(-0.667 + 1.784 \ln(D) + 0.207\{\ln(D)\}^2 - 0.0281\{\ln(D)\}^3)$	0.66	Chave et al. (2005)
Plumeria alba *	$AGB = p \times exp(-0.667 + 1.784 \ln(D) + 0.207\{\ln(D)\}^2 - 0.0281\{\ln(D)\}^3)$	0.5	Chave et al. (2005)
Polyalthia longifolia *	$AGB = p \times exp(-0.667 + 1.784 \ln(D) + 0.207\{\ln(D)\}^2 - 0.0281\{\ln(D)\}^3)$	0.54	Chave et al. (2005)
Pongamia pinnata	$V/D = -0.077965/D - 1.481043 + 9.797028D$	0.54	FSI 1996
Prosopis juliflora	$V/D = -0.00342/D - 0.0922/D + 2.28178 + 9.46641D$	0.85	FSI 1996
Psidium guajava *	$AGB = p \times exp(-0.667 + 1.784 \ln(D) + 0.207\{\ln(D)\}2 - 0.0281\{\ln(D)\}3)$	0.58	Chave et al. (2005)
Pterocarpus marsupium	$V/D^2 = -0.04659/D^2 + 8.06901$	0.67	FSI 1996
Sapindus mukorossi *	$AGB = p \times exp(-0.667 + 1.784 \ln(D) + 0.207\{\ln(D)\}^2 - 0.0281\{\ln(D)\}^3)$	0.58	Chave et al. (2005)
Schrebera swietenioides *	$AGB = p \times exp(-0.667 + 1.784 \ln(D) + 0.207\{\ln(D)\}^2 - 0.0281\{\ln(D)\}^3)$	0.82	Chave et al. (2005)
Soymida febrifuga *	$AGB = p \times exp(-0.667 + 1.784 \ln(D) + 0.207\{\ln(D)\}^2 - 0.0281\{\ln(D)\}^3)$	0.97	FSI 1996
Sterculia urens *	$AGB = p \times exp(-0.667 + 1.784 \ln(D) + 0.207\{\ln(D)\}^2 - 0.0281\{\ln(D)\}^3)$	0.67	Chave et al. (2005)
Tamarindus indica	$V = 0.046883 - 0.894379D + 7.220441D^2$	0.75	FSI 1996
Tecoma stans *	$AGB = p \times exp(-0.667 + 1.784 \ln(D) + 0.207\{\ln(D)\}^2 - 0.0281\{\ln(D)\}^3)$	0.46	Chave et al. (2005)
Tectona grandis	$V = -0.27773 + 3.10419D - 6.12739D^2 + 15.16993D^3$	0.58	FSI 1996
Terminalia arjuna	$V = 0.50603 - 6.64203D + 25.23882D^2 - 9.19797D^3$	0.68	FSI 1996
Terminalia catappa	$V = 0.50603 - 6.64203D + 25.23882D^2 - 9.19797D^3$	0.52	FSI 1996
Trichilia dregeana *	$AGB = p \times exp(-0.667 + 1.784 \ln(D) + 0.207\{\ln(D)\}^2 - 0.0281\{\ln(D)\}^3)$	0.48	Chave et al. (2005)
Un-identified	$V = exp(-2.289 + 2.649 \times \ln(D) - 0.021 \times \ln(D^2))$		Brown et al. (1989)
Vachellia farnesiana	$VV = -0.00142 + 2.619111D - 0.54703VD$	0.70	FSI 1996
Vachellia nilotica	$V = 0.043849 - 0.552735D + 2.952386D^2 + 0.334508\,D^2H$	0.67	FSI 1996
Ziziphus mauritiana	$V = 0.027354 + 4.663714\,D^2$	0.58	FSI 1996

Figure A1. Species specific volumetric equations.

References

1. Strohbach, M.W.; Haase, D. Above-ground carbon storage by urban trees in Leipzig, Germany: Analysis of patterns in a European city. *Landsc. Urban Plan.* **2012**, *104*, 95–104. [CrossRef]
2. Nowak, D.J.; Crane, D.E.; Stevens, J.C.; Hoehn, R.E.; Walton, J.T.; Bond, J. A Ground-Based Method of Assessing Urban Forest Structure and Ecosystem Services. *Arboric. Urban For.* **2008**, *34*, 347–358.
3. Nero, B.; Callo-Concha, D.; Anning, A.; Denich, M. Urban Green Spaces Enhance Climate Change Mitigation in Cities of the Global South: The Case of Kumasi, Ghana. *Procedia Eng.* **2017**, *198*, 69–83. [CrossRef]
4. Kabisch, N. Ecosystem service implementation and governance challenges in urban green space planning-The case of Berlin, Germany. *Land Use Policy* **2015**, *42*, 557–567. [CrossRef]

5. Nesshöver, C.; Assmuth, T.; Irvine, K.N.; Rusch, G.M.; Waylen, K.A.; Delbaere, B.; Krauze, K. The science, policy and practice of nature-based solutions: An interdisciplinary perspective. *Sci. Total Environ.* **2017**, *579*, 1215–1227. [CrossRef]

6. Raymond, C.M.; Frantzeskaki, N.; Kabisch, N.; Berry, P.; Breil, M.; Nita, M.R.; Calfapietra, C. A framework for assessing and implementing the co-benefits of nature-based solutions in urban areas. *Environ. Sci. Policy* **2017**, *77*, 15–24. [CrossRef]

7. Frantzeskaki, N. Seven lessons for planning nature-based solutions in cities. *Environ. Sci. Policy* **2019**, *93*, 101–111. [CrossRef]

8. Lafortezza, R.; Chen, J.; Bosch, C.K.; Thomas, B.R. Nature-based solutions for resilient landscapes and cities. *Environ. Res.* **2018**, *165*, 431–441. [CrossRef]

9. Novara, A.; Pulido, M.; Rodrigo-Comino, J.; Di Prima, S.; Smith, P.; Gristina, L.; Keesstra, S. Long-term organic farming on a citrus plantation results in soil organic matter recovery. *Cuad. De Investig. Geográfica* **2019**, *45*, 271–286. [CrossRef]

10. Jenkins, J.C.; Chojnacky, D.C.; Heath, L.S.; Birdsey, R.A. National-scale biomass estimators for United States tree species. *For. Sci.* **2003**, *49*, 12–35.

11. Nowak, D.J.; Crane, D.E. Carbon storage and sequestration by urban trees in the USA. *Environ. Pollut.* **2002**, *116*, 381–389. [CrossRef]

12. Nagendra, H.; Gopal, D. Street trees in Bangalore: Density, diversity, composition and distribution. *Urban For. Urban Green.* **2010**, *9*, 129–137. [CrossRef]

13. Imam, A.U.K.; Banerjee, U.K. Urbanisation and greening of Indian cities: Problems, practices, and policies. *Ambio* **2016**, *45*, 442–457. [CrossRef] [PubMed]

14. Govindarajulu, D. Urban green space planning for climate adaptation in Indian cities. *Urban Climate* **2014**, *10*, 35–41. [CrossRef]

15. Alberti, M.; Marzluff, J.M.; Shulenberger, E.; Bradley, G.; Ryan, C.; Zumbrunnen, C. Integrating humans into ecology: Opportunities and challenges for studying urban ecosystems. *BioScience* **2003**, *53*, 1169–1179. [CrossRef]

16. Rao, P.; Puntambekar, K. Evaluating the Urban Green Space benefits and functions at macro, meso and micro level: Case of Bhopal City. *Int. J. Eng. Res. Technol.* **2014**, *3*, 359–369.

17. Anguluri, R.; Narayanan, P. Role of green space in urban planning: Outlook towards smart cities. *Urban For. Urban Green.* **2017**, *25*, 58–65. [CrossRef]

18. Vittal, I.; Dobbs, R.; Mohan, A.; Gulati, A.; Ablett, J.; Gupta, S.; Kim, A.; Paul, S.; Sanghvi, A.; Sethy, G. India's Urban Awakening: Building Inclusive Cities, Sustaining Economic Growth. Available online: https://www.mckinsey.com/featured-insights/urbanization/urban-awakening-in-india (accessed on 15 January 2018).

19. Kumar, V. *Coping with Climate Change: An Analysis of India's State Action Plans on Climate Change*; Centre for Science and Environment: 2018. Available online: http://cdn.cseindia.org/attachments/0.40897700_1519110602_coping-climate-change-volII.pdf (accessed on 15 January 2019).

20. Chaudhary, P.; Tewari, V.P. Managing urban parks and garden in developing countries, a case from an Indian cities. *Int. J. Leis. Tour. Mark.* **2010**, *1*, 248–256. [CrossRef]

21. Tripathi, N.G.; Bedi, P. IOP Conference Series: Earth and Environmental Science 18. In *Digital Earth for Manipulating Urban Greens towards Achieving a Low Carbon Urban Society*; IOP Publishing: Bristol, UK, 2014; pp. 1–5. [CrossRef]

22. Davies, Z.G.; Edmondson, J.L.; Heinemeyer, A.; Leake, J.R.; Gaston, K.J. Mapping an urban ecosystem service: Quantifying above-ground carbon storage at a city-wide scale. *J. Appl. Ecol.* **2011**, *48*, 1125–1134. [CrossRef]

23. Ugle, P.; Rao, S.; Ramachandra, T. Carbon Sequestration Potential of Urban Trees. In Proceedings of the Lake 2010: Wetlands, Biodiversity and Climate Change, Bangalore, India, 22–24 December 2010.

24. Sudha, P.; Ravindranath, N.H. A study of Bangalore urban forest. *Landsc. Urban Plan.* **2000**, *47*, 47–63. [CrossRef]

25. Meyer, B.; Grabaum, R. MULBO—model framework for multi critieria landscape assessment and optimisation. A support system for spatial land use decisions. *Landsc. Res.* **2008**, *33*, 155–179. [CrossRef]

26. Mitra, S. Some aspects of ecology of walls at Vishakhapatnam. Ph.D. Thesis, Andhara University, Waltair, India, 1993.

27. Madan, M.S. Composition of the ground vegetation of Visakhapatnam. *J. Natcon* **1993**, *5*, 77–82.

28. Chaudhary, P. Valuing recreational benefits of urban forestry- A case study of Chandigarh city. Ph.D. Thesis, FRI Deemed University, Dehradun, India, 2006.

29. Dwivedi, P.; Rathore, C.S.; Dubey, Y. Ecological benefits of urban forestry: The case of Kerwa Forest Area (KFA), Bhopal, India. *Appl. Geogr.* **2009**, *29*, 194–200. [CrossRef]

30. Khera, N.; Mehta, V.; Sabata, B.C. Interrelationship of birds and habitat features in urban greenspaces in Delhi, India. *Urban For. Urban Green.* **2009**, *8*, 187–196. [CrossRef]

31. Waran, A.; Patwardhan, A. Urban Carbon Burden of Pune City: A Case Study from India. Master's Thesis, Department of Environmental Sciences, University of Pune, Pune, IN, USA, 2001.

32. Gill, S.E.; Handley, J.F.; Ennos, A.R.; Pauleit, S. Characterising the urban environment of UK cities and towns: A tem- plate for landscape planning. *Landsc. Urban Plan.* **2008**, *87*, 210–222. [CrossRef]

33. Whitford, V.; Ennos, A.R.; Handley, J.F. "City form and natural process"–indicators for the ecological performance of urban areas and their application to Merseyside, UK. *Landsc. Urban Plan.* **2001**, *57*, 91–103. [CrossRef]

34. Gupta, R.B.; Chaudhari, P.; Wate, S. Floristic diversity in urban forest area of NEERI Campus, Nagpur, Maharashtra (India). *J. Environ. Sci. Eng.* **2008**, *50*, 55–62. [PubMed]

35. Surawar, M.; Kotharkar, R. Assessment of Urban Heat Island through Remote Sensing in Nagpur Urban Area Using Landsat 7 ETM+. *Int. J. Civ. Environ. Struct. Constr. Archit. Eng.* **2017**, *11*, 851–857.

36. Dhyani, S.; Lahoti, S.; Khare, S.; Pujari, P. Verma; Ecosystem based Disaster Risk Reduction approaches (EbDRR) as a prerequisite for inclusive urban transformation of Nagpur City, India. *Int. J. Disaster Risk Reduct.* **2018**, *32*, 95–105. [CrossRef]

37. Ugemuge, N.R. *Flora of Nagpur District, Maharashtra State*; Shree Prakashan: Nagpur, India, 1986.

38. Chaturvedi, A.; Kamble, R.; Patil, N.G.; Chaturvedi, A. City–forest relationship in Nagpur: One of the greenest cities of India. *Urban For. Urban Green.* **2013**, *12*, 79–87. [CrossRef]

39. Lahoti, S.; Kefi, M.; Lahoti, A.; Saito, O. Mapping Methodology of Public Urban Green Spaces Using GIS: An Example of Nagpur City, India. *Sustainability* **2019**, *11*, 2166. [CrossRef]

40. Threlfall, C.G.; Ossola, A.; Hahs, A.K.; Williams NS, G.; Wilson, L.; Livesley, S.J. Variation in Vegetation Structure and Composition across Urban Green Space Types. *Front. Ecol. Evol.* **2016**, *4*, 1–12. [CrossRef]

41. Lahoti, S.; Lahoti, A.; Saito, O. Benchmark Assessment of Recreational Public Urban Green Space Provisions: A Case of Typical Urbanizing Indian City, Nagpur. *Urban For. Urban Green.* **2019**, *44*, 126424. [CrossRef]

42. Tree. In *I-Tree Eco User's Manual v.6.0*; Available online: https://www.itreetools.org/resources/manuals/Ecov6_ManualsGuides/Ecov6_UsersManual.pdf (accessed on 4 January 2019).

43. Curtis, J.; McIntosh, R.P. An upland forest continuumin the prairie forest border re gion of Wisconsin. *Ecology* **1951**, *32*, 476–496. [CrossRef]

44. Lu, D.S. The potential and challenge of remote sensing based biomass estimation. *Int. J. Remote Sens.* **2006**, *27*, 1297–1328. [CrossRef]

45. Brown, S.; Gillespie, A.; Lugo, A. Biomass estimation methods for tropical forests with applications to forest inventory data. *For. Sci.* **1989**, *35*, 881–902.

46. Hughes, R.F.; Kauffman, J.B.; Jaramillo, V.J. Biomass, carbon, and nutrient dynamics of secondary forests in a humid tropical region of Mexico. *Ecology* **1999**, *80*, 1897–1907.

47. Henry, M.; Bombelli, A.; Trotta, C.; Alessandrini, A.; Birigazzi, L.; Sola, G.; Santenoise, P.E. GlobAllomeTree: International platform for tree allometric equations to support volume, biomass and carbon assessment. *Iforest Biogeosci. For.* **2013**, *6*, 326–330. [CrossRef]

48. Nowak, D.J. Air pollution removal by Chicago's Urban Forest. In *Chicago's Urban Forest Ecosystem: Results of the Chicago Urban Forest Climate Project*; US Department of Agriculture, Forest Service, Northeastern Forest Experiment Station: Upper Darby, PA, USA, 1994; pp. 63–83.

49. Ravindranath, N.H.; Ostwald, M. Carbon Inventory Methods. 2008. Available online: https://www.springer.com/cn/book/9781402065460 (accessed on 1 May 2018).

50. Chave, J.; Andalo, C.; Brown, S.; Cairns, M.A.; Chambers, J.Q.; Eamus, D.; Yamakura, T. Tree allometry and improved estimation of carbon stocks and balance in tropical forests. *Oecologia* **2005**, *145*, 87–99. [CrossRef]

51. Pandya, I.Y.; Salvi, H.; Chahar, O.; Vaghela, N. Quantitative analysis on carbon storage of 25 valuable tree species of Gujarat, India. *Indian J. Sci. Res.* **2013**, *4*, 137–141.

52. Forest Survey of India (FSI). *Volume Equations for Forests of India, Nepal and Bhutan. Forest Survey of India*; Ministry of Environment and Forests, Govt. of India: Dehradun, India, 1996.

53. Aguaron, E.; McPherson, E.G. Comparison of Methods for Estimating Carbon Dioxide Storage by Sacramento's Urban Forest. In *Carbon Sequestration in Urban Ecosystems*; Lal, R., Augustin, B., Eds.; Springer Science: Dordrecht, The Netherlands, 2012. [CrossRef]

54. Ngo, K.M.; Lum, S. Aboveground biomass estimation of tropical street trees. *J. Urban Ecol.* **2018**, *4*, jux020. [CrossRef]

55. Chow, P.; Rolfe, G.L. Carbon and hydrogen contents of short-rotation biomass of five hardwood species. *Wood Fiber Sci.* **1989**, *21*, 30–36.

56. Magnussen, S.; Reed, D. Modelling for Estimation and Monitoring. *FAO-IUFRO* **2004**, *1*, 111–136.

57. Brack, C.L. Pollution Mitigation and Carbon Sequestration by an Urban Forest. *Environ. Pollut.* **2002**, *116*, 195–200. [CrossRef]

58. Kent, M. *Vegetation Description and Data Analysis: A Practical Approach*, 2nd ed.; Wiley-Blackwell: Hoboken, NJ, USA, 2011; ISBN 978-0-471-49093-7.

59. Mani, S.; Parthasarathy, N. Above-ground biomass estimation in ten tropical dry evergreen forest sites of peninsular India. *Biomass Bioenergy* **2007**, *31*, 284–290. [CrossRef]

60. Murali, K.S.; Bhat, D.M.; Ravindranath, N.H. Biomass estimation equations for tropical and evergreen forests. *Int. J. Agri.-Cult. Resour. Gov. Ecol.* **2005**, *4*, 81–92. [CrossRef]

61. Kumar, A.; Sharma, M.P. Estimation of carbon stocks of Balganga Reserved Forest, Uttarakhand, India. *For. Sci. Technol.* **2015**, *11*, 177–181. [CrossRef]

62. Borah, N.; Nath, A.J.; Das, A.K. Aboveground Biomass and Carbon Stocks of Tree Species in Tropical Forests of Cachar District, Assam, Northeast India. *Int. J. Ecol. Environ. Sci.* **2013**, *39*, 97–106.

63. Rahman, M.M.; Kabir, M.E.; Jahir Uddin Akon AS, M.; Ando, K. High carbon stocks in roadside plantations under participatory management in Bangladesh. *Glob. Ecol. Conserv.* **2015**, *3*, 412–423. [CrossRef]

64. Joshi, R.K.; Dhyani, S. Biomass, carbon density and diversity of tree species in tropical dry deciduous forests in Central India. *Acta Ecol. Sin.* **2018**, *39*, 289–299. [CrossRef]

65. Gandhi, S.D.; Sundarapandian, S. Large-scale carbon stock assessment of woody veg- etation in tropical dry deciduous forest of Sathanur reserve forest, Eastern Ghats India Environ. *Monit. Assess.* **2017**, *189*, 187–196. [CrossRef] [PubMed]

66. Meena, A.; Bidalia, A.; Hanief, M.; Dinakaran, J.; Rao, K. Assessment of above- and belowground carbon pools in a semi-arid forest ecosystem of Delhi, India. *Ecol. Process.* **2019**, *8*, 8. [CrossRef]

67. Sallustio, L.; Quatrini, V.; Geneletti, D.; Corona, P.; Marchetti, M. Assessing land take by urban development and its impact on carbon storage: Findings from two case studies in Italy. *Environ. Impact Assess. Rev.* **2015**, *54*, 80–90. [CrossRef]

68. Liu, Y.; Meng, Q.; Zhang, J.; Zhang, L.; Jancso, T.; Vatseva, R. An effective Building Neighborhood Green Index model for measuring urban green space. In *International Journal of Digital Earth*; Taylor and Francis: Milton Park, UK, 2015. [CrossRef]

69. Devi, R. Carbon storage by trees in urban parks: A case study of Jammu, Jammu and Kashmir, India. *Int. J. Adv. Res. Dev.* **2017**, *2*, 250–253.

70. Ministry of Urban Development (MoUD). *City Development Plan for Nagpur, 2041*; Government of India, 2015. Available online: http://www.metrorailnagpur.com/pdf/FinalCDP_Nagpur-Mar15ofNMC.pdf (accessed on 1 December 2018).

71. Profous, G.V.; Rowntree, R.A.; Loeb, R.E. The urban forest land- scape of Athens, Greece: Aspects of structure, planning and management. *Arboric. J.* **1998**, *12*, 83–108. [CrossRef]

72. Mishra, A.K.; Sharma, M.P.; Singh, H.B. Addition to the Flora of Delhi. *Indian J. Plant Sci.* **2015**, *4*, 1–6.

73. Pandey, R.K.; Kumar, H. Tree Species Diversity And Composition In Urban Green Spaces Of Allahabad City (U.P). *Plant Arch.* **2018**, *18*, 2687–2692.

74. Bhalla, P.; Bhattacharya, P. Urban Biodiversity and Green Spaces in Delhi: A Case Study of New Settlement and Lutyens' Delhi. *J. Hum. Ecol.* **2017**, *52*, 83–96. [CrossRef]

75. Magurran, A.E. *Measuring Biological Diversity*; Blackwell Science Ltd.: Cornwall, UK, 2004.

76. Chacalo, A.; Aldama, A.; Grabinsky, J. Street tree inventory in Mexico City. *J. Arboric.* **1994**, *20*, 222–226.

77. Thaiutsa, B.; Puangchit, L.; Kjelgren, R.; Arunpraparut, W. Urban green space, street tree and heritage large tree assessment in Bangkok, Thailand. *Urban For. Urban Green.* **2008**, *7*, 219–229. [CrossRef]

78. Sudhira, H.S.; Ramachandra, T.V.; Subrahmanya, M.H.B. City profile Bangalore. *Cities* **2007**, *24*, 379–390. [CrossRef]

79. Nair, J. *The Promise of the Metropolis: Bangalore's Twentieth Century*; Oxford University Press: Oxford, UK, 2005.
80. Baishya, R.; Barik, S.K.; Upadhaya, K. Distribution patter of aboveground biomass in natural and plantation forests of humid tropics in north-east India. *Trop. Ecol.* **2009**, *50*, 295–304.
81. Mansell, M.G. *Rural and Urban Hydrology*; Thomas Telford Ltd.: London, UK, 2003.
82. Watson, G.W.; Kelsey, P. The impact of soil compaction on soil aera- tion and fine root density of Quercus palustris. *Urban For. Urban Green.* **2006**, *4*, 69–74. [CrossRef]
83. Jim, C. Sustainable urban greening strategies for compact cities in developing & developed economies. *Urban Ecosyst.* **2012**, *16*, 741–761.
84. McPherson, E.G.; Nowak, D.; Heisler, G.; Grimmond, S.; Souch, C.; Grant, R.; Rowntree, R. Quantifying urban forest structure, function and value: The Chicago Urban Forest Climate Project. *Urban Ecosyst.* **1997**, *1*, 49–61. [CrossRef]
85. Pauleit, S.; Duhme, F. Assessing the environmental performance of landcover types for urban planning. *Landsc. Urban Plan.* **2000**, *52*, 1–20. [CrossRef]
86. Anamika, A.; Pradeep, C. Urban Vegetation and Air Pollution Mitigation: Some Issues from India. *Chin. J. Urban Environ. Stud.* **2016**, *4*, 1650001. [CrossRef]
87. Chaudhry, P.; Bagra, K.; Singh, B. Urban Greenery Status of Some Indian Cities: A Short Communication. *Int. J. Environ. Sci. Dev.* **2011**, *2*, 98–101. [CrossRef]
88. Tengö, M.; Gopal, D. Nagendra H (NA) Sacred Trees in the Urban Landscape of Bangalore, India. Current Conservation, Issue 8.1. Available online: https://www.currentconservation.org/?s=issue+8.1 (accessed on 1 April 2019).
89. Keesstra, S.D.; Bouma, J.; Wallinga, J.; Tittonell, P.; Smith, P.; Bardgett, R.D. The significance of soils and soil science towards realization of the United Nations Sustainable Development Goals. *Soil* **2016**, *2*, 111–128. [CrossRef]
90. Keesstra, S.; Mol, G.; de Leeuw, J.; Okx, J.; de Cleen, M.; Visser, S. Soil-related sustainable development goals: Four concepts to make land degradation neutrality and restoration work. *Land* **2018**, *7*, 133. [CrossRef]
91. Lahoti, S.; Lahoti, A.; Saito, O. Application of Unmanned Aerial Vehicle (UAV) for Urban Green Space Mapping in Urbanizing Indian Cities. In *Unmanned Aerial Vehicle: Applications in Agriculture and Environment*; Ram, A., Teiji, W., Eds.; Springer: Basel, Switzerland, 2020; ISBN 978-3-030-27156-5. [CrossRef]

MDPI
St. Alban-Anlage 66
4052 Basel
Switzerland
Tel. +41 61 683 77 34
Fax +41 61 302 89 18
www.mdpi.com

Land Editorial Office
E-mail: land@mdpi.com
www.mdpi.com/journal/land